Studies on China's High-Speed Rail New Town Planning and Development

Lan Wang · Hao Gu

Studies on China's High-Speed Rail New Town Planning and Development

Lan Wang
Department of Urban Planning
College of Architecture
and Urban Planning
Shanghai, China

Hao Gu
College of Architecture
and Urban Planning
Tongji University
Shanghai, China

ISBN 978-981-13-6915-5 ISBN 978-981-13-6916-2 (eBook)
https://doi.org/10.1007/978-981-13-6916-2

Jointly published with Tongji University Press, Shanghai, China.
The print edition is not for sale in China Mainland. Customers from China Mainland please order the print book from: Tongji University Press, Shanghai, China.

This Springer imprint is published by the registered company Springer Nature Singapore Pte Ltd.
The registered company address is: 152 Beach Road, #21-01/04 Gateway East, Singapore 189721, Singapore

Foreword

Studies on China's High-Speed Rail New Town Planning and Development is a book on high-speed rail (HSR) and Metro planning but is not limited to these. The basic consensus reached by the urban planning society in China is that urban planning responds to and guides social and economic changes by regulating and intervening in the arrangements for various resources that make up the land space. Contemporary urban planners are not only concerned with the planning project itself but also committed to promoting human progress, which is manifested in social equity, economic prosperity and ecological health and as the ultimate goal of human development, through such construction projects as transportation and new town development. These goals are undoubtedly addressed in various social fields and spatial dimensions. This book begins the discussion of the HSR new town from three aspects: regional range as the macrolevel, urban range as the mesolevel and the surrounding area of the station as the microlevel. The main concern at the regional level is the issue of economic cooperation both inside and outside the region, and at the urban level, HSR will be regarded as a new economic growth point, while the planning of the surrounding area of the station will place the main objective of improving the life quality of residents in space nodes through diversified land use functions.

This book has innovative significance in three aspects. First is the research topic. HSR new town planning itself is a new topic in the urban planning society of China. Although railway construction has been ongoing in the world for nearly two centuries, HSR is a new phenomenon that emerged only 50 years ago. HSR in China started in the mid-1990s and just matured so far. However, since the beginning, HSR construction in China has not only aimed to solve traffic problems but also has been entrusted with the tasks of stimulating economic growth, promoting regional development and rejuvenating the nation. It is not only a traffic or economic issue but also a political issue, guided by the government, with the full support of the country. This is different from HSR construction in some developed countries where relatively mature cities already existed, and how to speed up travel between those cities by HSR construction is mainly a traffic problem. Therefore, the alignment of HSR in China and the selection of HSR stations are not only to facilitate traffic flow, and more consideration must be given to the balanced development of regional economies, the economic transformation opportunities of local cities, etc. As a result, economic and political considerations together with traffic considerations have become the cornerstone of site selection for HSR in China. The site selection of HSR will inevitably affect the future of a region, a city, the surrounding area of a station and its inhabitants with economic, social and ecological impacts. In practice, we also observe another tendency: Due to too much emphasis on economic and local political balance, including unrealistic optimism regarding future development and constraints like investment, land availability and others, some HSR stations are far away from downtown and cannot fully meet the traffic planning requirements for convenience, thus deviating from the original intention of HSR construction. These positive and negative experiences make research on China's HSR new town different from traditional traffic planning research on transportation methods. This book not only provides valuable information for the development of more HSR towns in the future for China and other countries, especially developing countries, where HSR construction is also the stakeholder of high hopes of promoting the country's economic development and an important pillar of the government's performance.

The second innovative aspect is the research method. The book uses a combination of quantitative and qualitative methods to analyze the cases. Although limited by the availability of data, the quantitative research methods currently used in urban planning studies in China have become extremely common and effective. The book focuses on the 22 cities along the Beijing-Shanghai HSR line and provides a good quantitative analysis based on theories such as location theory, "traffic-land" coupling theory and bus-oriented development. With theories to guide the construction of quantitative models and the choice of variables, the quantitative analysis of the empirical results returns to the qualitative theoretical framework to explain and demonstrate in combination with foreign cases. These kinds of "theoretical-empirical" multiple interactive research approaches, as the author explains, provide a good mode of research pathways to the methods of "linking urban theory with planning theory."

The third innovative aspect is the urban planning theory. It involves two aspects: First is the relationship between theory and practice, one of the focuses of traditional planning theory. Traffic planning and new town locations have obvious technical content and technical characteristics of the rational. This way of putting technical rationality into complex practice is similar to Donald Schon's proposal for transforming technical rationality into reflection-in-action.[1] From the experience of the construction of HSR new towns, existing theories were tested and examined, such as the "traffic-land" coupling theory, to amend, enrich and perfect them. The second aspect is the application of western planning theory. Most planning and construction projects in the contemporary world are in developing countries, but the theory of guiding planning is still based on the experience of developed Western

[1]See Schon, 1983, "From Technical Rationality to Reflection-in-Action," in Jean Hillier and Patsy Healey (edit.) 2008, *Political Economy, Diversity and Pragmatism*, published by Ashgate.

countries. How to summarize the developing countries' own planning practice, compare the experience of similar planning projects in developed countries, focus on the different institutional structures and different historical contexts in the promotion (or constraint) of planning and implementation of the mechanisms and motivations and make a theoretical upgrade to form a planning theory based on the experience of developing countries are the major issues. China's large-scale HSR construction and new town planning projects provide an extremely rich case.

Since the reform and opening up, China has sent many students to study abroad in developed countries. At present, a considerable number of these students have returned to China to serve in various fields and have become the backbone of various academic disciplines. Professor Lan Wang, the author of the book, is such a returnee. Her Ph.D. research experience and work experience in the Metropolitan Area Planning Council of Chicago (CMAP) in the USA have laid a solid theoretical and methodological foundation for her academic research. This book is also the test and affirmation of her academic research results. As her doctoral supervisor, I congratulate her and hope more returnees will make great contributions to the rejuvenation of China.

Chicago, US

Tingwei Zhang
Emeritus Professor
Urban Planning Department
University of Illinois at Chicago

Preface

As a regional transport infrastructure with high passenger capacity, the HSR carries an important mission of linking many cities and towns on the vast land area of China. It is also regarded as a public investment carrier to stimulate economic growth. At the same time, planning and gradually building the surrounding area of the HSR have become important parts of China's urbanization process.

The HSR station is the core of the HSR new town, usually during the agricultural land re-planning and construction of a comprehensive new town. Because HSR has led to the flow of people and not to the flow of goods, HSR service is mainly targeted at high-mobility communities, including investors, regional commuters and entrepreneurs. Each new town plan incorporates the demand of local government for HSR stations to bring investment opportunities and economic growth. However, does the development of the new town follow the original intention of planning to achieve the expectations of local governments? What factors have affected the development of HSR? This book analyzes the development impact elements of China's first HSR line—the Beijing-Shanghai HSR line—and strives to provide similar references for the development of new towns.

Studying the planning and development of the area surrounding the HSR station started with the first strategic planning consulting project conducted by the author after her return to China: a medium-sized city in northern Anhui Province. Due to the restriction of conventional railways and expressways to the east and west, the city shows a pattern of north–south expansion: The central part is the old town, the south side planned to be an industrial park, and the north side is a public investment park, with the green space providing a value-added space for the development of residential real estate. Approximately in 2006, a station of Beijing-Shanghai HSR line sited in this city. The city government believes this station is a potential strategic growth point and invited planners to plan the areas surrounding the HSR station and their interaction with the development of the downtown areas. When the author's research team started this planning and study, the site location had been identified and was under construction; it is located in the eastern part of the city, approximately 20 km from the city center. For a medium-sized city, this distance creates a dilemma: On the one hand, the radiation capability of the city is limited and cannot bring about the development of the new town at such a distance; on the other hand, it may be difficult for the resources and human flow brought by the HSR station to have a positive interaction with the downtown because of the distance. The reason for this site selection by the Ministry of Railways is allegedly to ensure the smooth and speedy linearity of the HSR. In this case, the default conditions are insufficient due to the convergence of government departments, and the construction of an expressway that can connect the downtown area and the HSR station has encountered the problem of placing a ramp, which will be located 6 km more than originally from the HSR station, thus further reducing the accessibility of the HSR station. In this case, subsequent planning is limited to improvement, as the transport accessibility optimization and space allocation can be carried out only according to the city's existing resources and industrial characteristics. The reality of the city's development has inspired the author a lot. Authority barriers and lack of communication may result in heavy sunk costs that are difficult to recover; urban development conditions and resources are changing, resulting in the planning which is hard to ensure long-term development benefits. The HSR station construction has developed rapidly, but the site selection, site surrounding area planning, the development of the HSR new town and other issues need further study to better achieve the integration of regional transport facilities and urban development. With such a real-life research issue, combined with study and research experience in the USA, the author applied for a grant from the National Natural Science Foundation of China for a youth fund project and started to conduct urbanization studies around the HSR stations.

The basic idea of this research is to build a quantitative model based on the case of 22 cities along the Beijing-Shanghai HSR line and analyze the core elements that affect the development around the HSR line. Given the limited data, the research takes the expansion of the built-up area as the dependent variable of the development model of HSR new town. The independent variable is selected according to the theory of traffic-land coupling and TOD theory, etc. Using cluster analysis and regression models, the authors discuss factors such as site selection, scale, land use and the relationship with other transport facilities. The final analysis of the site location (the relationship of site and the existing urban built-up area) and the city's financial capacity have significant impact on the development of HSR. Half of the stations along the Beijing-Shanghai HSR line are outside the current built-up areas, limiting the interaction of surrounding area of stations and the station city; therefore, it is difficult for the regional transport facilities to lead the city development, and it is difficult to rely on the existing urban facilities for resource integration and development within the areas surrounding the regional transport facilities. The city's financial ability as a significant influence factor shows that the current development of HSR stations surrounding area is leading by the local government, economy-developed cities develop fast, while private investment focus on residential real estate projects. Only extra investment or obvious economic development can improve the current situation and develop the surrounding areas of station in cities with weak economy.

The study did not end with the original quantitative model, case studies for planning and development started accordingly. According to cluster analysis, it is found that there are great differences in the scale of sites for development along the Beijing-Shanghai HSR line. The comparative analysis of the subgroups shows that planning has a guiding role, especially in cases where the planned urban development goal coincides with that of the HSR sites, and the development shows a positive interaction. In the planning of the area surrounding the station, the initial goal is to build an integrated transport hub, and most cities have determined this location to be a new economic growth area. In the comprehensive development of the new district or new town, planning covers various functions such as business, office and leisure facilities. It is similar in positioning and functional configuration. According to the field survey, in a better-developed HSR new town area, most of land use is residential and commercial (including commercial service facilities and commercial facilities) supplemented by a certain amount of public management and public service land (including land for administrative offices, cultural facilities, science education and sports); green land and public facility land are mainly based on the original natural conditions or functional needs. The new plan of the HSR new town carries out the planning and layout of the urban characteristics of the station based on the HSR characteristics.

To further analyze the planning and development of HSR new town, a comparative case study was subsequently implementation. The study selected two cities with similar site locations and urban financial capacities but different development situation: Wuxi and Changzhou. Based on the theory of spatial political economy—"growth machine theory"—the framework of exchange value and use value for comparative analysis, this study analyzes the reasons for the change of planning content, describes the main decision-making stakeholders and its relations, processes and key decisions and discusses the important factors leading to different outcomes. The comparative study reveals that the planning and development of the HSR new town are rooted in planning and decision-making processes that are complicated, and the determination of the mechanism determines the effects of development and implementation. Finally, using the planning research on the development opportunities carried by Wuhu HSR station optimization to its surrounding area as a case study, this book exemplifies the space-by-layer analysis ideas and methods for the planning of HSR surrounding areas.

Based on the research and technical methods, the book is divided into six chapters. Chapter 1 is an outline of the domestic and foreign HSR development and analyze the characteristics of HSR. Based on current studies, explore the influence of HSR lines and stations on the development of the surrounding areas and propose the theoretical framework and empirical research on planning and development of HSR new town. In Chap. 2, 22 stations along the Beijing-Shanghai HSR line as cases to discuss the relationship between the station location and development priorities of city, as well as the domestic and foreign-related cases. Chapter 3 introduces the HSR new town planning along the Beijing-Shanghai HSR line in detail and, through cluster analysis, defines the development types of the new towns along the Beijing-Shanghai HSR line. By a quantitative model with multiple factors, we analyzed the key development factors of HSR new towns. Chapter 4 selects two cities along the Beijing-Shanghai HSR line in similar development stage, urban scales and industrial structures, which are Wuxi and Changzhou, analyzes the reasons for different developed states and the main stakeholders, planning process and important decision-making points of them. Chapter 5 compares the planning and development of HSR in Wuxi and Changzhou to propose the principles and key points for HSR new town planning. Chapter 6 proposes the analytical framework of the planning of HSR new town and its surrounding areas and continues the demonstration with cases.

This book is based on the studies of China's HSR new town planning and development led by Prof. Lan Wang. The specific contributors in Chinese version are as follows: Chap. 1, Lan Wang; Chap. 2, Lan Wang, Chang Hu, Hao Gu, and Sheng Cao; Chap. 3, Lan Wang, Can Wang, Chen Chen, and Hao Gu; Chap. 4, Lan Wang and Hao Gu; Chap. 5, Lan Wang and Hao Gu; and Chap. 6, Lan Wang. Among the contributors, Can Wang, Chen Chen and Sheng Cao were Ph.D. candidates at the College of Architecture and Urban Planning (CAUP) at Tongji University. Hao Gu and Chang Hu were graduate students in CAUP at Tongji University. Shuwen Liao, Su Li, Wenyao Sun, Fangfang Jiang and Sile Hu, who helped to create the charts and diagrams in this book, were also graduate students in CAUP at Tongji University. For the English version, Owl Xu and Luyun Shao helped translate the draft version, and Esther Morán Flores, Xiji Jiang, Yinglu Luo, Yirui Du, and Weiyang Yang helped to proofreading the text and figures, who are graduate students in CAUP at Tongji University. Sihan Chen, a current master student at University Hong Kong, also helped with proofreading and coordination. Some sections are based on papers published in the *Journal of Urban Planning* (Issue 4, 2014), *Planners* (Issue 7, 2011) and *Chinese Science Papers* (Issue 4, 2015). The case study in Chap. 2 is based on the report prepared for the project—"Urban Planning of HSR New Town in Tongling City, Anhui Province." This project was led by Prof. Zilai Tang of CAUP at Tongji University, who also provided guidance on the writing of the case study. Wanrong Wang, Zhi Sun, Xiaoyong Wan, Huiquan Cao and Yujie Tao of the Wuhu City Urban and Rural Construction Strategy and Planning Research Center provided data support for the analysis and study in Chap. 6.

Quantitative and qualitative research combination is one of the characteristics of this study. The research uses mixed-method analysis, uses quantitative analysis as the main method to establish the model, modified the quantitative model through qualitative analysis and provides in-depth logical explanations. The formation of an interactive process between quantitative analysis and qualitative analysis will increase the sophistication of the quantitative research and the representativeness of the qualitative research. The quantitative model is applied to the 22 station cities along the Beijing-Shanghai HSR line as a sample. The analysis results provide a basis for further qualitative selection of case studies. Qualitative research is conducted through in-depth interviews and planning documents analysis to support and deepen the quantitative analysis of the results. Such a progressive research design provides a rigorous research way for planning and developing research.

At the same time, this study attempts to integrate planning theory and urban theory by exploring the working objects of urban planners under a specific planning method. Planning theory needs to consider under what conditions human activity can create a better city for all citizens. In answering this question, we must focus on the interaction between the planning process and the outcome; at the same time, it is necessary to explore the characteristics of what can be called a better city, what kind of strategy can achieve a better city, and what are the main obstacles. There is a need to understand the history of a specific city's development, the relationship between urban economic fundamentals and social structure, and the relational systems around which policy decisions are made. An empirical study of the planning and development of the HSR new town and its surrounding areas provides a mirror; it reflects how the planners influence the city-related resource allocation and spatial development within the existing institutional framework. Thus, the book begins with the HSR new town and attempts to explore the reality of planning and development in China through this particular phenomenon of urban development.

Current relations between different levels of government in China, departmental relations within the same level of government and the relationship between the market and the government define and constrain the design and development of space for planners and implementers in multiple dimensions. In the system, individuals rely on their own literacy, competence and authority to interact with other stakeholders to gain the exchange value and use value brought by space production to jointly promote urban development. This analysis draws on the core points of the growth machine theory and political theory in the process of urban development and redevelopment in the USA. To a certain extent, this approach can also explain the current urban planning and development in China. However, the relationship between the market and the government in China is still in a stalemate phase. The border between the two is not clear, and the relationship is also different from the pattern in which the USA elects its members from the bottom up to govern urban development in a constituency. This difference makes the relational pedigree based on the theory frame of the growth machine different, and there are also differences in the development level of the new town in different cities, which brings about differences in the planning and development process.

From the perspective of planning ontology, the comparative study of the HSR new town highlights the importance of the consistency and stability of planning. Development and implementation are obviously affected by administrative systems and decision-making processes. The consistency that is mainly reflected by the decision-making people at all levels is basically the same for the planning of core decisions, including site selection, the overall planning of urban development direction, the HSR Metro space, positioning judgment and functional assumptions. Stability shows that the decision-making process of planning and implementation remain relatively stable, including the implementation of the main amendments to the planning details. This stability depends on the constant growth of machines of urban space production. Coordination within the growth machinery can ensure that the development resources invested by decision-making bodies at all levels within a specific space are effectively coordinated; stability can be based on spatial planning, at a certain space level and within the scope of the progressive implementation of the implementation, to ensure that before and after the planning, investment and construction will not conflict with each other to avoid duplication of investment. The planner must understand the limitations of the decision-making people and the administrative framework and try to include stakeholders outside the growth machine in the process. This is a key and difficult point in the transformation of the planning paradigm for China. However, property law, civil society, urban renewal, etc. have formed the external conditions and pressures to promote changes in planning, and how to address these problems requires optimizing the understanding of the current planning decisions, the implementation of the logic and problems.

In retrospect, in terms of theory, as one of the urban development categories in China, the new HSR has become an important case for analyzing China's planning and development because of the large investment, plenty stakeholders and high expectations that are involved. In practice, new and upgraded HSR stations will change the surrounding areas. New development opportunities will require new planning to address. This is an important part of urban development and planning in the future. The book combines quantitative model construction and qualitative analysis of comparative cases, with the specific project of an empirical analysis of planning and development in China, and proposes planning principles and policy recommendations. As the construction of HSR changes from public investment to marketization, the planning and development of HSR stations and their surrounding areas must be optimized in the following aspects: site selection, ownership of land and establishment of institutions. More empirical research should be conducted to determine how to revise and improve the process; we hope this book could be one of significant studies to provide a basis for future research.

Shanghai, China

Lan Wang
Hao Gu

Contents

List of Figures

List of Tables

Chapter 1
High-Speed Rail (HSR) and Urban Development

In 2015, the total length of the world's HSRs reached 29,792 km. A total of 3603 HSR trains were put into operation, with a high speed of 320 km/h. The highest speed record for HSR, which is the highest speed data for the French HSR in 2007, is 574.8 km/h. Each year, the total number of HSR passengers is 1.6 billion, with 800 million in China, 355 million in Japan, 130 million in France and 315 million in other countries and regions. When the travel time is less than 2.5 h, 80% of travelers prefer to choose HSR rather than air travel.

HSR will increasingly become people's primary travel option for a certain distance and period, which makes it even more significant for the economic and social development of cities and towns.

1.1 Development Overview of HSR

As a large-capacity and fast regional transportation infrastructure, HSR can enhance the accessibility of an entire area and the economic attractiveness of individual cities and promote the development of the areas surrounding the stations. In human history, the railroad was invented in the UK in the early 19th century and is an important carrier for land transportation. The development of aviation and road traffic has impacted the railway. The former is more dominant in speed, and the latter has advantages in door-to-door convenience. The emergence of HSR has increased the attractiveness and competitiveness of railways and affected the development of the cities and towns associated with them.

1.1.1 The Worldwide Development of HSR

The earliest HSR appeared in Japan. The industrial relations between Japanese cities require the support of large-capacity transport facilities, and the belt-shaped urban

L. Wang and H. Gu, *Studies on China's High-Speed Rail New Town Planning and Development*, https://doi.org/10.1007/978-981-13-6916-2_1

agglomerations in Japan support the efficiency of rail transport. From the development of HSR to the actual construction, the Olympic Games were a major event to promote the realization of the infrastructure in the region. The Tokaido Shinkansen was approved after the successful bid for the Olympic Games in Tokyo in 1958 and began to construct in 1959. It was officially opened in 1964 before the opening of the Tokyo Olympic Games and became the world's first HSR. With a top speed of 210 km/h and a total length approximately 515 km, it greatly reduced the travel time between Tokyo, Osaka and other major cities. With the planning and construction of several routes, such as the Sanyo Shinkansen and Nagano Shinkansen, an HSR network covering most urbanized areas was formed. At the same time, HSR was well organized with other transport modes to achieve a seamless connection between various modes of transport and to support population mobility in Japan's high-density urban agglomerations. Japan's HSR reached 2176 km by the end of 2010.

HSRs have also enjoyed rapid growth in Europe. France, Germany and Italy successively established HSRs after the opening of the Shinkansen in Japan. The Netherlands, Belgium, Spain and other countries subsequently started construction. Currently, Europe has almost formed an HSR network to achieve cross-border railway lines and convenient multiline transfer sites. The first French HSR (Train à Grande Vitesse; TGV) opened in 1981, connecting Paris with Lyon and raising the highest speed to over 300 km/h. Subsequently, France built 282 km of the Atlantic line, 333 km of the Nordic lines and other HSR convergences. The first German HSR (intercity express, ICE) was served in 1987 from Hanover to Würzburg, and the highest speed reached 400 km/h, setting a new record for HSR. At present, the German HSR covers the major German cities and is easily connected to all medium- and low-speed railways, forming a railway system that reaches all of the German towns and cities as the preferred way of traveling in Germany. By the end of 2010, the total length of the French HSR was 1884 km, and that of the German HSR was 1443 km. Currently, they are 2023.6 and 2331 km, respectively. Europe plans to build approximately 20,000 km of HSR for a European HSR network that is a total of 35,000 km long (Beijing-Shanghai HSR Ltd. 2011).

The US statewide interstate highway plan, started by the federal government, is a plan to build a large-scale highway project linking the states. The massive popularity of automobiles, low-cost gasoline and suburbanization have prompted the development of road traffic in the USA, and rail transport (especially passenger transport) has been declining. Currently, travelers in the USA still use cars as their main mode of travel, including self-driving, car rental and bus (coach) travel, and fewer choose to take the train. Although the USA plans to promote the construction of HSR, budget plans, land ownership problems and other issues make it difficult to implement the plan. For example, California plans to build an HSR link from San Francisco to Los Angeles but has difficulty selecting routes. The specific problems are as follows: The western location of the line, near the coastline, may have an ecological impact and be opposed by environmentalists. An inland eastern location of the line may encounter land ownership issues. As a result, the construction of the line has run aground. At present, the only HSR under construction in the USA provides a link between Washington and New York.

The construction of HSR in Japan, Europe and the USA has a different course in each country, but the pursuit of speed is the focus of railway development, and reducing travel time is one of the core competitive advantages of this transport mode. High speed has changed people's choice of their mode of travel, affecting the development opportunities and prospects of cities.

1.1.2 HSR Development in China

China's railway system developed slowly before the 1980s. After the reform and opening up, the demand for transportation increased rapidly, and both passenger and freight transport exceeded capacity, which brought pressure and impetus to railway development. At the same time, competition emerged from air transport and the development of an expressway transport system, especially the realization of the expressway from door to door. For example, the market share of passenger traffic for railways, civil aviation and highways in 1990 was 12.3%, 0.2% and 83.9%, respectively (Beijing-Shanghai HSR Ltd.). Against such a development background, the Ministry of Railways decided to use "increasing speed" as a strategic direction for reversing the decline of railway transportation. In December 1994, China's first quasi-HSR, the Guangzhou–Shenzhen line, was officially opened, with a total length of 147 km and a speed of 160 km/h, thus starting a comprehensive increase in the speed of China's railways. From 1997 to 2007, the speed has increased six times, focusing first on the three already busy main lines of the Beijing-Shanghai, Beijing-Guangzhou and Beijing-Harbin areas. The total distance is 5046 km, accounting for 9.5% of the total distance of the entire national railroad system, and passenger and freight traffic accounted for 39.4 and 34.4% of the total traffic volume, respectively. Changing the speed and transport capacity of these trunk lines reflected the strategic intention of developing the HSR at the national level. The Beijing-Shanghai, Beijing-Guangzhou and Beijing-Harbin trunk lines, from the past speeds of approximately 90 km/h, rose to approximately 200 km/h, and by 2007, an HSR that reached 350 km/h appeared on the traffic map of China.

In January 2004, the State Council considered and approved the first *Medium- and Long-term Railway Network Plan* in the history of China's railways and proposed the construction of a high-speed passenger dedicated railway line with a total distance of 13,000 km. In 2005, construction officially started on China's first HSR passenger dedicated line, the Beijing-Tianjin intercity HSR, and in 2008, the Beijing-Tianjin intercity HSR officially opened with a high speed of 350 km/h. Afterward, the construction of HSR in China was fully and rapidly promoted. The construction of the Beijing-Shanghai HSR line, the Wuhan-Guangzhou HSR line and the Shanghai-Nanjing intercity HSR line began one after another, and the construction of the HSR network was in full swing. In early 2013, Mr. Sheng, former minister of the Ministry of Railways, noted at the National Railway Work Conference that the total length of China's HSR operation would reach 9356 km by the end of 2012, and the total number of travelers boarding would reach more than 1.3 million a day, with the total distance and passenger traffic ranked first in the world. According to the *three-year railway construction plan* after the 12th Five-Year Plan, in 2013, the investment in fixed assets in the national railway amounted to 650 billion yuan, of which 520 billion yuan was for capital construction, and more than 5200 km of new line was put into operation.

The construction and development of HSR will gradually affect the development pattern of regions and cities in China. According to the latest *National New Urbanization Plan (2014–2020)*, China will perfect its comprehensive transport corridor and the backbone of the inter-regional transportation network. Strengthening the traffic links among urban agglomerations, accelerating the integrated planning and construction of urban agglomerations and improving the external traffic of small- and medium-sized cities and small towns will provide support and guidance for an integrated transport network with an urbanization pattern. By 2020, the ordinary railway network will cover all the cities which have a population of over 200,000 people, and the rapid rail network will basically cover these cities which have a population of over 500,000 people. The synergy between HSR construction and urban development has become a key issue. As an important strategic deployment at the national level, the construction of the HSR network has many important characteristics such as large investment, wide scope and a long construction period. Cities directly or indirectly affected by the development of the HSR or along the HSR network are facing different development opportunities and challenges. How cities combine their own development conditions and stages of development to make full use of the opportunities brought by the construction of HSRs to achieve healthy and rapid development is inevitably the issue of greatest concern among all stations and cities and the core link for realizing the strategic value of HSRs.

1.1.3 Definition and Characteristics of HSR

The speed of HSR was 200 km/h from the earliest time, reaching a top speed of 500 km/h. The speed is the main standard for the current definition of HSR. Japan's definition of HSR in 1970 was the railway with the highest operating speed of 200 km/h. The USA defines a railway that operates at speeds above 160 km/h as a HSR. According to the requirements of the International Union of Railways (UIC) and the European Union (EU), the speed of newly built HSRs should be over 250 km/h, and that of retrofitted HSRs should be over 200 km/h (Dai et al. 2011). China complies with international standards and defines a HSR as a railway with a top speed of approximately 250–350 km/h.

HSR is characterized by the transport of people as the main object of service; it does not transport goods. According to specific travel times, HSR passengers can be divided into three categories: (1) short-range passengers traveling for 1 h or less, (2) midrange passengers traveling for 2–3 h, and (3) long-haul passengers exceeding 3 h (Greengauge 21 2006). The first two types of passengers are commuters on the same day. Commercial traders and commuters use high-speed trains mainly because HSR is more convenient than other modes of transport, and their travel is time-sensitive. Most of the third type of passengers is those who stay for more than one day at the place of origin or at the end of the trip instead of traveling on the same day, mainly tourists and business travelers. HSR travel needs are mainly related to a large number of business needs, to commuters and to the short-distance travel needs of major cities. Regional resource needs vary slightly over time; for example, short-range passengers within a one-hour distance do not need accommodation, and long-haul passengers over three hours do need it.

The primary effect of building an HSR line is that the travel time between cities will increase from the lower level to the upper level, guiding the flow of capital and resources between cities from major regional cities to HSR-accessible cities and thus creating a rapid traffic corridor. There are two key elements of the HSR corridor area: the overall size of the regional passenger market and the quality of HSR services. The demand for travel in the market depends on the economic strength of the city. The stronger the economy is, the higher the demand. However, the low frequency of stops in small cities along the route will affect the output of resources from major cities to small cities. This finding is a contradiction. The quality of service offered by the HSR line also determines the travel needs of the market; for example, the French TGV system has become a brand for its speed, frequency, operational energy and toll structure.

According to the current basic operating speed of HSR, which is 250–300 km/h, the suitable operating distance for HSR is 500–600 km. That is, the most suitable time is about two hours. This is the most suitable travel time for business. It is convenient for business people to have a day-return trip, e.g., from Tokyo to Osaka (in Japan) or Paris to Lyon (in France). Therefore, the impact of HSR on urban development is reflected in the accessibility of the service economy. Simply providing HSR services may not necessarily enhance the economic viability of a city. Its service efficiency depends on its economic activities and development mode. HSR supports the service economy, including business, public administration, recreation, commerce and tourism. The characteristics of HSR listed above determine the development of HSR stations in the city and surrounding areas.

Numerous studies have demonstrated that large-scale transport facilities often promote urban development (Knaap et al. 2001; Handy 2005; Hess and Almeida 2007; Cervero and Murakami 2009). The construction of large transport facilities involves the introduction of a large amount of public investment, which has a profound impact on the development of urban areas. Against this realistic development background, there is an urgent need to deepen the research on the settings of railway stations and their surrounding areas and to propose planning ideas and principles for site selection, development scale and functional allocation.

HSR is a regional large-capacity rapid transportation infrastructure with passenger flow at its core, shortening the space-time distance and changing the spatial relationship of regional urban networks, which changes the basic assumptions of urban development. HSR stations that serve as hubs for crowds of people may bring rapid accumulations of capital, talent and goods for the cities to be built, thereby changing the urbanization process in the surrounding areas (Hirota 2004; Sands 1993; Priemus

2006). In the current analysis of the development of HSR and urban areas, there are three main spatial scales: (1) the macrolevel is a regional area where reachability and integration benefits are emphasized; (2) the mesolevel is the urban area with HSR as a new economic growth point; and (3) the microlevel is the centralized analysis of the area around the site (Wang 2011). Domestic studies focus on the quantitative analysis of the impact on regional accessibility of railways, HSRs or urban rail transit lines and stations (Luo et al. 2004; Ma et al. 2008) and case studies on the changes of land use (He and Gu 1998; Liu and Zeng 2004; Li et al. 2007). Against this research background, research on the area surrounding the HSR station must be strengthened to provide a theoretical basis for the planning of similar areas.

1.2 Research on the Impact of HSR on Urban Development

According to the research domestically and abroad, HSR shortens the space-time distance between cities in a region and changes the spatial relationship of the city network at the regional level. In particular, it should link the areas with weaker economic growth to the economically developed areas more rapidly and promote regional integration and development. At the urban level, HSR has brought population inflows and outflows, including investors, employers, tourists and occupants who play important roles in urban development. Population migration has promoted urbanization. In the area around the station, high-speed trains promote the development of high density and mixed use, forming a new urban development node.

Research on the impact of transportation facilities on urban development domestically and abroad is conducted at three levels: the region, the city and the area surrounding the station. According to the different spatial scales, the research focus is also different: when focusing on the regional level, the reachability and regional integration brought by the rail transit will be analyzed; at the urban level, the impetus for urbanization brought by rail transit and new economic growth opportunities are emphasized; and the research on the area surrounding the site mainly includes theoretical and empirical studies of circle-style development and the theory of the transit-oriented development (TOD) of the new urbanism.

1.2.1 Regional Level: Accessibility and Urban Areas

At the regional level, the current study focuses on the impact of HSR lines and stations on regional accessibility (Luo et al. 2004; Ma et al. 2008; Greengauge 21 2006). Traffic facilities usually improve regional accessibility by reducing transportation costs in terms of currency and time, bringing about major changes in the modes of travel of residents along the route and resulting in the rapid increase of large-scale population movements. In these studies, cities are nodes in the regional network, and regional accessibility is the main content of the study. For example, three parallel indicators, weighted average travel time, economic potential and daily accessibility, are used to measure the improvement of accessibility between cities in the region (Luo et al. 2004).

With the improvement of accessibility, the basic assumptions of individual urban development have been changed as urban agglomerations participate in regional and global competition, forming a global city region that can be linked to the world economy at each urban scale. Such research focuses on how enterprise networks and industry chains extend from regional central cities to small- and medium-sized cities, increase employment opportunities, optimize industrial structures and open up markets (Chen 2007; Brito and Correia 2010). Rapid regional transport links cities that lack large-scale external transport facilities (deep-water ports and airports) with cities that own these facilities to connect with the global economy. Such studies focus on the analysis of industrial links among cities and the spatial requirements of enterprises and investors in cities in the region.

In research at the regional level, HSR enhances the reachability of the whole region, changes the industrial relations within the region and enhances the interaction between the city and the region. The impact on urban space is reflected in the spatial needs of external resources, including businesses, investors and tourists. There is also a slight difference in regional resource needs over time, such as a demand for accommodation space. The link between HSR and regional transport facilities also affects the spatial structure of cities. It mainly reflects the conditions of urban development and the links with a wider market and industrial chain. For example, the Beijing-Shanghai HSR line will likely promote the linkage between the two major urban agglomerations in Beijing, Tianjin and the Yangtze River Delta, resulting in major changes in the spatial scope and industrial structure. For the urban agglomerations in the Yangtze River Delta, the Beijing-Shanghai HSR line is the path and channel for industrial relocation to the outlying urban agglomerations that have Shanghai as their core. Cities along the line undertake the transfer of the focus of industrial development and layout. *The Regional Planning in the Yangtze River Delta Region* and *The Planning of the Demonstration Zone for Undertaking Industry Transfer in the City of Wanjiang* were issued in response to the regional integration and industrial spillover brought about by the HSR construction. The spatial development of the city where the site is located must absorb the regional industrial transfer.

1.2.2 Urban Level: Urbanization and Economic Development Node

The connection and interaction between transport facilities and urban development have recently been important topics in the urban planning field both in China and worldwide. Numerous studies have demonstrated that large-scale transport facilities are often catalysts for urban development (Hirota 2004; Knaap et al. 2001; Handy 2005; Priemus 2006; Hess and Almeida 2007; Cervero and Murakami 2009). For example, statistics on changes in the population of major Japanese jurisdictions between 1975 and 1988 show that cities with Shinkansen sites have significantly higher population growth than other cities (Sands 1993). This type of research focuses on how rail transit changes the land use, types of industries and real estate prices in the surrounding area (He and Gu 1998; Liu and Zeng 2004; Li et al. 2007). The travel demand for HSR includes business requirements, commuting requirements for the regional dominant city and neighboring towns and short-distance travel and leisure requirements. Serving the business and financial industry, leisure and entertainment industry and tourism industry are also requirements for HSR. The development of these industries is the main driving force of urbanization.

At present, research on the impact of the transportation system on urban space is mostly qualitative case studies (Shu and Shi 2008; Yang et al. 2009; Ai 2010), including the development and analysis of sites around Suzhou, the relationship between Beijing's spatial diffusion and traffic and the impact of HSR on the urban spatial development in Tianjin. Such studies consider the area around the rail transit station as a new economic development node in the city that may form the center of gravity of the current spatial structure. Whether HSR can form an economic growth point that affects the development of urban space is related to many factors. The service level of HSR, for example, the stop frequency of a line, is one of the important factors. The market demand for HSR depends on the economic strength of the city. The stronger the economy is and the higher the demand, the more stop frequency a line will have and the more resources it will bring. As a result, the frequency of stops in smaller cities on the route is usually low, affecting the resources and opportunities that high-speed trains bring to smaller cities. Existing studies have illustrated the limitations of HSR impacts through studies of small cities along TGV lines in France: Cities with low or no service-based stops often struggle to secure positive urban development from HSR lines and station setups (Greengauge 21 2006). Therefore, for a single city, the city must set more stops to improve the service quality of the HSR line and integrate the service industry and the station function. The relationship between the HSR station sites and the built-up areas affects whether the HSR station sites become economic nodes in the development of urban space. The indicators describing the relationship include the distance from the urban center and whether it is within the built-up area. HSR and urban transport system integration are equally important. HSR should be built with a variety of transport hubs so that all parts of the city can more quickly reach the site, increasing the service needs of HSR and thereby enhancing the service level of HSR and promoting urban economic development.

1.2.3 Surrounding Sites: Spatial Function Layout and Development

In the theoretical model of the development of the area surrounding the traffic facilities, the "three-layer" structure model (Schutz 1998; Pol 2002) has a greater impact.

In the three-circle structure, the first circle is the core area, approximately 5–10 min from the station, focusing on high-level high-density business office functions; the second circle is the affected area, approximately 10–15 min from the station, mainly developing commercial office and supporting functions with a slightly lower building density and height than the core area; and the third circle is the peripheral area of influence, where the specific functions of development may include housing, research and development and industry (Zheng and Du 2007). There is a slight difference between the theoretical model and the actual situation of land development around the site in the empirical study.

An empirical study of the impact of transport facilities on the area around the site (1/4–3 mile, or 400–4800 m) focused on the impact of transportation systems on land use (Cervero and Landis 1997; Polzin 1999) and real estate prices (Ryan 1999; Knaap et al. 2001; Hess and Almeida 2007; Immergluck 2009) in the area around the site. The research mainly focuses on quantitative analysis and tests the changes in land use and real estate prices in the surrounding areas due to the site settings. Studies have shown that different factors change with different distances from the site. For example, in the immediate vicinity of the site, real estate prices decline; with a rapid increase in distance, rapidly reaching the peak price, and then rapidly decreasing again. Thus, there are a few differences from the theoretical model of the three circles. These studies provide a variety of empirical models, including multi-agent models, pricing models and hedonic regression models. The explanatory variables involved in such research mainly include the size and quality of the houses, the distance to transportation facilities, the distance to the downtown area and public service facilities, etc. At the same time, foreign countries should gradually conduct research to explore the social impact of transport facilities on the surrounding communities and analyze changes in social life, demographic structure and ethnic characteristics.

The planning and design theories of the area surrounding the site are mainly "transit-oriented development" (TOD) in new urbanism. The TOD planning paradigm includes relatively high-density development, with mixed use and pedestrian-friendly public spaces in the site area, and provides principles for site planning and environmental design in the site area. The design of the HSR site area should be in line with the principles of bus-oriented development.

1.3 Research Ideas and Methods

At the regional level, HSR will shorten the space-time distance between cities in the region and change the spatial relationship of regional city networks. In particular, HSR links the regions with weaker economic growth to the economically developed regions more quickly and promotes regional integration and development. At the urban level, high-speed trains bring in large numbers of people, including investors, employers, tourists and residents, who play important roles in urban development. Such population movements will promote the process of urbanization. In the area around the station, HSR stations promote the development of high density and mixed use in the surrounding areas, forming a new development node in the city. The research on the impact of domestic and international transport facilities on urban development is conducted in these three aspects: the region, the city and the surrounding area.

Through the combination of related theories on these three levels, this book defines the important spatial elements and constructs an analytical framework of the impact of HSR on urban space at the theoretical level. Taking the Beijing-Shanghai HSR line as an example, it conducts an empirical analysis under the framework and discusses the principles of spatial planning around the site.

1.3.1 The Overall Research Ideas

HSR is characterized by the transport of people, rapid circulation and an elevated setting in a large area, differentiating its impact on urban space from the impact of traditional railways and urban rail transit. Based on the above theories, we can analyze the impacts of HSR development on urban space from three aspects: the region, the city and the area surrounding the station (Fig. 1.1). Each level of analysis focuses on different content, divided into spatial relations and spatial needs, and includes both empirical and paradigmatic types of research.

The construction of an HSR system will change the reachability of cities in the region and enhance the interdependence between cities, forming a cluster of urban areas in large, medium-sized and small cities. The density of cities in the region will further increase, attracting various mobile resources through HSRs in a networked form. Analysis of the impact of HSR urban space should first identify the regional level of the flow of resources into and out of the trend. The setting of HSR stations will shift the center of gravity of the city space and tilt it toward the HSR site area. This shift will enable the HSR to reach the external economy, attract external resources and talents and open up external markets. The analysis of the three spatial scales from the area to the area surrounding the site constitutes the research framework for the impact of HSR on urban space.

In regional-level analysis, cities are nodes in the region. HSR has changed the attractiveness of cities to outside investors, tourists and talent through the improvement of regional accessibility; therefore, the spatial analysis of urban space at the regional level should explore the space requirements of these resources in view of the availability of resources within one hour, two hours or more than three hours in the region. The connection between HSR sites and other transport facilities (airports, ports, etc.) in the larger area has a certain impact on the development of urban space; there is a need to analyze how to establish quick links between multiple transport nodes through spatial planning and to configure functions and specific industries based on this analysis. The spatial relationship analysis includes both the regional enterprise network and its industrial chain to make links between HSR cities.

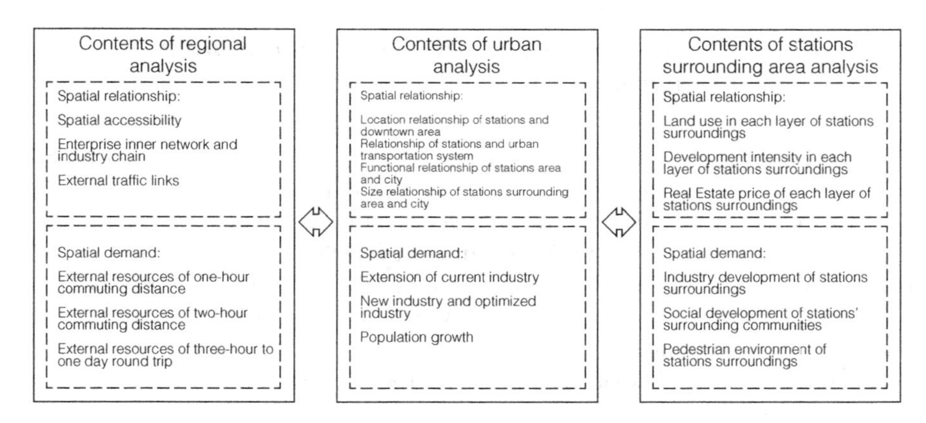

Fig. 1.1 Analysis framework of the impact of HSR on urban space

In the spatial analysis at the urban level, the process of urbanization triggered by HSR stations is important research content. The spatial relationship analysis includes the relationship between the HSR line and the built-up area (including the distance between the site and the city center and whether the site is within the built-up area), linkages between HSR and other transport facilities (provision of other transport services such as highways, intercity railways, ports and bus lines). Based on distance and traffic integration analysis, the function and scale of the facilities around the HSR stations and the whole city should be explored. The main body of space requirements includes the status quo industries in the cities that may be expanded by HSRs, the industries that may be optimized and increased by HSRs and the population brought in by HSRs.

In the analysis of the area surrounding the site, the spatial relations formed by elements such as land use, development intensity and real estate price can be discussed in a circle. Industrial development around the site is one of the main needs of the space. At the same time, the analysis of the social space impact of transport facilities deployment is on the rise in the American research field, with studies mainly analyzing the social development of the surrounding communities (including family income, race, community interaction, etc.). TOD around HSR station sites must consider the space needs of pedestrians.

The three-level analysis framework covers many aspects of the impact of HSR on urban spatial development. However, this book is not devoted to establishing a complete analysis but is based on the theory to determine the important aspects of analysis; it focuses on the analysis of the relationship between Beijing-Shanghai HSR line stations in the main urban areas and the relationship between the site size, positioning and function, mainly in terms of HSR station planning and development.

1.3.2 Research Framework of HSR New Town Planning and Development

This book explores the multifaceted impact of HSR sites on urbanization in the surrounding areas. The setting of transportation facilities includes the relationship between the site and the city, the site size, the current situation around the site and the functions of the planned land. Surrounding areas include the area within a certain range (formed by a specific radius) near the site and the city where the site is located. The research questions to be answered are: What is the impact of regional transport facilities such as HSR stations on the cities in which they are located? What is the impact on the process of urbanization in the surrounding area and what are the main influential elements?

Based on the related theory of regional traffic facilities, traffic-land linkage development and other related theories, this paper studies and analyzes the mechanism of how HSR stations influence the development of their cities and surrounding areas, establishes the analytical framework and conducts empirical research. The empirical research includes two parts: (1) a quantitative analysis of 22 cities along the Beijing-Shanghai HSR line and (2) a comparative analysis of two cities on the Beijing-Shanghai HSR line. The first part is the empirical part of the second empirical basis for the selection of cases.

For the first part of the empirical study, 22 cities along the Beijing-Shanghai HSR line were selected as a case: Beijing, Langfang, Tianjin, Cangzhou, Dezhou, Jinan, Taian, Qufu, Tengzhou, Zaozhuang, Xuzhou, Suzhou, Bengbu, Chuzhou, Nanjing, Zhenjiang, Danyang, Changzhou, Wuxi, Suzhou, Kunshan and Shanghai. There are 24 stations in these 22 cities, with Tianjin and Chuzhou each having two HSR stations. In Tianjin, including the South Station and the West Railway Station, in the built-up area, the original sites were expanded for the multiline shared site, and the sites were built earlier, so the surrounding development and construction is more mature and therefore is not suitable as an HSR case analysis. In Chuzhou, Dingyuan Station is located in the village of Qinggang, Chihe Town, Dingyuan County, approximately 50 km from the city of Chuzhou. To establish a quantitative model to analyze the impact of sites on urban development, this study chooses a site in each city for further analysis and forms 22 analytical models corresponding to 22 cities. Therefore, Tianjin South Railway Station and Chuzhou South Railway Station and Tianjin West Railway Station and Chuzhou Dingyuan Railway Station are not included in this quantitative model construction. After selecting the case studies, this book studies the impact of HSR deployment on urban development indicators and urbanization around the site and then establishes a multiple regression model to explain the impact of site setting on the surrounding areas.

Based on the quantitative model analysis of the core elements of the impact of the development of the HSR new town, the second part of the empirical research is based on a comparative analysis of selected cities along the Beijing-Shanghai HSR line, Wuxi and Changzhou, as a further in-depth analysis of a group of cases. Through a deep interview and planning meta-analysis, this study conducts a comparative analysis of the planning process, development process and decision-making relationship between the Wuxi and Changzhou HSR new towns to clarify the key principles for the planning and development of the HSR new town.

The specific research contents are as follows:

1. Theoretical Study on the Mechanism of Urbanization by Regional Transport Facilities

The study will explore the mechanism of the impact of regional transport facilities on the development of the surrounding areas, including location theory, traffic-land coupling theory, the transit-oriented development (TOD) model and so on, and then explicitly explains the explanatory variables in the model. The theoretical model of urbanization in the area surrounding the station is proposed, and theoretical hypotheses are formed to provide a basis for the subsequent series of empirical studies.

2. Analysis of the Impact of HSR Stations on the Station Cities

The impact analysis of HSR stations in cities mainly focuses on the shift of the urban space center of gravity and the key indicators of urban development.

HSR has an impact on the city from the announcement of the project establishment and planning for its construction and use. It is both an event and a process for the city. The study explored the changes before and after the two major points in time after the route was confirmed and formally announced and determined the role of the HSR site setting in the town development.

3. Study on the Impact of HSR Site Settings on Urbanization in the Surrounding Areas

The study intends to measure urbanization through economic development and population employment in the area surrounding the station. However, due to the statistical caliber failing to support the data in the specific surrounding area, the size of the built-up area is taken as the index to measure the process of urbanization. Based on satellite images, a measured analysis of the expansion of the built-up area at different time points is conducted using AutoCAD and GIS software. In this book, the site is taken as the center of the circle, and the data of the built-up area within a certain radius are measured. Clustering analysis is conducted to determine the type of expansion in the area surrounding 22 stations. At the same time, a comparison analysis of the HSR stations' surrounding area planning, the status quo and development planning similarities and differences is conducted.

4. Empirical study on the Mechanism of Urbanization of HSR Sites

The study initially considered that the distance between the station and the original town, the current situation of the surrounding communities, the current status of the HSR station, the predicted passenger volume and the urban public investment ability theoretically determine the impact of transportation facilities on urbanization in the surrounding areas (Hess and Almeida 2007; Immergluck 2009) as independent variables to influence the model. Based on the data of 22 HSR stations in cities and surrounding areas, a multilevel regression model was developed to explain the impact of site setting on the surrounding areas. The initial idea of this impact model is as follows:

$$U = f(D, V, G, P)$$

where the dependent variable U is the expansion area of the built-up area as a symbol of urbanization, D is the distance between the site and the original town center, V is the site status/forecast passenger volume, G indicates the integration of HSRs and other transport facilities in the station city and P expresses the government fiscal capacity to guide the situation. G is an independent variable group with multiple variables. G, which indicates the integration of highways and other transport facilities in the stations, will include the provision of other transport services such as expressways, intercity railways, ports and bus lines. Each indicator will be expressed by means of per capita change or individual dummy variable. At the same time, the scale, population, GDP and other factors of the city will serve as covariates. The model aims to clearly define the key elements of HSR development.

5. A Case Study of Planning and Development

On the basis of clarifying the key influence factors of the development of HSR, the study selects two cases according to a specific comparative analysis method and conducts in-depth analysis. Through in-depth interviews, government officials, real estate investors and entrepreneurs study and compare the similarities and differences between the two cases to further clarify the impact of the site on urbanization in the surrounding areas and improve the model. The study combines the planning and development process, explores the interactive relationship among decision-makers, deepens the understanding of the planning and development of the HSR new town and proposes relevant ideas and principles.

The book follows the "theoretical–empirical" interactive research ideas. Based on the research questions, through theoretical analysis, a theoretical model of the impact of regional transport facilities on urbanization in the surrounding areas is established. Based on the theoretical assumptions, the empirical research is mainly based on quantitative analysis. Combined with qualitative analysis, a multiple regression model is established to correct and perfect the theoretical model. On the basis of clarifying the core influencing factors of the development around the HSR, a comparative study on selecting specific cases, planning and development is conducted, and finally, planning suggestions are proposed (Fig. 1.2).

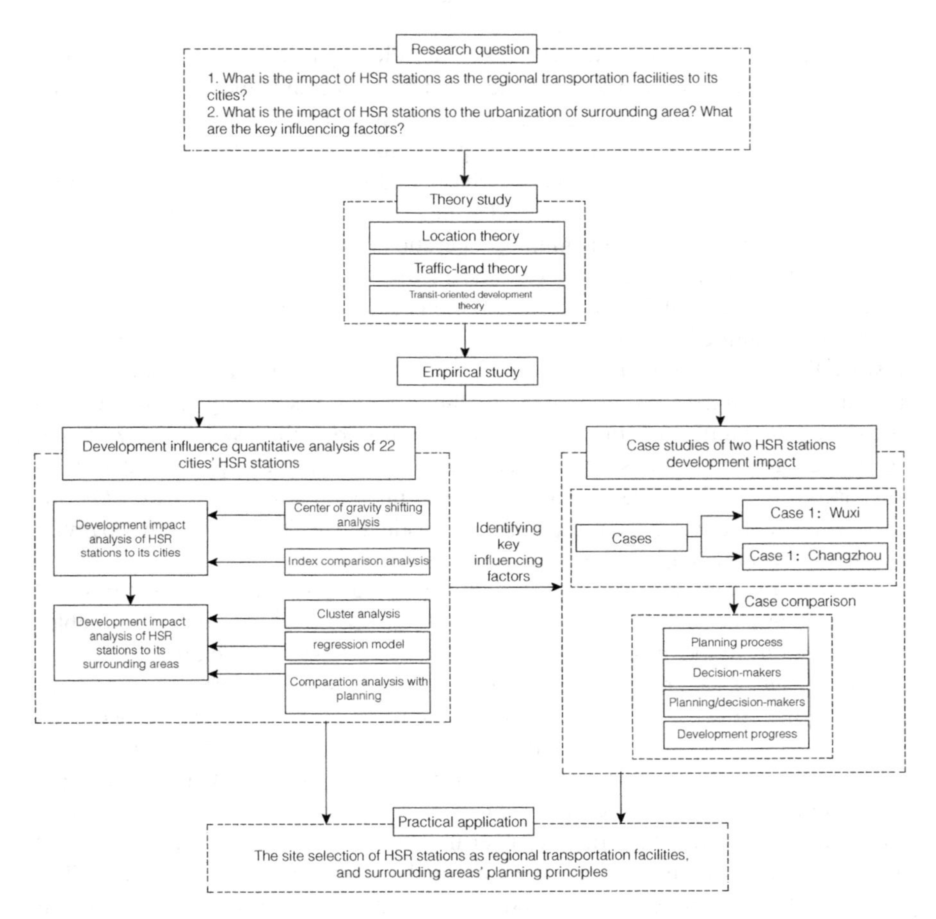

Fig. 1.2 Project research framework

1.3.3 Research Methods

This study is based on quantitative analysis methods combined with qualitative analysis methods.

Quantitative research methods mainly use cluster analysis and quantitative model analysis based on the theoretical model of the impact of HSR stations on urbanization in the surrounding areas. Cluster analysis aims to clarify the different types of HSR around the development and its characteristics. The quantitative model uses structural equation modeling and multiple stepwise linear regression models to analyze the relationship between different variables and the expansion of built-up areas around the site to clearly identify the significant elements that influence the expansion.

Qualitative data collection methods are based primarily on in-depth interviews conducted in individual cities. This method involves one-on-one conversations with important people to obtain answers to a list of questions, gain an insightful perspective on the interviewer's questions about the research issues and study the process and details of the case. This study selected government officials of the cities along the route, high-level decision-makers in the planning and construction of HSRs and real estate investors and entrepreneurs who intend to invest in areas near the HSR sites. The interviews mainly focus on the impact and impact factors of HSR from planning and construction to beginning operation. Qualitative data will aid in the logical interpretation of quantitative data. In this study, quantitative data will be combined with the establishment of a common support model for qualitative data.

The key scientific issue of the study is how to accurately reflect the complex mechanism of the impact of regional transport facilities on urbanization in the surrounding areas through empirical models. To solve the problem of selecting empirical models to accurately reflect the impact of regional transport facilities on urbanization in the surrounding areas, the study will conduct a preliminary computational simulation based on empirical data to determine the significant influencing factors. On this basis, using qualitative analysis methods, we conducted in-depth interviews to revise and supplement the empirical models.

The research data include various statistical data, government documents and planning materials of cities and towns along the line (Table 1.1). Among the variables of the model, the population employment status and industrial structure are obtained from the *China City Statistical Yearbook* and the statistical yearbooks of various provinces and cities. Data for the sites and the original town center distance, land use and urban infrastructure are obtained from the *Statistical Yearbook of China's Urban Construction*, Yearbook of China's Urban Construction in Cities, status quo and planning text of overall urban planning in various cities and satellite images. Site status/detailed forecasts of passenger traffic and site setting data are obtained from the railway administrations in China and various provinces and cities. The provision of transport services, including highways, intercity railways, ports and bus lines near HSR stations, are mainly provided by the transport bureau of each city while integrating data from urban master plans and urban transport planning. Public investment data, focusing on foreign direct investment, investment in fixed assets and fiscal revenue and expenditure, are obtained from *China Urban Statistical Yearbook* and *China Urban Construction Yearbook*. Relevant plans include all plans and designs related to the city master plan and the area surrounding the site.

Based on the above research ideas and methods, the book takes 22 cities of the Beijing-Shanghai HSR line as an example to discuss the relationship between HSR stations and their cities and analyzes the relationship between the station location and the development direction of the cities. Using cluster analysis to identify the type of development of the new towns of the Beijing-Shanghai HSR line, it constructs a quantitative model with multiple influential elements and analyzes the key elements influencing the development of the HSR new town. On this basis, the author chooses two cities with similar stages of development from the cities along the Beijing-Shanghai HSR line, namely Wuxi and Changzhou, with similar urban size and industrial structure, and analyzes why the HSR development of the two similar cities presents different states. The analysis traces and analyzes the main body

Table 1.1 Variables and data sources

Variable	Specific indicators	Recourse
U	Built-up areas' extended areas	Satellite images
Covariate	Population	***China City Statistical Yearbook***, statistical yearbook of all provinces and cities
	Industrial structure	***China City Statistical Yearbook***, statistical yearbook of all provinces and cities
	GDP	***China City Statistical Yearbook***, statistical yearbook of all provinces and cities
D	Stations and original town center distance	Satellite images, overall planning text
V	Stations current situation/predicted passenger capacity	China Railway Bureau and bureaus of all provinces and cities
G	Express railway and integration of station cities' other transportation facilities	Communications Bureau of all cities, urban overall planning and urban transport planning data
P	Direct investment by foreign merchants (number of signed projects of the year, money spent during the year)	***China City Statistical Yearbook***, statistical yearbook of all provinces and cities
	Investment in fixed assets	***China City Statistical Yearbook***, statistical yearbook of all provinces and cities, ***Yearbook of Urban Construction in China***
	Fiscal revenue, financial expenditure	***China City Statistical Yearbook***, statistical yearbook of all provinces and cities
Planning concerned	Urban overall planning	Urban Planning Bureau, City Design Study Institute
	Specific control planning of stations, urban design, etc.	Urban Planning Bureau, City Design Study Institute, HSR new town management committee

structure, planning process and important decision-making points of the two cities and compares them. Through case studies, the study finally explores the ideas and methods of planning for the new HSR and its surrounding areas. This book will study and analyze the process and results, follow the data and maps and explore the role of HSR in urbanization.

References

Ai BT (2010) Study on the impacts of high-speed railway hub on Tianjin urban spatial dvelopment. City (4):12–14
Beijing-Shanghai HSR Ltd. (2011) On the Beijing-Shanghai HSR line. China Railway Press, Beijing
Brito CH, Correia R (2010) Regions as networks: towards a conceptual framework of territorial dynamics. Fep Working Papers
Cervero R, Landis J (1997) Twenty years of the bay area rapid transit system: land use and development impacts. Transp Res Part A Policy Pract 31(4):309–333
Cervero R, Murakami J (2009) Rail and property development in Hong Kong: experiences and extensions. Urban Stud 46(10):2019–2043
Chen X (2007) A tale of two regions in China rapid economic development and slow industrial upgrading in the Pearl River and the Yangtze River Deltas. Int J Comp Sociol 48(2–3):167–201
Dai S, Cheng Y, Sheng ZQ (2011) Urban transport planning in HSR era. China Architecture & Building Press
Greengauge 21 (2006) High speed trains and the development and regeneration of cities
Handy S (2005) Smart growth and the transportation-land use connection: what does the research tell us?. Int Reg Sci Rev 28(2):146–167
He N, Gu BN (1998) Analysis of the role of urban rail transit in land use. Urban Mass Transit 1(4):32–36
Hess DB, Almeida TM (2007) Impact of proximity to light rail rapid transit on station-area property values in Buffalo, New York. Urban Stud 44(5):1041–1068
Hirota R (2004) Air-rail links in Japan: present situation and future trends. Jpn Railway Transp Rev (39)
Immergluck D (2009) Large redevelopment initiatives, housing values and gentrification: the case of the Atlanta Beltline. Urban Stud 46:1723
Knaap GJ, Ding CR, Hopkins LD (2011) Do plans matter?: The effects of light rail plans on land values in station areas. J Plann Educ Res 21(1):32–39
Li LL, Zhang GH, Cao YL (2007) Technical method for planning and adjusting land use around rail transit stations: a case study of Suzhou. UTC J 5(1):30–36
Liu JL, Zeng XG (2004) Urban rail transit and land use integrated planning based on quantitative analysis. J China Railway Soc 26(3):13–19
Luo PF, Xu YL, Zhang NN (2004) Study on the impacts of regional accessibility of high speed rail: a case study of Nanjing to Shanghai region. Econ Geogr 24(3):407–411
Ma AK, Cao RL, Zhang PG et al (2008) Impacts of the expansion of the railway network on reginal accessibility: a case study of the urban agglomeration along with the Jiaoji Railway. J Shandong Normal Univ (Natural Sci) 23(2):89–93
Pol PMJ (2002) A renaissance of stations, railways and cities: economic effects, development strategies and organisational issues of European high-speed-train stations. J Virol 72(6):5046–5055
Polzin SE (1999) Transportation/Land-use relationship: public transit's impact on land use. J Urban Plann Dev 125(4):135–151
Priemus H (2006) HST-railway stations as dynamic nodes in urban networks
Ryan S (1999) Property values and transportation facilities: finding the transportation-land use connection. J Plann Lit 13:412–427
Sands B (1993) The development effects of high-speed rail stations and implications for California. Built Environ. University of California Transportation Center Working Papers 19(3/4):257–284
Schütz E (1998) Stadtentwicklung durch Hochgeschwindigkeitsverkehr (Urban development by High-Speed Traffic), Konzeptionelle und Methodische Abs tzezum Umgangmit den Raumwirkungen des schienengebunden PersonenHochgeschwind igkeitsverkehr(HGV)als Beitrag zur L sung von Problemen der Stadtentwicklung. Informationen zur Raumentwicklungs (6):369–383
Shu HQ, Shi XF (2008) The effect Tokyo Metropolis circle's railway system to urban spatial structure development. Urban Plann Int 23(3):105–109

Wang L (2011) Research framework and empirical study of the impacts of HSR on urban space. Planners 27(7):13–19

Yang SH, Ma L, Chen S (2009) Interaction of spatial structure evolution and urban transportation. UTC J 7(5):45–48

Zheng DG, Du BD (2007) Looking for the balance between transport value of node and functional value of city: discussing theory and practice in the development of airport area and high speed rail station area. Urban Plann Int 22(01):72–76

Chapter 2
Site Selection and Urban Development of HSR

The previous chapter introduced the development of HSR in China and worldwide. From a theoretical perspective, the impact of HSR on three different spatial scales and levels of development—the region, city and area surrounding the station—is discussed to explain the research ideas, and methods, and to provide the overall framework for the book's development. This chapter focuses on the relationship between station site selection and the urban development of HSR by analyzing Beijing-Shanghai HSR line as the core case. Station site selection is a complex technical demonstration and political consultation process involving all parties' interests and various considerations. The final location of the station site is very important for the development of the towns.

2.1 HSR Stations Site Selection

As an important transport infrastructure for cities and regions, HSR stations are of great significance to urban development. In theory, both paradigmatic and empirical studies point out that important transport infrastructure is a catalyst for urban development (Handy 2005; Hess and Almeida 2007; Cervero and Murakami 2009). Its setting has an important influence on the spatial structure, development direction and functional layout of the city where it is located. Transport facilities promote the flow of capital, population and commodities in the surrounding areas by reducing transport costs in terms of currency and time. It also increases accessibility which thereby changes the land use types of industries and real estate prices in the surrounding areas. In addition, the facilities of the site change the spatial layout of the city as a whole by providing a new driving force in urban development. In January 2006, the State Council executive meeting adopted the *Long-term Railway Network Planning*, which promotes rapid rail network construction. As a result, HSR station site selection in the studied cities will affect the development of urban areas in the long term. As China's HSR technology matures, in-depth study and analysis on the

L. Wang and H. Gu, *Studies on China's High-Speed Rail New Town Planning and Development*, https://doi.org/10.1007/978-981-13-6916-2_2

construction of HSR and site planning will be needed in other countries and regions worldwide. This section focuses on the impact of HSR station site selection on urban development and notes the key elements that should be considered.

At present, domestic studies mainly focus on the influence of railway lines, such as HSR or urban rail transit lines and stations on regional accessibility (Luo et al. 2004; Ma et al. 2008) and land use changes (He and Gu 1998; Liu and Zeng 2004; Li et al. 2007). These studies explore a single dimension (e.g., accessibility) of the impact of transport facilities on urban areas. Simultaneously, an empirical study on the change of urban spatial structure based on a geographic information system (GIS) was conducted in China (Li et al. 2007). In addition, a theoretical study on the dynamics of spatial structure evolution (Zhang 2001; Wei and Yan 2006) and a case study on the impact of a transportation system on urban space (Shu and Shi 2008; Yang et al. 2009) were conducted. However, most of these studies are qualitative analyses such as the development of sites around the city of Suzhou, the spatial diffusion and transportation in Beijing and the development of a metropolitan rail transit station in Tokyo. A small number of studies supplemented by local quantitative analysis (e.g., accessibility or a change in the spatial structure) failed to establish an empirical model on the impact of transport facilities on overall urban spatial structure.

Foreign studies on the impact of transport facilities on urban development focus on the impact of transport systems on land use (Cervero and Landis 1997; Polzin 1999) and real estate prices (Ryan 1999; Knaap et al. 2001; Hess and Almeida 2007; Immergluck 2009) and are mainly based on quantitative analysis. They test the intensity of changes in land use and real estate prices in the surrounding areas due to changes in site settings, quantifying changes as distances vary from site to site. These studies provide a variety of empirical models, including multi-agent models, pricing models and hedonic regression models. The explanatory variables involved in such research mainly include the area and quality of houses, the distance to transportation facilities, the distance to the downtown area, the facilities nearby, the characteristics of the communities in which the facilities are located, and the characteristics of the population. These studies mainly analyze the impact of transportation sites on residential land and housing, but do not address changes in urban space. At the same time, research is concentrated in the range of ¼–3 miles (400–4800 m) around the site, with very few studies on the impact of site setting on overall urban development.

Against this practical and theoretical background. It is necessary to select typical cases, conduct empirical studies and analyze China's HSR station site selection and the relationship between cities. In 2006, the Ministry of Railways approved the plan to adopt the wheel-track technology for the Beijing-Shanghai HSR line, the first HSR in China with a world-class advanced level. Therefore, this chapter selects 22 stations on this line and the surrounding area as a case study. By means of satellite image interpretation, cross-section comparisons were made between the 22 stations of the Beijing-Shanghai HSR line in 2006 and 2012. The changes of the built-up areas between the two dates were analyzed with particular attention paid to the relationship between the transformation of the city center and the HSR stations. Based on the reconnaissance of the station cities and the surrounding areas, this chapter analyzes the relationship between the HSR stations and city development. The chapter notes the challenges of the development around the HSR stations, providing a basis for the site selection of HSR stations in the future to promote sustainable development of HSR and urban areas.

2.1.1 Beijing-Shanghai HSR Site Location

The total length of the Beijing-Shanghai HSR line is approximately 1318 km, which is roughly parallel to the existing trend of the Beijing-Shanghai railway. The entire line is a new elevated double track with a design speed of 350 km/h. The Beijing-Shanghai HSR runs through three municipalities—Beijing, Tianjin and Shanghai—and four provinces—Hebei, Shandong, Anhui and Jiangsu—passing through 22 cities with a total of 24 stations. Five locations are terminals for arrivals and departures, including three municipalities and two provincial capitals—Jinan and Nanjing. There are seven hub sites: the terminals for arrivals and departure plus Xuzhou and Bengbu. In addition, another 15 sites, including Tianjin South station and Chuzhou Dingyuan station, are stop sites. Since the completion of the Beijing-Shanghai HSR line, it takes just 5 hours to travel from Beijing to Shanghai. A functional division between the Beijing-Shanghai HSR line and the existing Beijing-Shanghai railway has been achieved with HSR for passengers and the traditional line for freight. HSR is generally an elevated line while the traditional line blocks circulation on the ground. These characteristics determine that the HSR line's relationship with the stations and the cities is different than that of the traditional railway line. Based on the information collected from Google maps and government Web sites, the author analyzes the relationship between the station sites of the Beijing-Shanghai HSR line and the main urban areas with the help of GIS.

The connection between the HSR station site and the existing main urban area mainly involves the distance to the city center and whether the site is within the scope of the Built up area. In the studied sites, located in 22 cities, there is a great distance between the station site and the city center (in this study, we use the city government's location as the city center) (Fig. 2.1), and there is no correlation with the city scale. The Langfang HSR station (3.3 km) is the closest to the city center. The Suzhou HSR station is approximately 24 km away from the city center. The Chuzhou Dingyuan HSR station, located in Dingyuan County, approximately 50 km away from the city center, is not included in this study. According to the distance analysis, there are four stations 0–4.9 km away from the city center: Langfang, Zaozhuang, Bengbu and Kunshan. At a distance of 10–14.9 km, there are two stations: Tianjin South Railway Station and Chuzhou South Station. There are four stations 15–20 km away from the city center: Dezhou, Wuxi, Suzhou and Shanghai. The stations in the remaining 12 cities are at a distance of 5–9.9 km. Among the 24 stations in the 22 cities, 12 stations are located in 11 built-up areas, while the other 12 stations are outside the built-up areas.

The connection between the station and the urban traffic system affects the location relationship of the site and the main city area. HSR stations in Beijing, Shanghai,

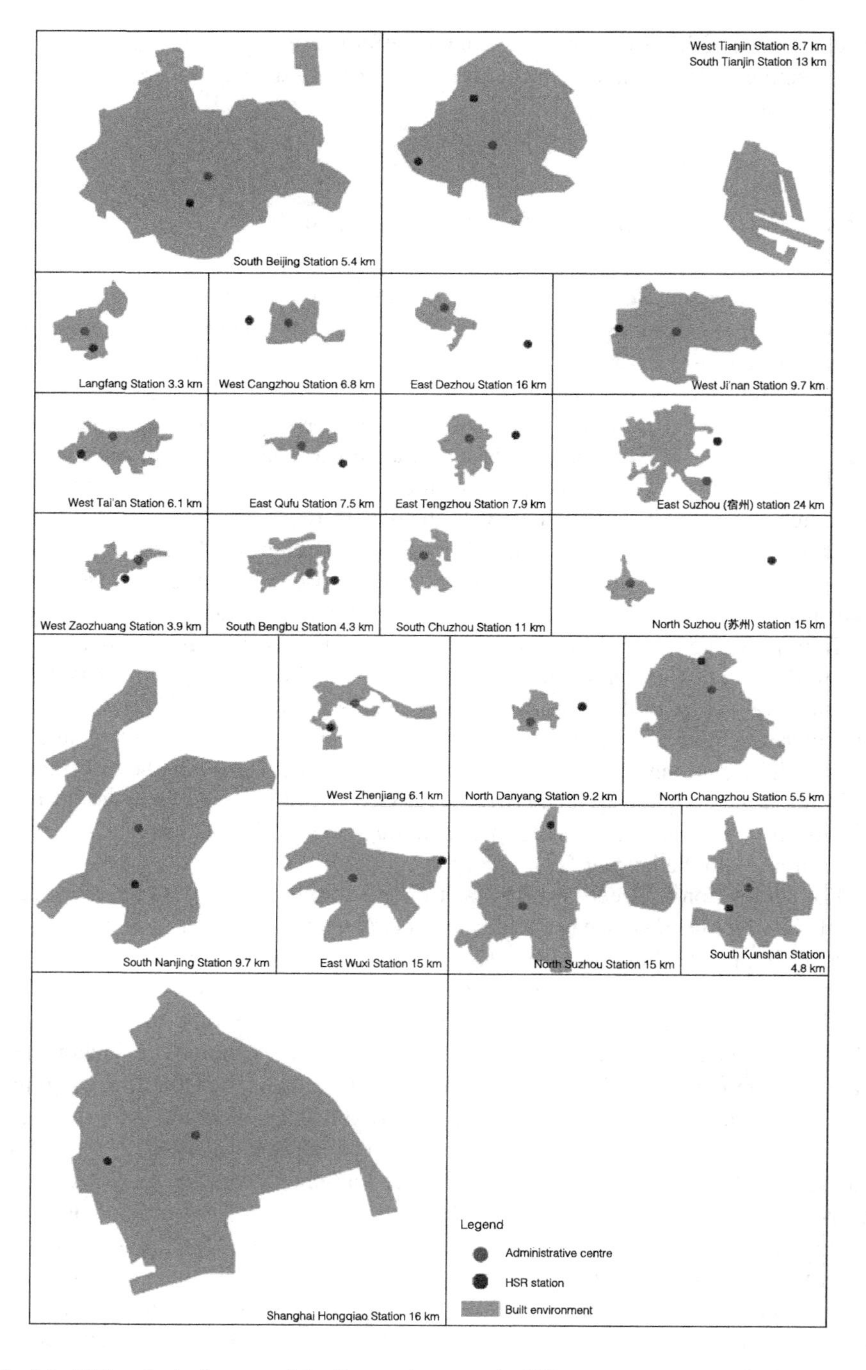

Fig. 2.1 GIS analysis diagram of the distance between the HSR stations and the main city area

Tianjin and Nanjing are well connected with other modes of urban transportation, especially with urban rail transit. For example, Shanghai Hongqiao Railway Station has an integrated transportation hub that contains the HSR station, the original Hongqiao Airport and the planned magnetic levitation stations. In addition, it also connects Metro Line 2 and Metro Line 10. Beyond that, three more metro lines are planned to connect to the HSR. Some stations do not have good connections with the city. For example, in Suzhou (Anhui Province), the Si-Xu highway is connected to the main city, but the exit is 6 km away from the HSR station which means a pedestrian must travel 6 km east from the main city area and then turn back toward the HSR station. It further increases the distance, which is already substantial. In addition, it reduces the driving role of the HSR in the city.

2.1.2 Comparison of the Beijing-Shanghai HSR Station Cities and the Surrounding Cities

There are 22 cities along the Beijing-Shanghai HSR line, including three municipalities—Beijing, Shanghai and Tianjin; two provincial capitals—Nanjing and Jinan; 13 prefecture-level cities—Suzhou, Wuxi, Changzhou, Zhenjiang, Chuzhou, Bengbu, Suzhou, Xuzhou, Zaozhuang, Tai'an, Dezhou, Cangzhou and Langfang; and four county-level cities—Kunshan, Danyang, Tengzhou and Qufu (Table 2.1, Fig. 2.2).

The Beijing-Shanghai railway line was completed in 1968, with a total length of 1462 km divided into three sections: the northern section is the Beijing-to-Tianjin

Table 2.1 Beijing-Shanghai HSR stations and City level

Station name	City level	Station name	City level
Shanghai Hongqiao	Municipality	Xuzhou East	Prefecture-level city
Kunshan South	County-level city	Zaozhuang West	Prefecture-level city
Suzhou North	Prefecture-level city	Tengzhou East	County-level city
Wuxi East	Prefecture-level city	Qufu East	County-level city
Changzhou North	Prefecture-level city	Tai'an West	Prefecture-level city
Danyang North	County-level city	Jinan West	Sub-provincial level city
Zhenjiang West	Prefecture-level city	Dezhou East	Prefecture-level city
Nanjing South	Deputy county-level city	Cangzhou West	Prefecture-level city
Chuzhou Dingyuan	County	Tianjin South	Municipality
Chuzhou South	Prefecture-level city	Tianjin West	Municipality
Bengbu South	Prefecture-level city	Langfang	Prefecture-level city
Suzhou East	Prefecture-level city	Beijing South	Municipality

Source Author

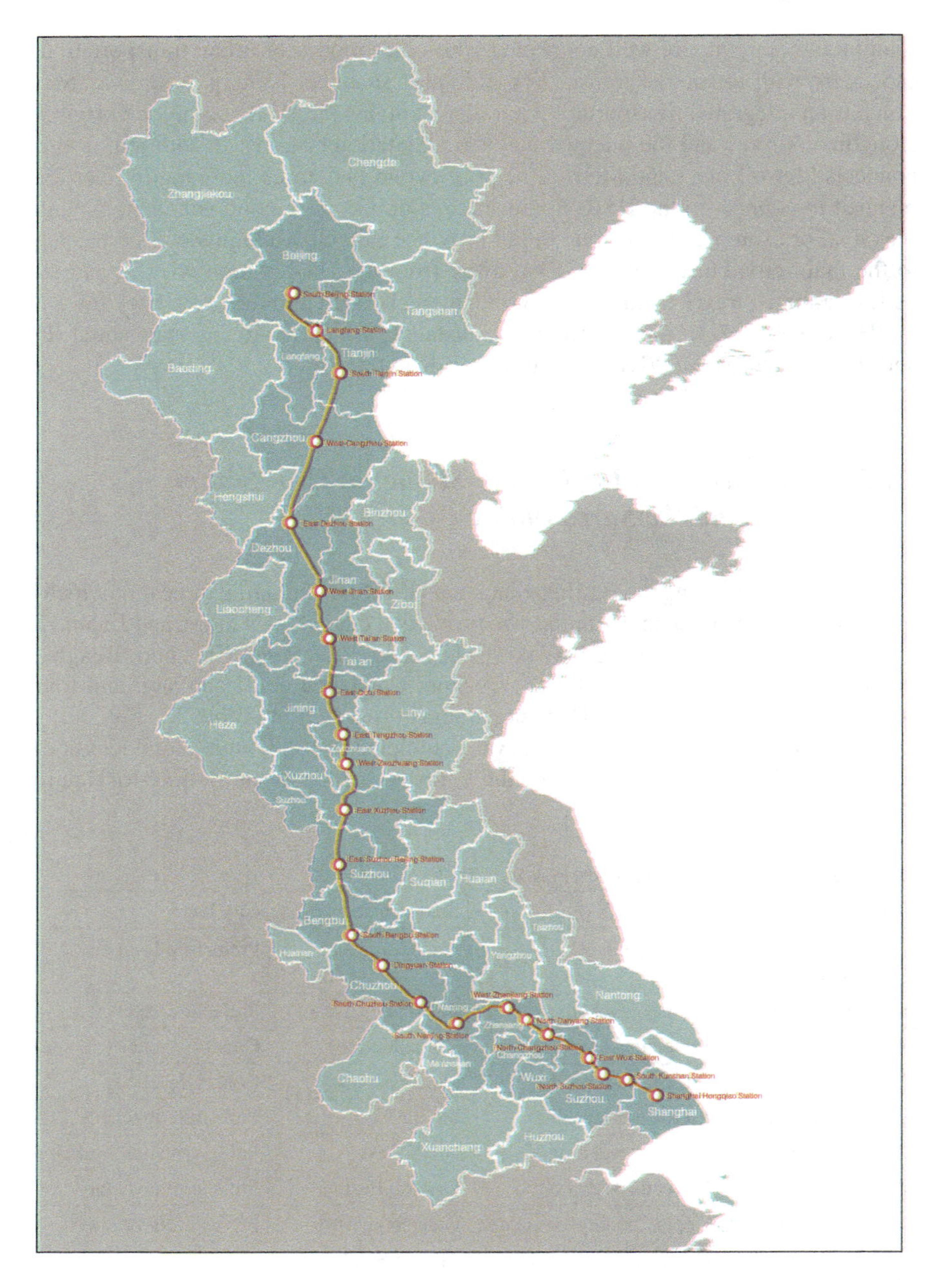

Fig. 2.2 Location map of station cities and surrounding cities along the Beijing-Shanghai HSR line

section of the Beijing-Shandong Railway line; the middle section is the Tianjing-Pukou railway line from Tianjin North Station to Nanjing Pukou Station; and the southern section is the Shanghai-Nanjing Railway line from Nanjing Pukou Station to Shanghai Station. There are 191 stations along the route, including prefecture-level cities such as Beijing, Beijing South, Langfang North, Tianjing West, Cangzhou, Dezhou, Yucheng, Jinan, Tai'an, Jining, Zoucheng, Tengzhou, Zaozhuang, Xuzhou, Suzhou, Bengbu, Chuzhou, Nanjing, Zhenjiang, Changzhou, Wuxi, Suzhou and Shanghai. Compared with the Beijing-Shanghai HSR line, the original line had more stop stations and more of those stations are close to the city center. However, for the HSR line, most of the stations are far from the city center. On the map, the two lines are basically matched, but the HSR line passes through the outer area of the city in most cases.

To analyze the influence of the HSR station on the city, we conducted a comparative study on the GDP, GDP per capita and tertiary industry in the cities along the Beijing-Shanghai HSR line and the surrounding cities. The surrounding cities include Zhangjiakou, Chengde, Tangshan, Baoding, Hengshui, Binzhou, Liaocheng, Zibo, Heze, Linyi, Suqian, Huaian, Huainan, Chaohu, Yangzhou, Taizhou, Ma'anshan, Xuancheng, Huzhou and Nantong.

Regarding total GDP in the cities along the Beijing-Shanghai HSR line (Figs. 2.3 and 2.4) from a city perspective, Beijing and Shanghai, the two terminals of the railway line accounted for an absolute predominance of GDP before the opening of the HSR line. From a regional perspective, the regional economic advantages of the Yangtze River Delta and the Beijing-Tianjin-Hebei region were obvious. In addition, the economic radiation areas of the surrounding cities to the surrounding areas were relatively small. The economic level of the peripheral cities had large gaps between the cities in which the stations were located. The economies of the station cities and surrounding cities were effectively improved with the opening of the HSR line. The rise and development of two hubs, Jinan and Nanjing, broke the polarization of Beijing and Shanghai. At the same time, the urban agglomerations in Shandong formed a new pattern of regional economic growth on the regional routes. From the perspective of incremental GDP, the stimulating effect of HSR on the surrounding cities is greater than that in the cities where the station is established, such as Zhangjiakou, Chengde, Heze, Suqian, Huaian and Taizhou, showing a broader radiation role.

In terms of per capita GDP of the cities with stations on the Beijing-Shanghai HSR line and the surrounding cities (Figs. 2.5 and 2.6), the per capita GDP and the total GDP has shown a spatial consistency by demonstrating a slightly higher level in the eastern part of the line than in the western part. The HSR line has played a significant role in driving the development of central cities along the Beijing-Shanghai line and has better balanced the spatial differences in urban economy between Beijing and Shanghai. Judging from the growth of per capita GDP, these cities also show three gradients. Some cities around the southern part have witnessed remarkable growth followed by cities in central China and finally Beijing, Shanghai and Suzhou-Wuxi-Changzhou. In addition, the growth rate of the southern section of the line is generally greater than the growth rate of the northern end of the line.

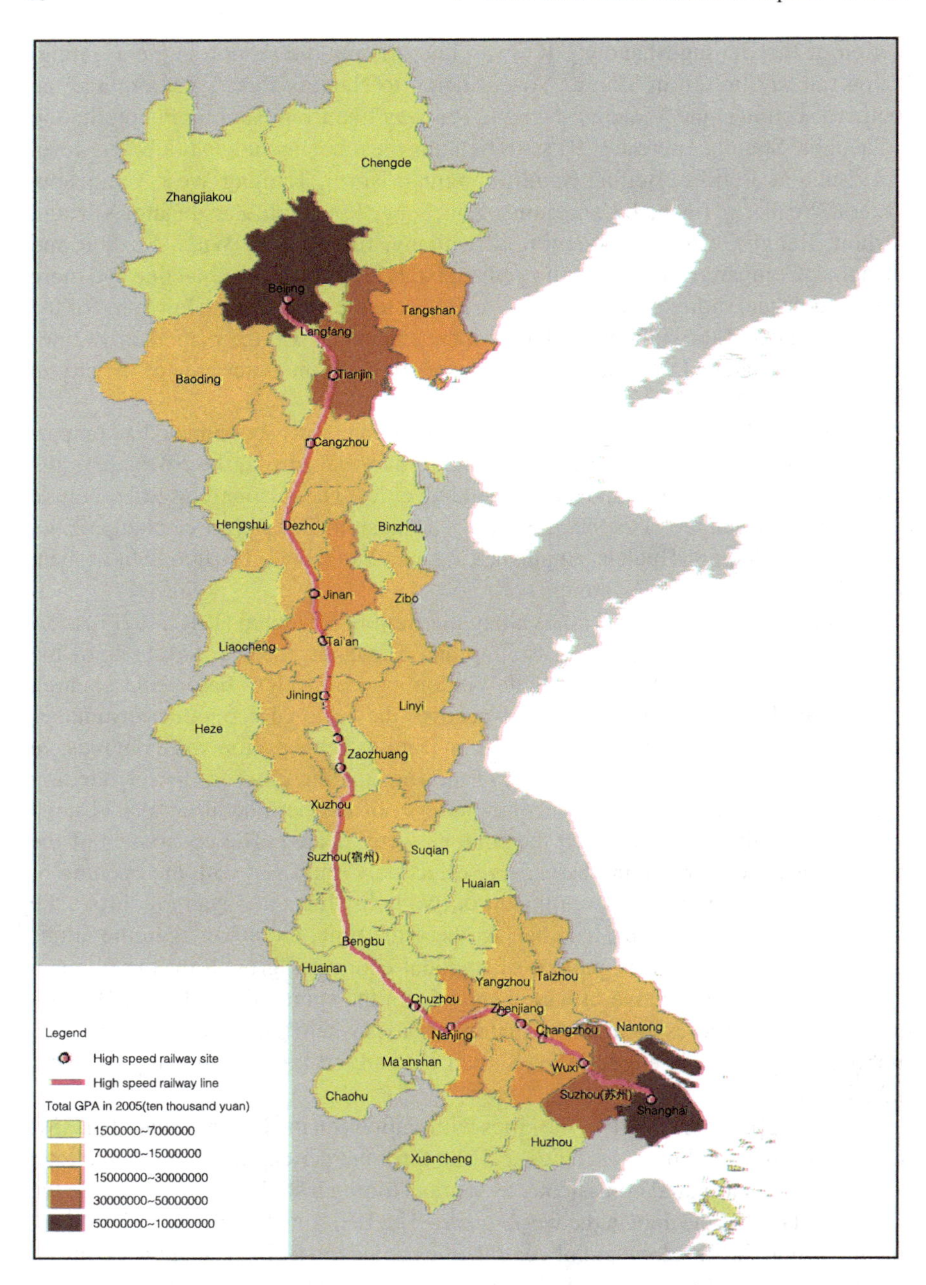

Fig. 2.3 Total GDP of Beijing-Shanghai HSR cities and surrounding cities (2005, 2010)

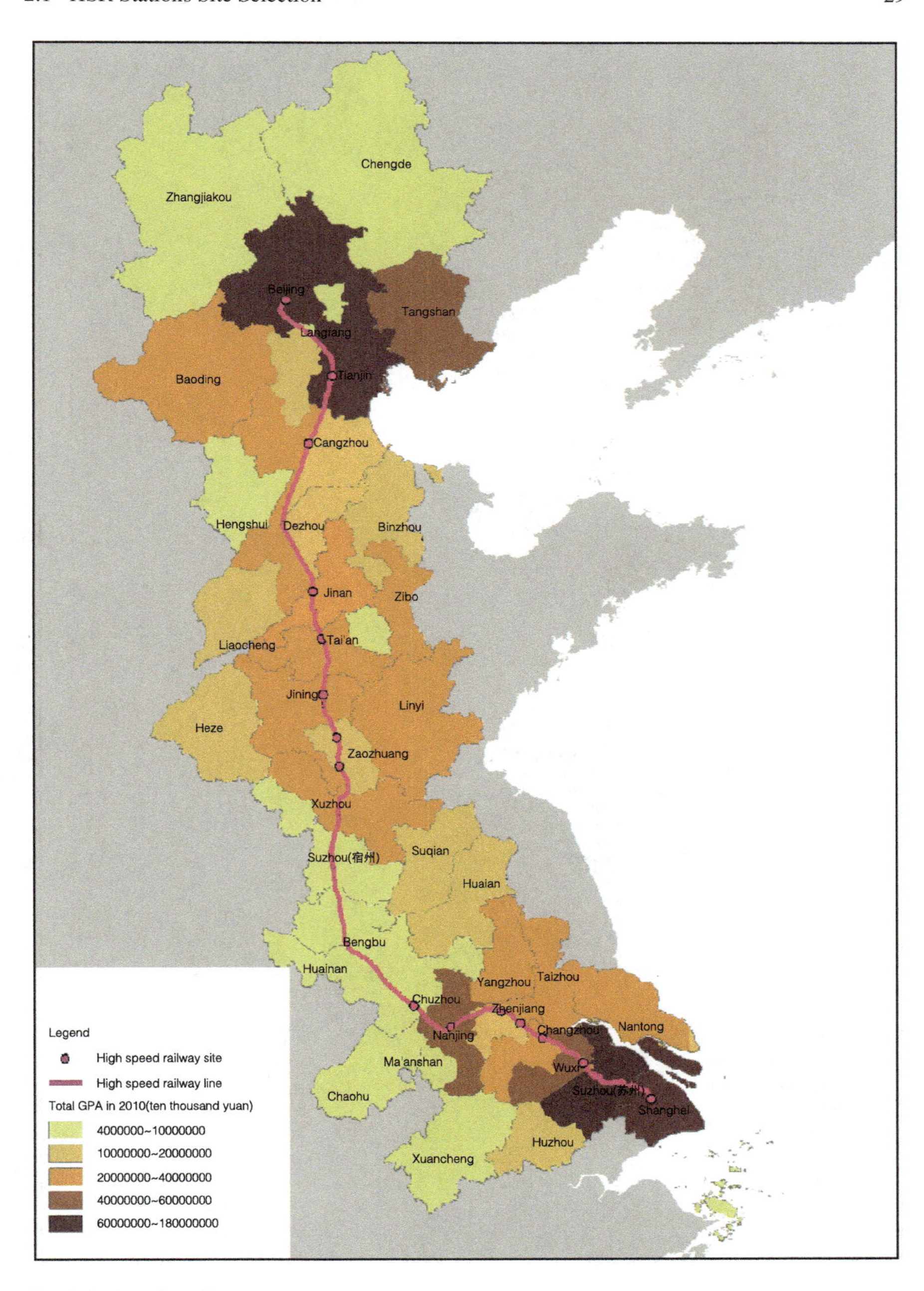

Fig. 2.3 (continued)

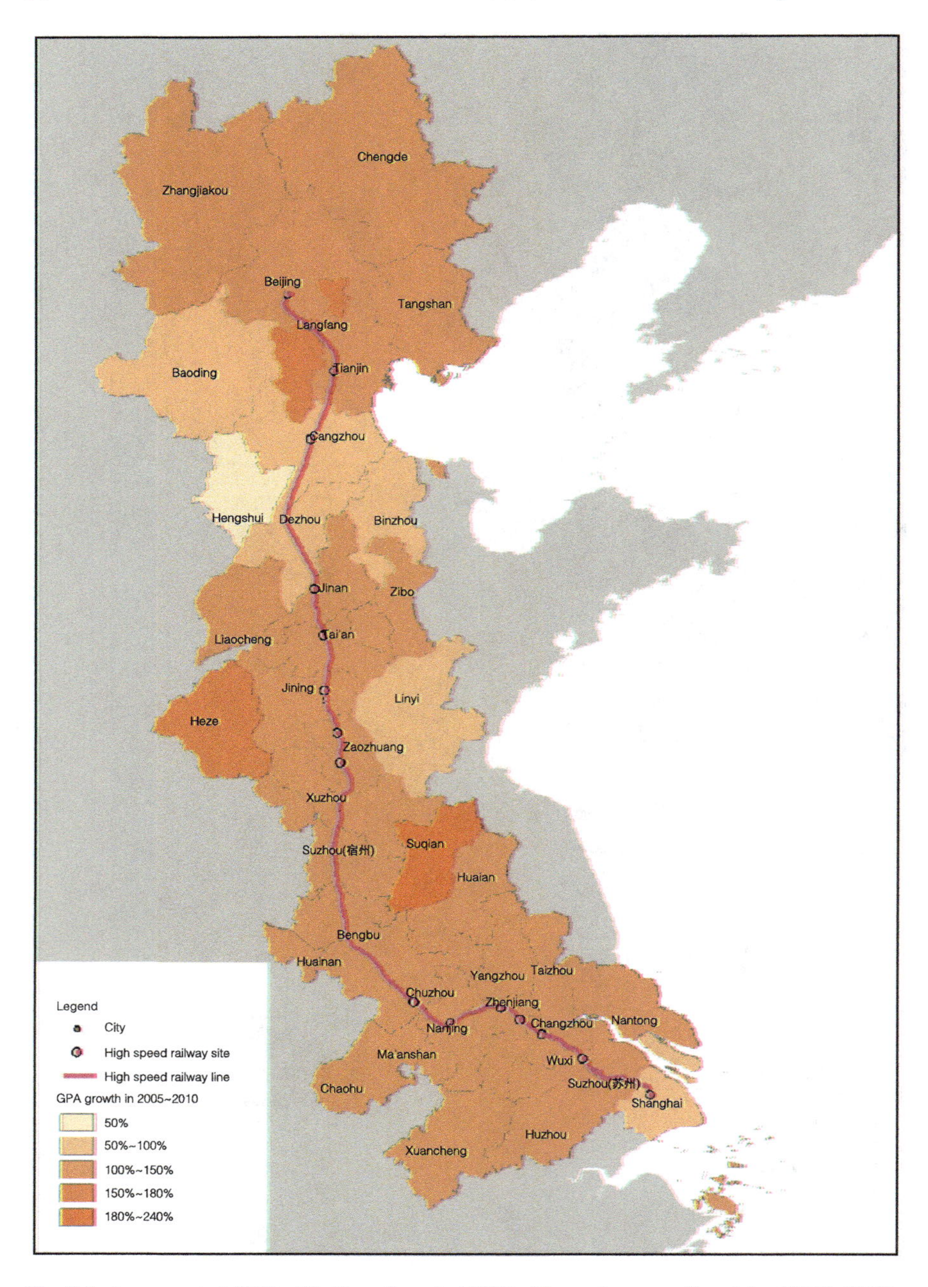

Fig. 2.4 Increment of GDP of Beijing-Shanghai HSR cities and surrounding cities (2005–2010)

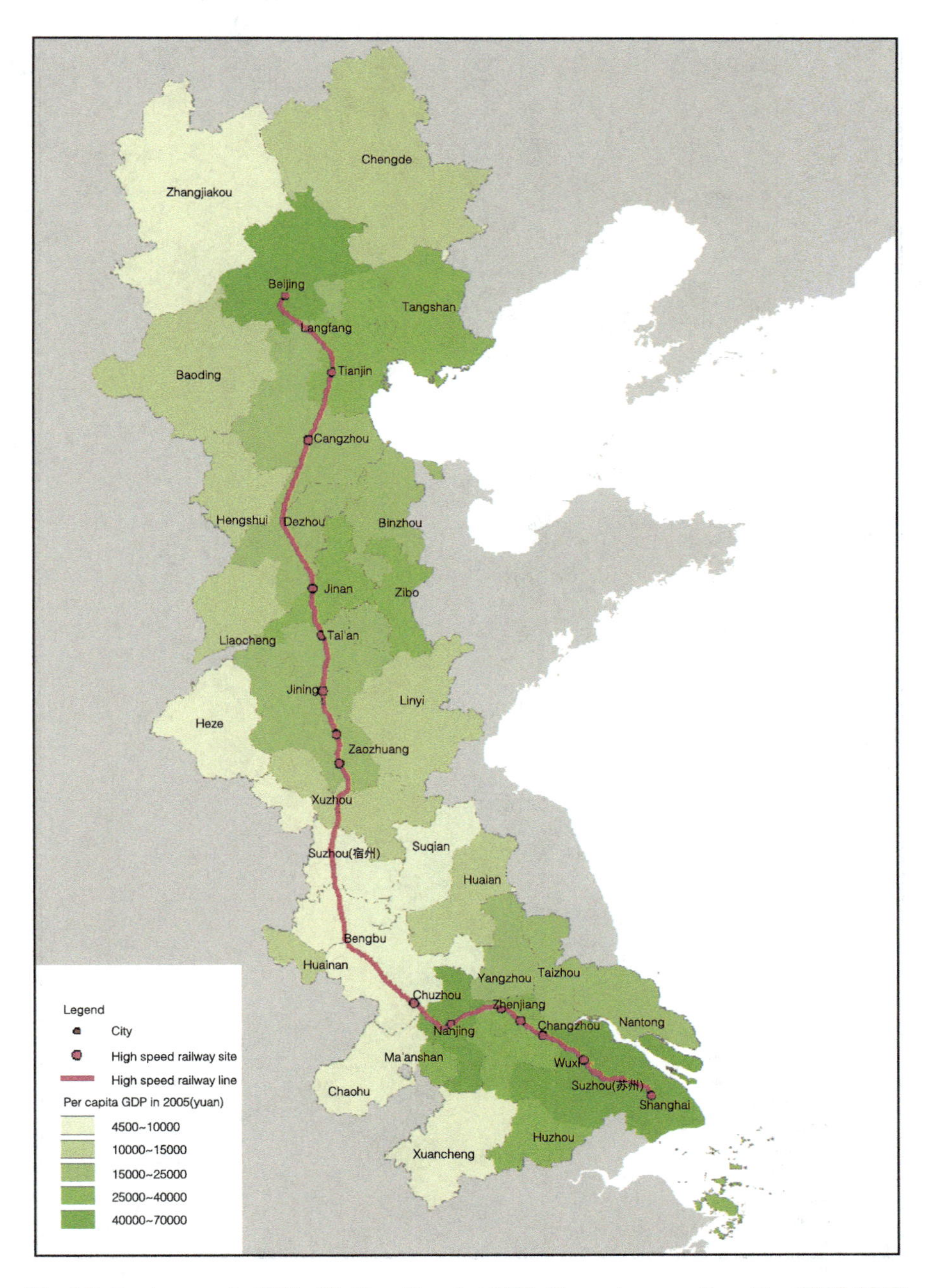

Fig. 2.5 Total per capita GDP of Beijing-Shanghai HSR cities and surrounding cities (2005, 2010)

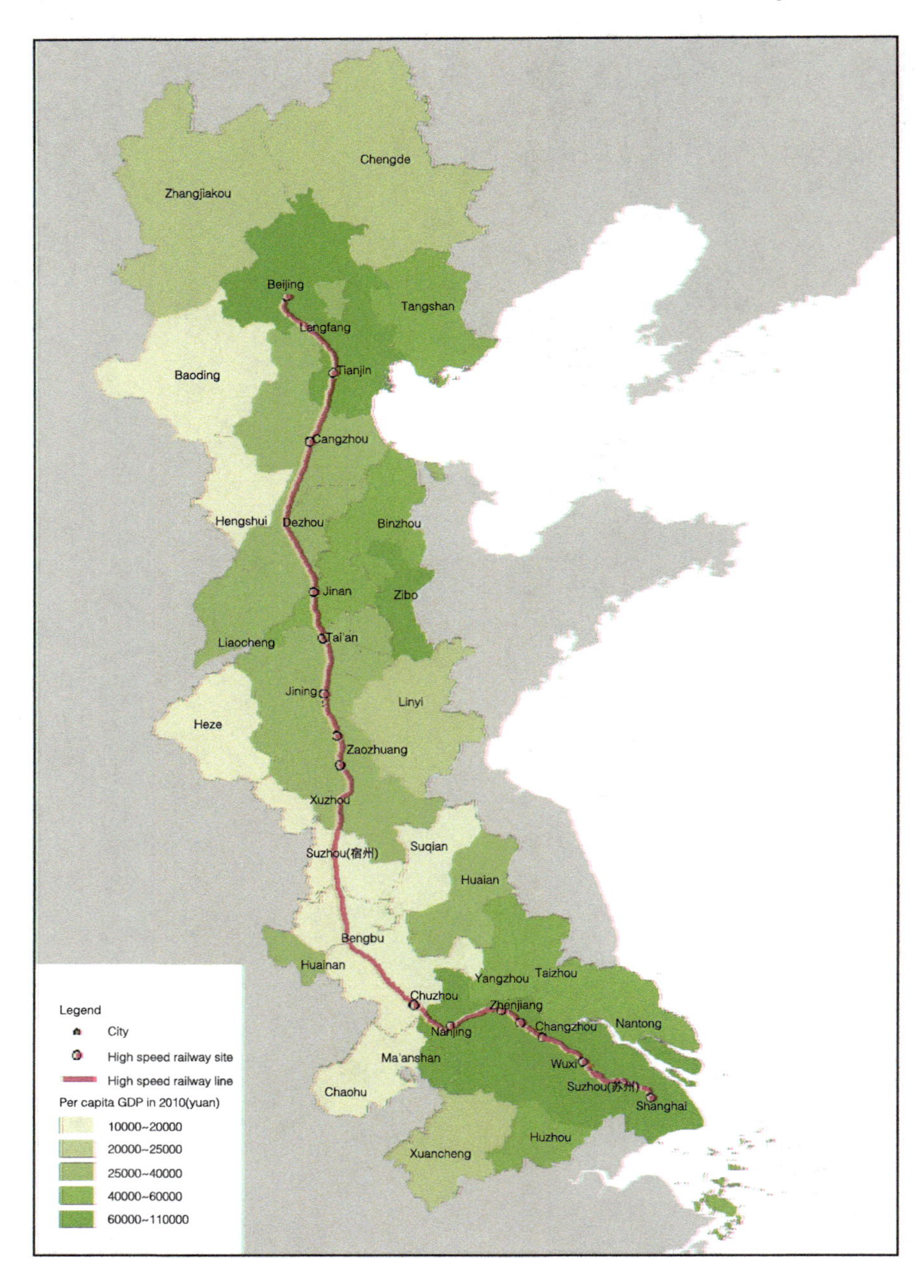

Fig. 2.5 (continued)

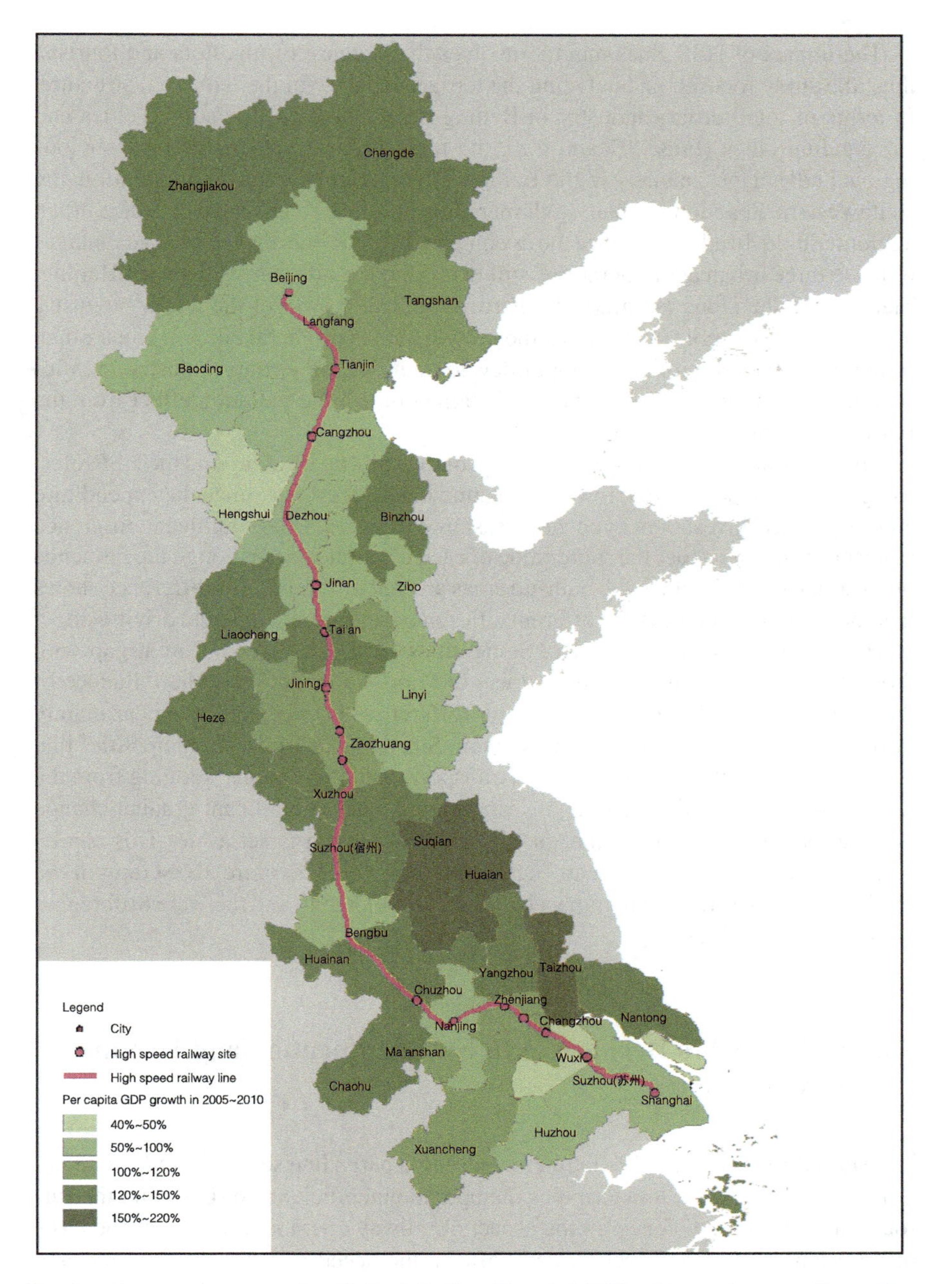

Fig. 2.6 Increase in per capita GDP of Beijing-Shanghai HSR cities and surrounding cities (2005–2010)

The impact of HSR lines mainly involves the increase of investors and tourists; thus, the study focused on analyzing the tertiary industry in the industrial structure. In terms of total tertiary industry in Beijing-Shanghai HSR line station cities and surrounding cities (Figs. 2.7 and 2.8), the total tertiary industry showed obvious regional advantages, namely in the Beijing-Tianjin-Hebei urban agglomeration, the southwestern Shandong urban agglomeration and the Yangtze River Delta urban agglomeration. In addition, since the opening of the HSR line, the regional advantages of these three urban agglomerations still exist, and the differences between Beijing, Tianjin and Shandong are gradually narrowing. The growth of the tertiary industry also shows the overall situation that the growth in the peripheral cities is greater than in the cities with stations. The rapid development of the tertiary industry in cities such as Heze, Liaocheng and Zhangjiakou may reflect the radiation effect from the HSR to the hinterland.

The construction of quantitative models of the built-up areas around the HSR sites, using GDP, total population, foreign investment, fiscal revenue, financial expenditure and fixed assets investment as covariates, shows that there is no significant difference in the built-up area under the same value of each covariate. Even though the influence of covariates is eliminated, the built-up areas are still different. This difference shows that the impact of the sites on different cities may be different, and the driving effect of HSR is obvious in some cities. Some cities have established a great gap with those cities without stations while others did not. After removing the influence of the economy and other factors, the major differences between cities appear mainly in Beijing, Tianjin, Nanjing, Jinan, Xuzhou, Shanghai and other municipalities that are provincial capitals, large cities and transportation destinations. Judging from the longitudinal data results, there is a clear insignificant to significant gradual change that can be attributed to whether or not the HSR station is set aside. This can be seen for example in cities such as Tianjin, Xuzhou, etc. Therefore, the setting of the HSR station has a significant impact on urban development, and there are differences among different cities.

2.2 Relationship Between HSR Station Location and Urban Development

The site selection of an HSR station includes two parts: line selection and site selection. At the same time, it determines the operational efficiency of HSR and the surrounding land use efficiency. In the aspect of railway operation, the line is the basis of the train movement which directly affects the speed and distance of the train. HSR stations are the nodes of the railway system, where the flow of people accesses and leave the HSR, with a direct impact on the site of the customer base range. The impact on the city include both positive and negative aspects, the negative impact is mainly reflected in the separation of urban space caused by railway lines. Urban separation is because a railroad requires a certain amount of space along the line, a safe

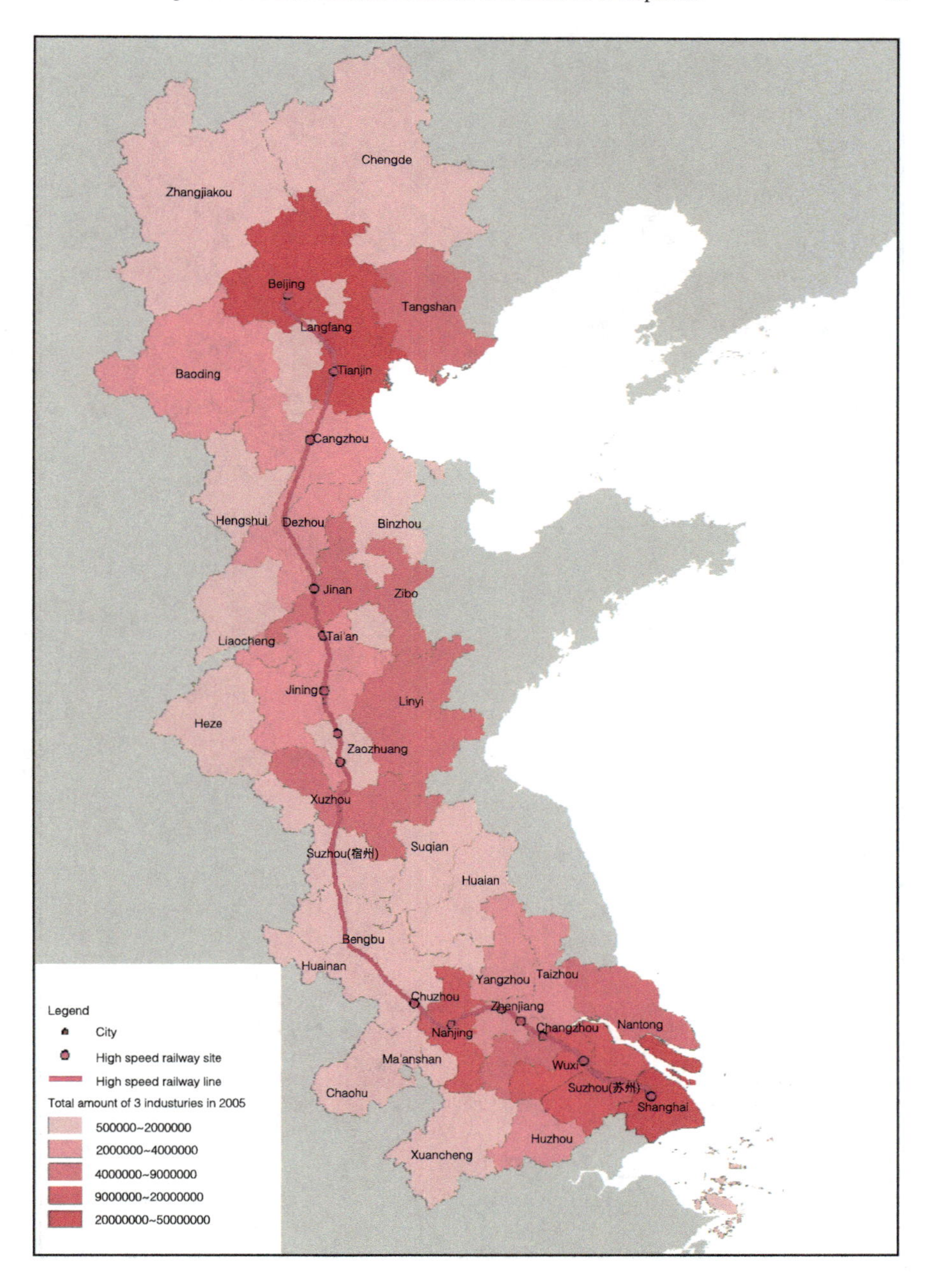

Fig. 2.7 Total tertiary industry of Beijing-Shanghai HSR cities and surrounding cities (2005, 2010)

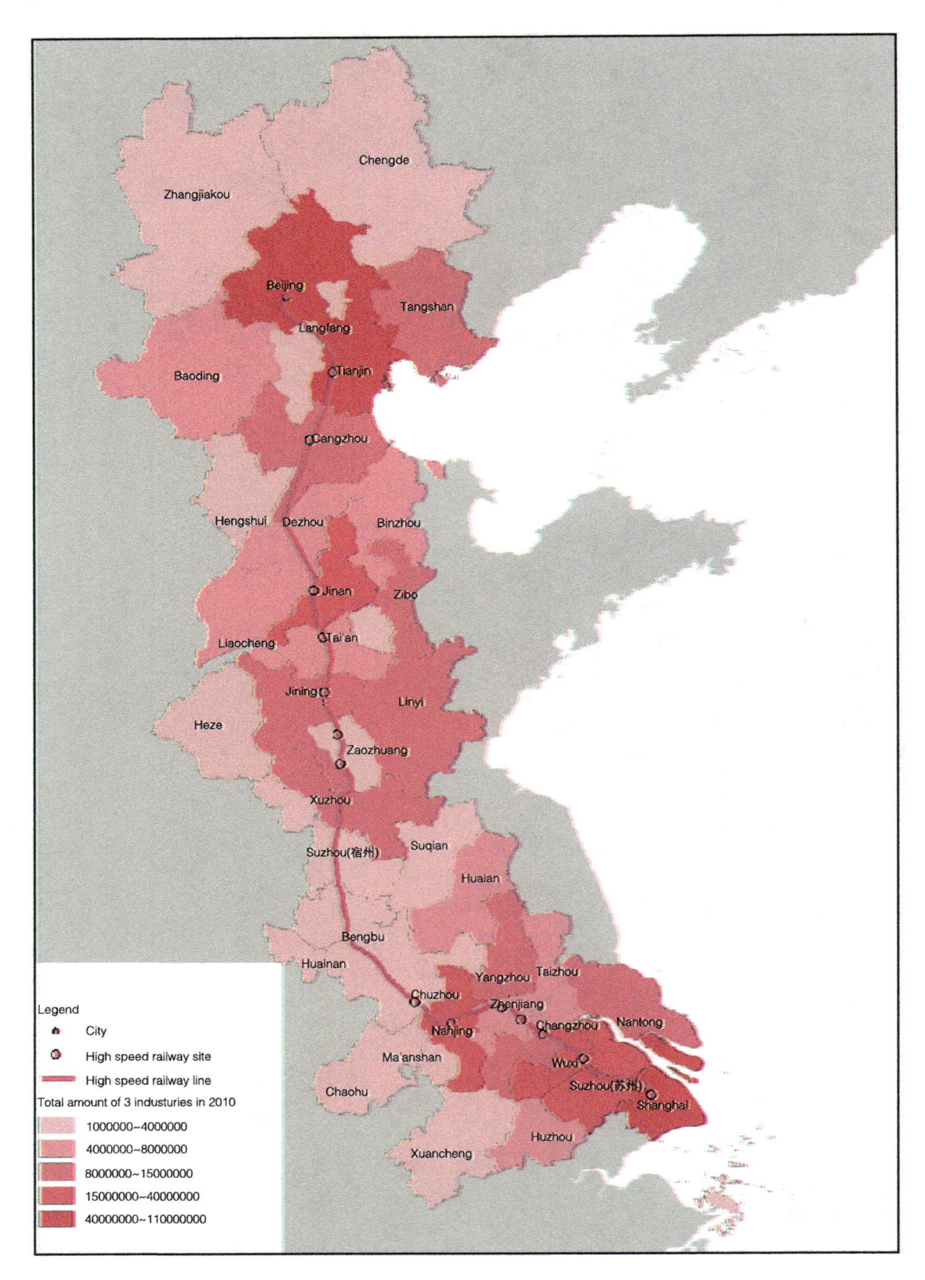

Fig. 2.7 (continued)

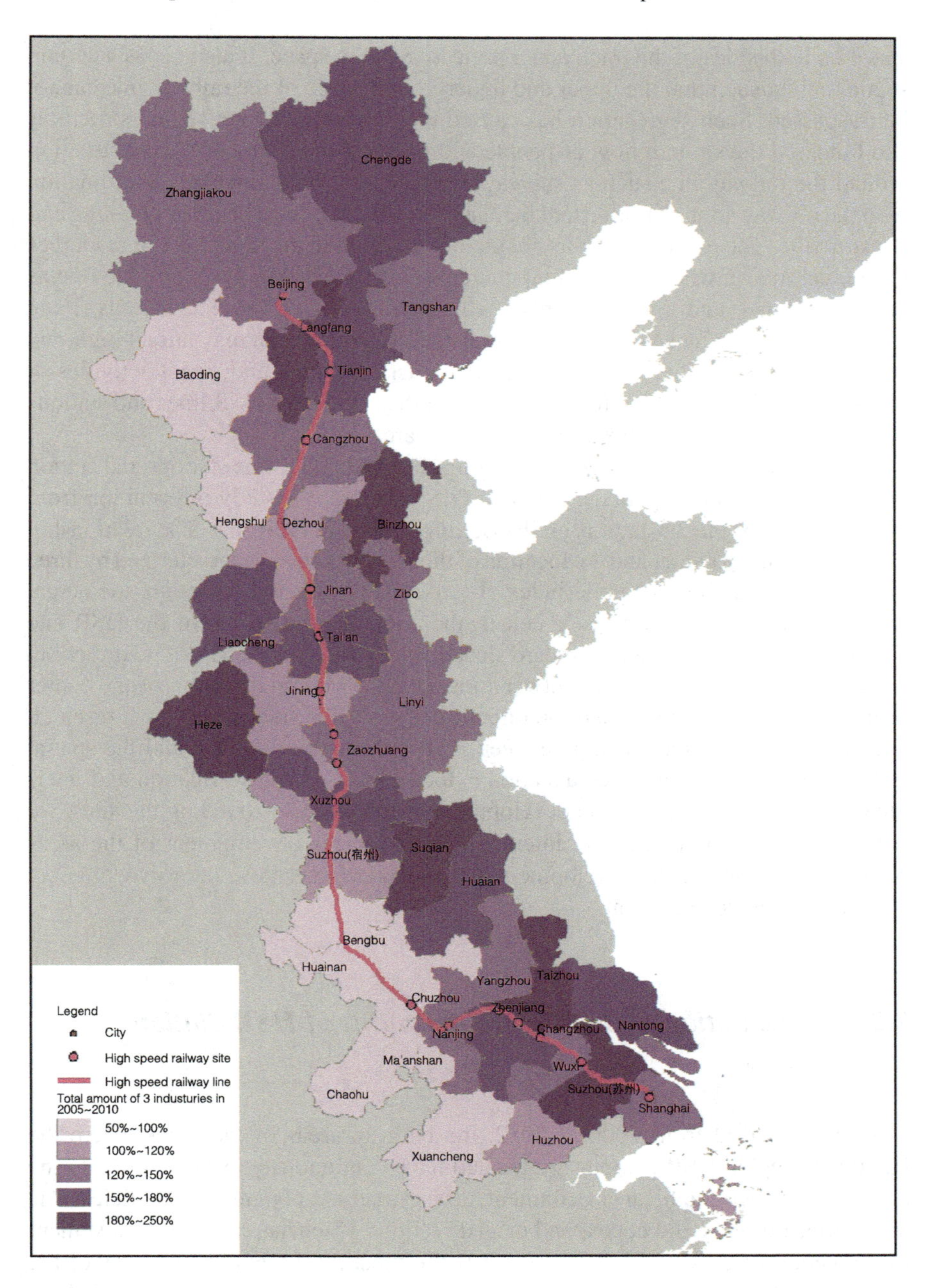

Fig. 2.8 Increment of tertiary industry of Beijing-Shanghai HSR cities and surrounding cities (2005–2010)

space on both sides of the track and a protective green space. It also needs a certain spatial height such that the upper and underground space of the railway line cannot be trespassed. Such segregation has caused the division of urban spatial structures and hindered the smooth flow of people and goods. At the same time, construction around the railway line such as subway overpasses or roads involves coordination with the railway administration and increased transaction costs in urban planning and construction. The positive benefits are mainly reflected in the rapid flow of resources in the region as a whole, which may increase investment in cities along the line as well as turnover and tourist numbers, which will benefit cities and regions (Feng et al. 2013). In the meantime, the proximity of the station area may attract high-end economic activities that directly affect the wider economic and social activities of cities (Poll and Zhou 2011). It can be seen that the location of HSR lines and stations is of great significance to the surrounding city areas.

The research mainly analyzes the Beijing-Shanghai HSR line stations and station cities. Based on satellite image data and GIS software, we study information from 2006 to 2012 for the 22 stations on the Beijing-Shanghai HSR line. The relationship between the urban center and the location of the HSR stations is characterized by three distance indexes and one angle index. The three distance indicators are the actual shift distance of the urban space center, the component distance of the HSR site direction and the component standard shift distance value of the HSR site direction.

An angle index is the angle between the actual deviation direction of the center of gravity of the city space and the direction of the HSR station. Through research and analysis of the relationship between HSR station site selection and the spatial expansion of the city in which a station is located and case investigation and interviews on the spot to clarify the development of the areas surrounding the sites, we will strive to promote effective interaction between the development of the areas surrounding HSR and the development of urban areas to achieve intensive, efficient and sustainable development.

2.2.1 Calculation of the City Center Shifts of HSR Station Cities

During the period from 2006 to 2012, the built-up areas of the cities where the Beijing-Shanghai HSR stations are located have continuously expanded due to the differences in geographical environment, urban structure, planning and policies. The urban area, however, did not expand in all directions which has caused the movement of the center of gravity. The city center is the most representative measure of the spatial distribution of urban space. It can be regarded as the average location of the city and the equilibrium point where the city is evenly distributed. The analysis of the relationship between the shift of the city center and the location of HSR stations helps to further understand the impact of HSR stations on urban space expansion. Based on satellite imagery, this section extracts GIS data in urban built-up areas in 2006 and 2012 to both calculate the center of gravity location in a city in two timelines and to determine the distance between the two points (Fig. 2.9). The light-colored

Fig. 2.9 City center shifting

area in the picture is the basic built-up area of Shanghai in 2006, and the dark area is the growth area of Shanghai's built-up area in 2012. The two green dots in the picture represent the centers of the urban built-up areas in 2006 and 2012. The shift between the two points can be calculated with GIS.

Based on the data, the study uses three distance indexes and one angle index to characterize the relationship between the urban space center and the location of the HSR stations. The three distance indicators are the actual shift distance of the urban space center, the component distance of the HSR site and the standard shift distance of the component distance from the HSR site (Fig. 2.10). The angle index is the angle between the actual shift direction of the center and the direction of the HSR station.

Figure 2.10 shows the actual shift of the city center, which is the measured distance from the city center in 2006 to the city center in 2012. The component distance of the HSR site direction is calculated by decomposing the actual city center along the

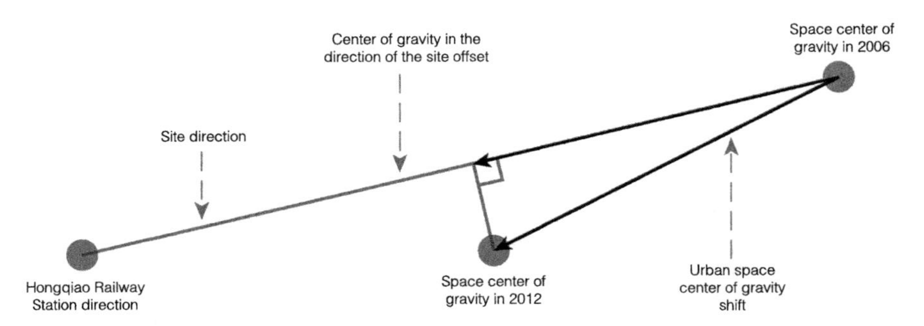

Fig. 2.10 City center shifts in the direction of the HRS station (Shanghai as an example)

line connecting the center and the location of the HSR station in 2006. Obtaining the shift distance of the city center in the direction of the site and combining that measurement with the angle between the direction of the station and the shift direction to reflect the relationship between the city center shift and the location of the station. Meanwhile, to more accurately reflect the relationship between the city center shift and the HSR stations, the spatial disparities of the HSR stations in different city sizes are standardized to avoid the impact on the size of urban built-up areas in order to make an effective comparison between cities. The standardization formula is $P' = P/\sqrt{S}$ where P' is the standard deviation value of the shift distance of the city center in the direction of the HSR station, P is the actual value (unit: meter) on the shift distance of the city center in the direction of the HSR station, and S is the size of the built-up area in 2012.

Based on the above calculation method, three spatial center of gravity shift indexes from 22 cities on the Beijing-Shanghai line are calculated to obtain the actual center of gravity offset distance of these cities. The three indexes include the HSR site component distance, the HSR site component standard distance and the angle between the direction of the actual center of gravity of the urban space and the direction of the HSR station (Table 2.2).

The shift of the city center reflects the change in the spatial distribution of the city. The actual values of the city center shifts between 2006 and 2012 were calculated by GIS software (Fig. 2.11). The vertical axis in the figure is the actual distance of the city center. A positive value indicates that the direction of the city center tends to be biased toward the HSR station overall, that is, the angle between the shift direction of the city center, and the HSR station is between 0° and 90°. A negative value means that the overall direction is away from the station site, that is, the angle of the actual shift direction and the direction of the HSR station is between 90° and 180°. Taking Shanghai as an example, from 2006 to 2012, the city center generally deviated to the southwest, with a distance of 2298 m to the Shanghai Hongqiao Railway Station. It can be found that in the 22 cities along the Beijing-Shanghai HSR line, half of the cities tended to develop in the direction of the HSR station, while the other half deviated from the station.

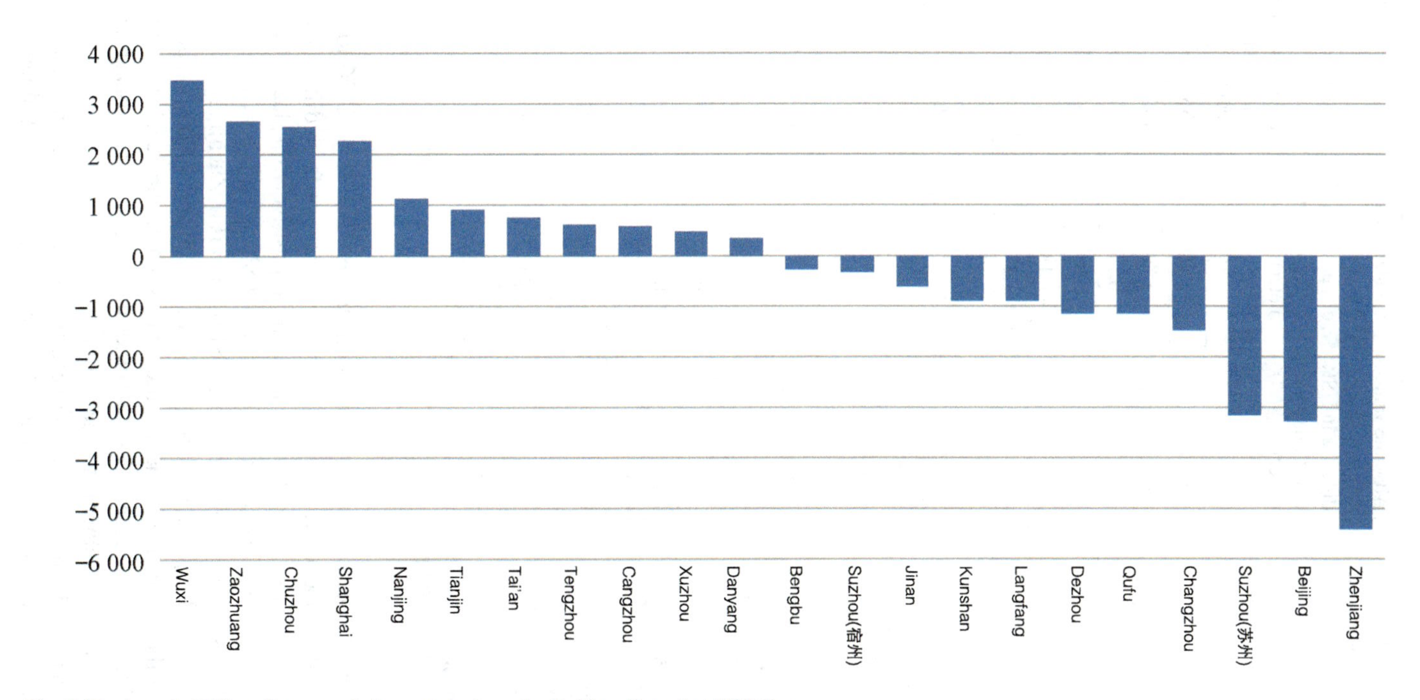

Fig. 2.11 Actual shifting distance of city center along the Beijing-Shanghai HSR line

Table 2.2 Index of city center shift

City	Actual shift distance (unit: meter)	Absolute value of component distance in HSR direction (unit: meter)	Standard value of component distance in HSR direction (unit: meter)	Angle between actual shift direction of city center and direction of HSR station (unit: degree)
Beijing	3275	−2340	−3.95	136
Langfang	915	−211	−1.80	103
Tianjin	900	47	0.14	87
Cangzhou	575	342	3.88	54
Dezhou	1139	−911	−8.13	143
Jinan	614	−496	−2.13	144
Tai'an	744	652	5.43	29
Qufu	1153	−675	−10.71	126
Tengzhou	617	224	2.31	69
Zaozhuang	2639	1248	10.85	62
Xuzhou	475	473	2.71	4
Suzhou	313	−251	−3.23	143
Bengbu	264	−222	−2.40	147
Chuzhou	2552	638	4.60	75
Nanjing	1147	242	0.76	78
Zhenjiang	5375	−1264	−10.88	104
Danyang	343	132	1.72	67
Changzhou	1472	−1038	−4.12	135
Wuxi	3453	2831	11.81	35
Suzhou	3155	−1107	−4.26	111
Kunshan	905	−171	−0.82	101
Shanghai	2298	2248	5.23	12

Due to different speeds of urban expansion in the cities from 2006 to 2012, there is a large difference in the distances of the city center shifts in different cities. Among them, Wuxi (more than 3000 m) is the most positively displaced from the HSR station, followed by Zaozhuang, Chuzhou and Shanghai (between 2000 and 3000 m). The largest negative shift from the HSR station is in Zhenjiang (more than 5000 m), followed by Beijing and Suzhou (between 3000 and 4000 m). The actual shift distance of the city center in the 22 cities shows that in the development of the cities where HSR stations are located, not all cities incorporate this regional transport facility into their space development considerations.

To further analyze the relationship between the city center shift and the station in this section, the space shift is calculated by vector decomposition along the line connecting the city center with the location of the station in 2006, and the actual city center shift is calculated in the direction of the HSR station (Fig. 2.12). In the direction of the HSR station, the city center shift is different from the actual shift distance, and the angle between the actual shift direction and the direction of the HSR station is considered; thus, different values are obtained. Wuxi, in the direction of the HSR station, still has the largest city center shift distance, and Shanghai surpassed Zaozhuang and Tai'an to become second in the city center shift distance. This finding shows that the city considered the location of the HSR station in its development to a certain extent. Beijing is the city with the most developed city center departure from the HSR network. Due to the reconstruction of the HSR station and expansion on the basis of the original railway stations in the built-up area of the city, the Beijing HSR station was not available to guide the development of the city.

In the meantime, to further consider the influence of the city scale on the actual shift distance (absolute value) of the city center, the study introduces the standardized distance. Through standardization, the relative distance between the city centers and the different scales are studied, making the cities more comparable. For example, if two cities with great differences in scale have HSR stations at the edge of the built-up area, the city center differs from the actual shift distance of the HSR station. After normalization, the normalized values of the two offset distances will be the same. The result is shown in Fig. 2.13.

2.2.2 Type Analysis Based on the City Center Shift and the Relationship with the HSR Station

Based on the measurement of the city center, the study is based on the angle between the city center and the direction of HSR. The study further subdivided the relationship between the HSR setting and city center and classifies the cities along the Beijing-Shanghai HSR line into three categories (Fig. 2.14, Table 2.3). If the included angle is between 0° and 60°, the direction of urban space expansion has obvious bias to the setting direction of HSR and belongs to the first type of city. If the angle is between 60° and 120°, the direction of urban space expansion is not obviously related to the setting direction of HSR and belongs to the second type of cities. If the included angle is between 120° and 180°, the development direction of the urban space is opposite to the setting direction of the HSR and belongs to the third type of city. For example, the deviations of the city center in Wuxi and Shanghai are biased toward the setting of HSR stations and their shift distances are large and belong to the first type of city. Although there is a larger expansion of Zaozhuang, the city center shifted significantly, but it is not obviously relationship with the HSR station, belongs to the second type of city.

According to this classification method, the 22 cities along the Beijing-Shanghai HSR line are divided into three types: (1) the urban development bias is toward the HSR station direction—there are 5 cities of this type, Shanghai, Wuxi, Xuzhou,

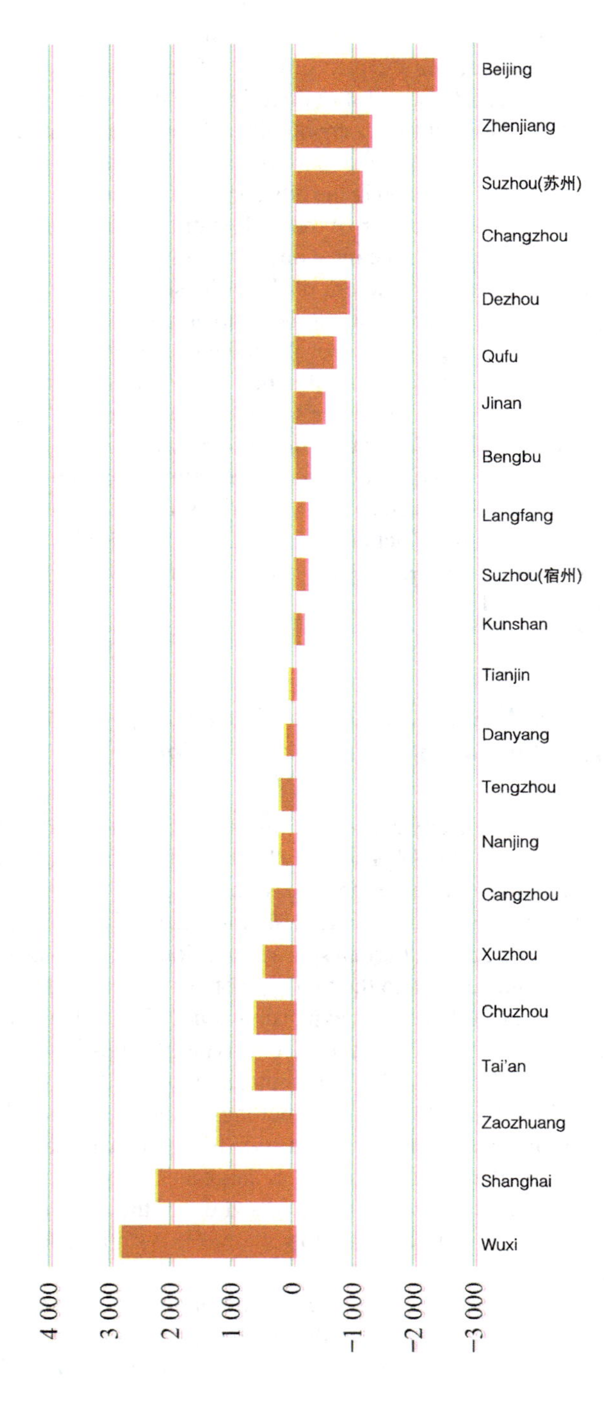

Fig. 2.12 Urban space center of gravity offset distance in the direction of the HSR station

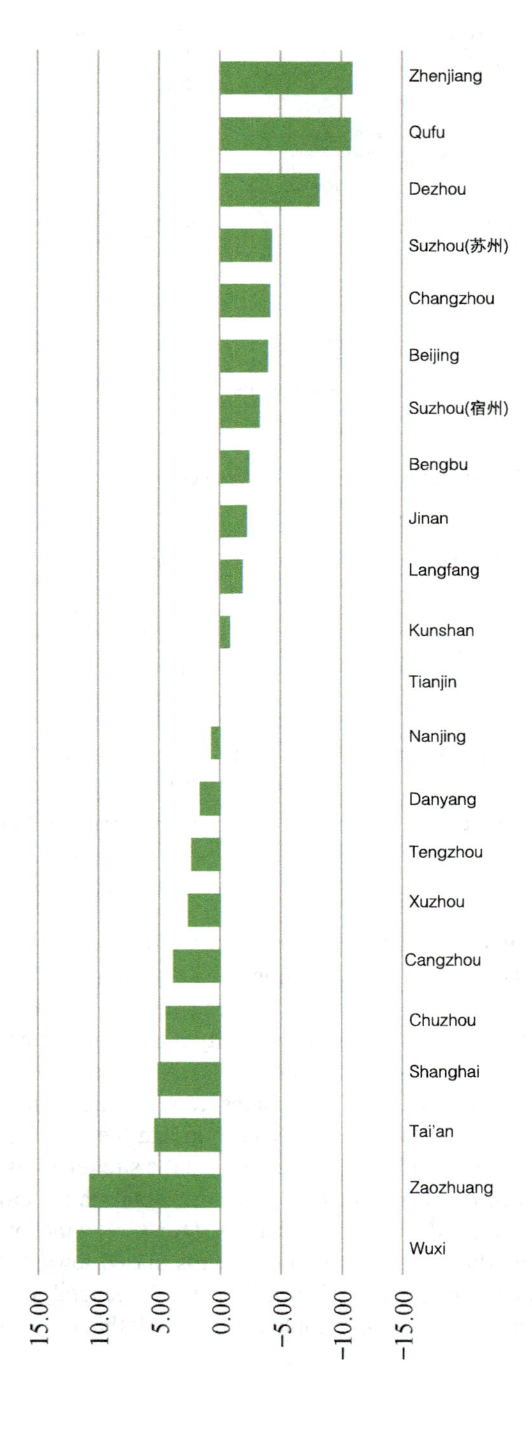

Fig. 2.13 Normalized value of the city center shift distance in the direction of the HSR station

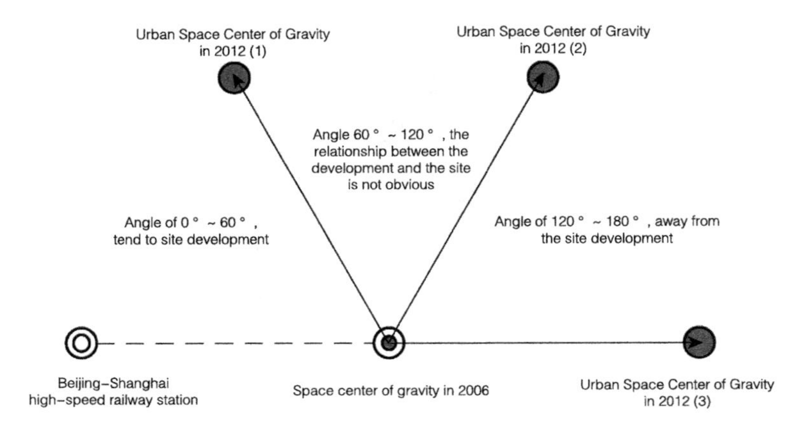

Fig. 2.14 Classification diagram of the relationship between city center and the location of HSR station

Table 2.3 Classification of urban space expansion and site relationships

Classification	City
Type I: Expansion close to station (five cities)	Xuzhou, Shanghai, Tai'an, Wuxi, Cangzhou
Type II: Relationship between expansion and station is not obvious (ten cities)	Zaozhuang, Danyang, Tengzhou, Chuzhou, Nanjing, Tianjin, Kunshan, Langfang, Zhenjiang, Suzhou
Type III: Expansion away from station (seven cities)	Qufu, Changzhou, Beijing, Dezhou, Suzhou, Jinan, Bengbu

Source Author, The order of the city is arranged from small to large according to the angle between the city center and the direction of HSR

Tai'an and Cangzhou where the average city center shift distance of this type is 1509 m and the shift distance toward the station is 1309 m (Fig. 2.15); (2) there are 10 cities of type II, in which the relationship between the spatial expansion direction and the station location is not obvious, the average city center shift distance is 1855 m and the shift distance from the station is 528 m; and (3) there are 7 cities of type III, in which the urban space expansion direction deviates from the direction of the station, the average distance of the city center shift is 1176 m and the deviation of the site direction is 848 m (Fig. 2.16). Urban development and the site relationship between the locations of the city centers accounted for 45% of the total station cities. The city center shift away from the site development accounted for 32% of the total. There are only 5 cities where the urban development direction is biased toward the direction of the HSR station which accounts for 23% of the total. In general, the facilities of the Beijing-Shanghai HSR stations have not led to the result that the development of those cities is around stations.

Fig. 2.15 City centers shift toward the Beijing-Shanghai HSR stations

Fig. 2.16 Cities in which the centers are far from the Beijing-Shanghai HSR stations

Through the above analysis combined with the field survey, the conclusions are as follows:

(1) There are individual differences in consistency between the location of the station and the basic development direction of the city. The expansion of urban space is affected by many factors. The HSR station, part of a large transportation infrastructure in the region, does not necessarily influence the city center shift toward or away from it. This means that the combination of site selection and city planning should be strengthened.
(2) The city centers in the three types of cities show different relations with the HSR stations. The main reasons are as follows: (1) the relationship between the site selection of the HSR stations and the existing built-up area, including the built-up area, the edge or the exterior; (2) whether the site selection is in line with the original development direction in the master plan of the city; and (3) the financial support ability of the city, etc.
(3) According to the change in absolute quantity and standardized value, cities such as Beijing and Shanghai have large absolute displacement of the city center due to their large scale and economic strength, the rapid expansion of built-up areas. However, after standardization, the changes are not as obvious as those in smaller cities, such as Zaozhuang.
(4) The urban space expansion is obviously affected by development policies. For example, Wuxi East Railway Station is located in the Xishan District. It belongs to the key development area of Wuxi. The goal is to build the city center at the district level. Therefore, the construction around the station is very rapid, and the city center shift is obvious. In 2006, Dezhou set up the Canal Economic Development Zone on the west side of the city with rapid construction at the provincial level. Dezhou East Station is not in the main expansion area of the city, and the city center shift is in the opposite direction from the location of the station.

2.3 Relationship Between Site Selection and Urban Development of HSR

Studies have shown that the distance between HSR stations and city centers significantly affects the urbanization of the surrounding area. In the 22 station cities along the Beijing-Shanghai HSR line, there are great differences in the distance between the stations and the city centers (in this study, we use city government to represent the city center). In addition, it has no significant correlation (that is, the distance from the station to the city center does not increase by the city scale). Half of the cities' HSR stations are outside the built-up area, and the other half are in the built-up area but mainly at the edge of it. Although the Ministry of Railways prefer a linear HSR line with cheap land access and minimal removals (meaning the speed and feasibility of implementation are the main principles), convenience of access for citizens should also be considered in the station setting. The existing urban development should also be combined with the HSR planning to maximize the impact of HSR in driving urban development. The two should lead to a comprehensive and balanced result. However, the current HSR stations do not play significant roles in urban development.

Among the five cities in type I in which the directions of urban expansion and the HSR station are the same, the relationship between the HSR stations and the city centers are as follows: Xuzhou—7.1 km, at the edge of the built-up area; Shanghai—16 km, at the edge of the built-up area; Tai'an—6.1 km, at the edge of the built-up area; Wuxi—15 km, at the edge of the built-up area; and Cangzhou—6.8 km, outside the built-up area. Therefore, with the exception of Cangzhou where the HSR station is outside the built-up area, these HSR stations have a close relationship with the urban development and the surrounding areas are well developed. Cangzhou is quite different because the overall direction of urban development shifted from the HSR direction from 2006 to 2012. However, it is difficult to develop the area surrounding the site in a short time because of the distance and limited financial support. The other four stations are built on the site edge, the distances are acceptable, and the cities' governments attach great importance to the development of the surrounding areas. These cities have taken the HSR stations into consideration in their development strategy. Based on the existing public construction investments, more private construction investments have started. This further promotes the role of HSR in urban development.

Most of the HSR stations in type II cities are located far from the city centers except those in Langfang, Zaozhuang and Kunshan; due to the small scale of these cities, the distance is approximately 3 km. In Tengzhou, Zhenjiang, Danyang, Nanjing and Chuzhou, the distance is more than 6 km. Tianjin South Railway Station and Suzhou Station are more than 10 km from the city centers. The urban development and HSR settings did not show a significant relationship, and the situations are relatively complicated. The main reasons include the location, the expectations of the city for HSR stations, the lack of impetus for the development of the city itself and inadequate investment of public funds. These cities should strengthen the links between the HSR station and the existing urban areas to attract investors, technicians and tourists to promote the development of the urban economy.

For the 7 cities in type III, where the direction of urban expansion is opposite to the HSR stations, the relationship between the station and the city center are as follows: Qufu—7.5 km, outside the built-up area; Changzhou—5.5 km, at the edge of the built-up area; Beijing—5.4 km, inside the built-up area; Dezhou—16 km, outside the built-up area; Suzhou—24 km, outside the built-up area; Jinan—9.7 km, at the edge of the built-up area; and Bengbu—4.3 km, outside the built-up area. The Beijing HSR station is inside the built-up area, so there is no new urban space in which to expand. The Bengbu station is regarded as a hub station, so there is a large planned area (approximately 23 km^2). The distance to the city center is short, the position is good, and the development scale is large (the plan is for a modern emerging area with multiple functions, such as finance, science, education, trade and housing). The development of the surrounding area is rapid, but it does not have a close relation to the built-up area. For the other cities in this type, the developments of the HSR new town are limited. This shows that inconsistency with the direction of urban development leads to the limited investment of cities in the areas surrounding the stations and hinders effective interaction between the areas surrounding the stations and the cities in which the stations are located. Regional transport facilities failed to promote the development of the city. However, it is difficult for the areas around the transport facilities to rely on existing urban facilities for resource integration and development. These cities should analyze and demonstrate the role of the HSR station in a new round of urban development strategic thinking, clarify the relationship between the major development directions of cities and HSR stations and better integrate their resources.

The case analysis further shows that the site selection of the HSR station should match the city's planning and development. Take Wuxi in type I as an example: Wuxi East Railway Station is located in Xishan District, which has two major agricultural areas. The urbanization rate is lagging behind that of other districts and requires the provision of public facilities to enhance public services. In terms of specific locations, the Anzhen-Yangjian area, where the station is located, is a high-tech development zone in Wuxi. There are industrial parks in the surrounding areas and the township and village enterprises are well developed. This provides a foundation and market for the development of producer services. Considering the information flow, capital flow and service flow that HSR stations can bring, Wuxi East Railway Station is located here, and the plan for the area surrounding the station is designated as "Wuxi High Speed Rail Business Area." In the process of construction, Wuxi first took advantage of the construction of HSR. The HSR was an opportunity to promote the demolition and construction of resettlement houses and established public facilities at key district-level hospitals and municipal key high schools to enhance the speed and quality of urbanization in the region. Second, Wuxi used the industrial base in the region to actively build the headquarters of regional enterprises to promote the development of the building economy. On this basis, combined with the advantages of the gateway location of HSR, it will further expand its commercial functions and build an integrated urban commercial complex. The guiding role of the site accords with the demands of regional development. The direction of urban space expansion is in line with that of the HSR stations, enabling rapid development in the area surrounding Wuxi East Railway Station and bringing about full vitality and employment.

Changzhou (type III) and Wuxi (type I) are both at a similar level of economic development and urbanization, but the development of Changzhou North Station has been relatively slow. Changzhou North Station, located in the north area of Changzhou Northern New Town, belongs to the Changzhou high-tech development zone (Xinbei). The east side of the station is close to the functional part of the Xinlong area. The south side is across the Shanghai-Nanjing expressway and the Beijing-Shanghai HSR line and connects with the industrial park. The southeast corner is in the core functional area of Xinbei District. It is where the district government and district-level businesses and commercial centers are located. Changzhou North Station is less than 10 min away from the government and commercial center of Xinbei District, and the linear distance is approximately 3 km. At present, the development of the northwest side of the HSR station is planned and positioned as "Xinlong International Business City" with the main purpose of developing business office functions. On one hand, the real estate industry as a whole has been in a downturn, and on the other hand, the proximity of the station to the original district center and continuous development near the original center have disadvantaged and slowed development around the site. At the same time, according to the research classification, Changzhou belongs to the third type of city. The main development direction of the city itself is in the southeast, away from the location of the HSR station; thus, the investment and development focuses of the city are not in the vicinity of the HSR station. Similar to Changzhou and with even more obvious features is Suzhou. Of the HSR stations in cities along the Beijing-Shanghai HSR, Suzhou's are located farthest from the city center and completely deviate from the original planned development direction of the city, resulting in the inefficient integration of resources.

The research shows that the large-scale transport infrastructure in the region plays a role in driving the development of the cities. However, this role must be matched with the original urban functional layout and planning. The site selection of an HSR station is very important. The combination of its location with the original expansion direction of the city will affect the effective integration of urban resources and determine the driving effect of HSR.

Based on the data processed by satellite imagery and GIS, this chapter analyzes the station site location and urban expansion direction of 22 cities along the Beijing-Shanghai HSR line. The empirical evidence shows that the relationship between site selection and the existing built-up areas and development directions of the existing cities affects the driving effect on the sites in the surrounding areas.

First, as a state-level transport infrastructure, HSR and their stations have improved accessibility at the regional level and have the positive effect of boosting regional development; specific benefits driven by the city level should be differentiated according to their specific circumstances. Overall, the location of the site and the distance from the city center to a certain extent determine the possibility and effectiveness of the interaction between the built-up area and the site. Although the HSR line selection should fully consider the speed needs of HSR along with the feasibility and cost of implementation, if the station is too far away from the city center, it will not only cause inconvenient access to the HSR, but also have a negative impact on HSR's role in urban development and may even lead to the areas surrounding HSR stations not being effectively developed. Therefore, the location of regional transport facilities should take full account of the status quo of the city built-up area and be arranged in an appropriate location. The HSR stations themselves can bring together popularity and activities. This requires good alignment and coordination between site selection and existing urban functions. Otherwise, it will be difficult for the effectiveness of the site to achieve full play.

Second, site selection and urban development should be coordinated. The localization of HSR stations is mostly based on a new urban growth point, which should be clearly defined for the specific stages and characteristics of each city. At the same time, it should be matched with the existing urban development direction and function orientation. Cities with the economic capability to expand outward can provide impetus for the development of HSR sites; HSR will likely bring investors, tourists and technicians to urban development. Such a benign interaction requires that the location of HSR stations be linked with the original urban master plan to analyze areas with development potential that are in line with the intended expansion. The type I cities mentioned above can further promote the established benign relations in development planning and take full advantage of the leading role of regional transport facilities. Type II and type III cities should strengthen the link between the site and the existing urban built-up area to demonstrate that the role of HSR is clear, and the integration and planning of urban development resources should be carried out.

Finally, the site selection is originally based on the Ministry of Railways plans, from the top down. If plans for the HSR line, land occupation and site location fail to fully consider the development needs of the city itself, the problem may lead to inconsistencies with the original plan of the city and sunk costs. At the same time, the original investment for the station comes from the Ministry of Railways. The initial investment in the surrounding area comes from the city government, including land acquisition and storage, roads and infrastructure construction. Cities with strong economic power usually develop rapidly and build more. Private investment is based on real estate projects. The site itself and the functionality of the surrounding area should be further improved. With the current institutional reform and HSR construction steadily advancing, careful consideration should be given to site selection in the future to effectively combine the development of HSR stations and the surrounding areas with urban development to enhance the efficient and orderly development of comprehensive utilization.

Appendix: Domestic and International Case Analysis

The selected cases in this section include HSR stations in Taiwan (China), Japan and France. The basic principles for selecting these cases are as follows: (1) the city where the HSR station is located in a medium-sized city or a prefecture-level city; (2) the station is located at the periphery of the city built-up area or at the edge of the city administrative region; and (3) the HSR stations are similar to most stations on the Beijing-Shanghai HSR line. The study will focus on the relationship between the site, area and city by analyzing the position, function and space shape of the area surrounding the HSR stations.

According to the relationship between HSR stations and cities, the cases are divided into three types: urban center type, urban edge type and emerging type. The cases in Japan and France are mainly the urban center type, including Lille HSR Station in France, Kagoshima HSR Station and Kumamoto HSR Station in Japan. The urban edge type cases include the Lyon Part-Dieu HSR Station in France and the Hsinchu HSR Station in Taiwan, China. The emerging type cases include Taoyuan HSR Station and Tainan HSR Station in Taiwan, China. Most of the stations on the Beijing-Shanghai HSR line and the Beijing-Fuzhou HSR line are of this type.

Cases in Taiwan, China

Taiwan HSR serves the western region in Taiwan, which is the most densely populated region. It forms a traffic corridor with a total length of 345 km. Based on the population scale and characteristics of the cities, the study selected Taoyuan, Hsinchu and Tainan to analyze the planning of HSR stations.

Case 1: Taiwan Taoyuan HSR Station

Taiwan Taoyuan HSR Station is located in northwestern Taoyuan, at a distance of 7 km from the city of Zhongli, 10 km from the Taoyuan city center and approximately 6 km north of Taoyuan International Airport. The station and Taoyuan are connected via a rail link and three highways. The station is the hub of north–south transit of Taoyuan International Airport and has become the preferred area for developing airport-related industries and playing a support role for the metropolitan area (Fig. 2.17).

Taoyuan Station is positioned as an "International Business City" and has taken advantage of the 4 h flights to major cities in East Asia to become a decision-making

Fig. 2.17 Location of Taoyuan HSR station, redrawn by the author, *original source* http://www.tycg.gov.tw/fckdowndoc?file=/aero.jpg&flag=pic

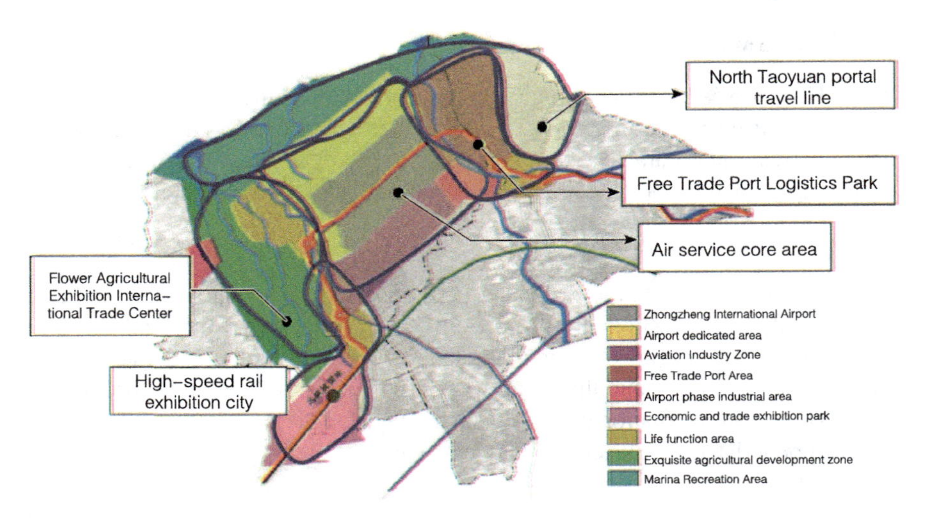

Fig. 2.18 Functional area around Taoyuan HSR station, redrawn by the author, *original source* http://1.bp.blogspot.com/-qtGc1-4Pb7c/T1rETzbhtI/AAAAAAAABH0/TtLoZ-UfTpM/s1600/eightfunction.png

and operational center for multinational corporations and Taiwanese businessmen to conduct business in the Asian markets as a pilot part of the Taoyuan Aviation City Project. The specific area of Taoyuan HSR Station is closely related to the development strategy of Taoyuan Aviation City. The major development and construction plans around it include the Taoyuan Aviation City Project, Far glory Free Trade Zone and Air Passenger Terminal Park (Fig. 2.18). Based on its position, its development strategies are as follows:

(1) To tie in with the promotion of the development policies of the Asia Pacific Operation Center and promote the common development of the surrounding areas around Taoyuan International Airport, establishing an overall development model that is interdependent with aviation plans;
(2) With the urban development characteristics of Zhongli and Taoyuan Gemini Satellite City, it is planned that the station area will be a multifunctional and interrelated industrial area that interacts with the airport; and
(3) The development of a composite area focusing on sightseeing, shopping, entertainment, international trade and business offices will attract domestic and foreign visitors, promote the development of tourism, raise the level of consumption and make the specific area of Taoyuan Station develop into a modern international airport city.

Taoyuan is an important industrial center in Taiwan because it is the location of several corporate headquarters. The planned area of the HSR station is 490 ha with a planned population of 60,000 and a residential density of 350 people per hectare. The planning area is divided into an exclusively industrial area, a subsidiary business area, a commercial area and a residential area as follows (Figs. 2.19, 2.20 and 2.21):

Fig. 2.19 Taoyuan HSR site planning

(1) The area of the special industrial zone is approximately 21.9 ha. In line with the development orientation of the Taoyuan Aviation City Project and with the advantage of its location at Taiwan's gateway, the International Enterprise Operational Planning Park will mainly provide office operation centers, commercial retail service facilities, international tourist hotels and other service facilities (such as cultural and leisure facilities, international medical services and other facilities) to multinational corporations and Taiwan businesses.
(2) The dedicated station area covers an area of approximately 8.5 ha and is adjacent to the Taoyuan HSR Station. It includes hotels and shops, shopping malls and office-related facilities to cope with the opening of the Taoyuan International Airport MRT and air transport pre-boarding, baggage and other transport services/transit benefits. In the long term, it will mainly focus on the development of complex functions such as transport services, shopping malls, leisure and entertainment, office-related facilities and hotels and other service facilities as the market matures.
(3) In response to the activities, crowds, and commercial services demand brought by the operation of Taoyuan International Baseball Stadium at the same time that the airport MRT opens to traffic, the business district close to the MRT station (A19) can develop leisure and entertainment, catering retail, hotels and other

Fig. 2.20 Land use surrounding the Taoyuan HSR station, redrawn by the author, *original source* http://www.hunhsin.com.tw/news/109-1.jpg

diversified business services. The remaining commercial districts are mainly engaged in providing business services in the neighborhood.

(4) Residential development planning in line with the development orientation of the Taoyuan Aviation City Project will serve local residential development needs. Centering on the core HSR functional areas, this area contains independent functional areas with a number of living communities as its mainstay.

The planned land is composed as shown in Table 2.4.

The road systems around the Taoyuan HSR station are divided between regional trunk roads and internal service loops that are externally linked. The areas within the circular service roads use new urbanism design techniques with linear twists and turns. External links exist to a light rail transit system and a number of expressways for fast-track use of corridor settings. Planning will be combined with the walking space of the integrated design of the station, with automobile parking outside. A complex function area will coordinate pedestrian activities and functions in order to

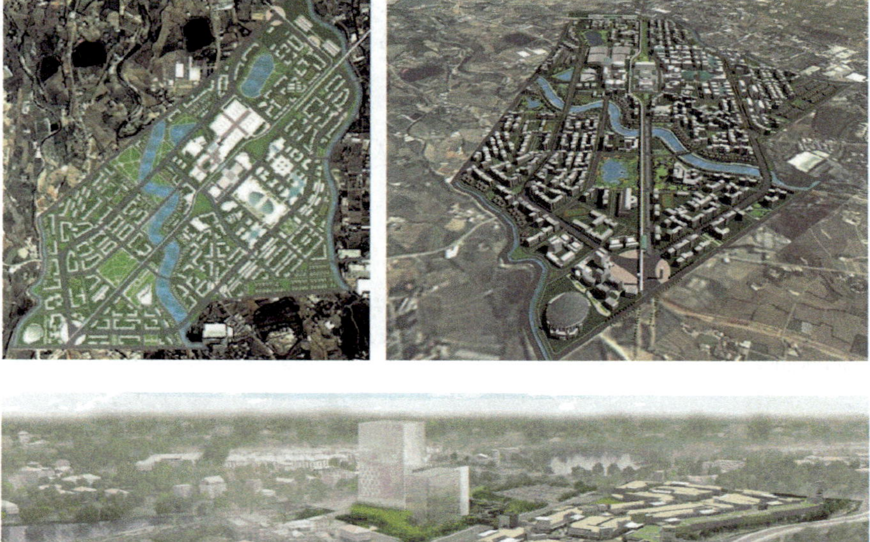

Fig. 2.21 Architectural layout of the areas surrounding Taoyuan HSR station, *source* http://blog.niceday.tw/wp-content/uploads/2014/06/【新聞相片】華泰名品城示意圖-3.jpg

provide transportation and public services at the same time. The external systems can be divided into the following two parts:

(1) MRT—The blue line in the MRT system of the Taoyuan metropolitan area, or the airport MRT line, passes through the HSR station and has three stations for transportation to and from the Taipei metropolitan area to Taoyuan Airport. The MRT line quickly connects to the Taipei MRT link.
(2) The road network includes the expressway system, such as highways 1, 2 and 3, the east–west expressway and HSR bridge roads with a high degree of convenience and accessibility.

Table 2.4 Planned land use surrounding Taoyuan HSR station

Land types	Area (unit: hectare)	Proportion of land use (%)
Station	20	4
Industrial	22	4
Commercial	54	11
Residential (including schools)	191	39
Green space (sports open space)	47	10
Roads (including parking space)	107	22
Others	49	10
Total land	490	100

Source Author

The spatial structure of the planning around the Taoyuan HSR station forms a T-shaped structure for the central park and the core areas of the station with high-density development of the business district on the T-shaped side and core buildings using large-scale highly integrated multifunctional development. At the same time, block-type development with high building density is adopted to form a small and elaborate axis sequence, and the maintenance of the original natural landscape is emphasized.

The station core area is approximately 80 ha, including one stop for four districts: HSR station area, mix use area, support service area, industry-specific area and community business service area. Some of the indicators for the development of some investment blocks are shown in Table 2.5, and the development intensity of major areas reached above 3.0.

Case 2: Taiwan Tainan HSR Station

Tainan HSR Station is approximately 9 km away from Tainan Airport and about 10 km away from the downtown area of Tainan, the southern science industrial park and the Kaohsiung Science Park. It is linked to Tainan through one rail link, two highways, and stands together with other railways.

Tainan Science Park is an important base for developing southern Taiwan's green energy industry. In recent years, Tainan has cooperated with the development of the southern science and technology industrial park, attracting many academic and research institutions, and the R&D production chain of the green energy industry in the city of Tainan has gradually taken shape. In the meantime, Tainan has another ecocultural village and four grass ecological protection zones. Therefore, Tainan HSR Station's has developed as a "Learning Eco City." With its advantage of being the largest hinterland of HSR transport nodes in the Kaohsiung area of Tainan, it will serve as a major base for Taiwan's R&D and the operations of the green energy industry and as a demonstration site for promoting green eco-communities.

The overall development strategy is to build a new city with ecological methods, concentrating on research and development institutions such as academic and

Table 2.5 Core area of Taoyuan HSR station

Land type	Area (hectare)	Architectural area (million square meter)	Gross volume ratio	Main function	Introduced facilities
Industrial core area	13.2	41.3	3.1	Business	–
Supporting service area	8.8	28	3.2	Shopping, residential, exhibition	Shopping center, hotel, exhibition facilities, themed recreational facilities
Composite functional area	10	33	3.3	Shopping, recreation, transportation	Business offices, hotel, food court, shopping mall
HSR area	8	–			
Community commercial service area	27				
Other areas	13				
Total	80				

Source Author

research institutions. To build green buildings, the surrounding area includes particular areas for a sustainable ecology research area, botanical gardens, a historical plant display area, industrial culture and the original eco-botanical garden planning to create both an academic and ecological green city. Specific development strategies include the following examples (Figs. 2.22, 2.23 and 2.24):

(1) To effectively diversify the existing over-intensive Tainan life circle with a multicore urban development model and pre-construct the commercial development needed for population industries, such as the future Tainan Science and Technology Industrial Park and the Tainan Science Park. To promote industrial upgrades in the southern region and move toward high-tech areas;
(2) To establish a scientific and technological industrial production function as the main demand as well as the preservation of tourism through the development of historic monuments, leisure and entertainment, shopping and living to strengthen group contacts and promote regional business development; and
(3) To cooperate with the lifestyle of the Tainan metropolitan area, planning a diversified leisure and entertainment center in the area and expanding the functions of combined sightseeing and recreation, consumer shopping, cultural and educational activities and tourist services.

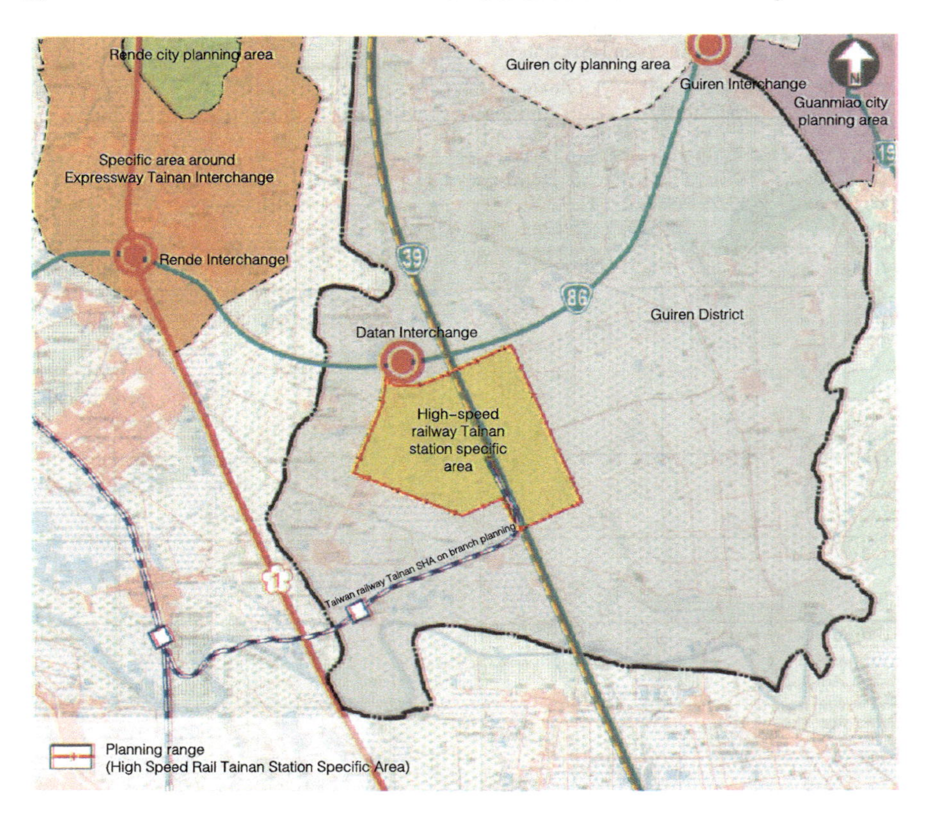

Fig. 2.22 Location of Tainan HSR station, *original source* Map of Taiwan, Tainan Government, *Proposal of Changing the Site of Tainan HSR Station*, April, 2014

The external road transport system of Tainan HSR Station includes the MRT network, road network and railway network:

(1) MRT station—Through the Red Line of Mass Rapid Transit (MRT) in the Tainan metropolitan area, the city of Tainan, and the Tainan railway station are connected. Travelers can also use the Extension Line to connect to the Science Park, coupled with the HSR station and Sharon Station Taichung, which in the future will connect to the Taiwan Railway Station in Nakasu and increase the convenience of HSR and the MRT-Taiwan Railway Interchange;
(2) Road network—There are liaison systems of highways 1 and 2, the east–west expressway, the Sharon Branch of Taiwan Railways and roads under HSR bridges to connect HSR stations and urban areas of Tainan in series with high traffic accessibility; and
(3) Railway network—HSR and Taiwan Iron and Steel Sharon Station are the same station, and Taiwan Iron and Steel Shalun Branch has the MRT function, so there is no need to transfer to an HSR station.

Fig. 2.23 Planning area of Tainan HSR station

The surrounding area is divided into an industrial area, a dedicated station area, a commercial area and residential areas (Figs. 2.22, 2.23 and 2.24):

(1) The industrial area covers an area of 47.1 ha. It will take Tainan National Chiao Tung University as the core of R&D and strive to attract optical institutes and laboratories. At the same time, it will guide green energy-related industries, such as solar photovoltaic, LED lighting and optoelectronics industries, wind power and biomass fuel, with its upstream and downstream service stations. It will provide R&D units and commercial service facilities (conference centers, hotels and business services), including business exhibitions, corporate operations, research and development, training, logistics centers and international certification laboratories.
(2) The dedicated station area is approximately 16.7 ha, of which the supporting facilities cover an area of approximately 4 ha and are located in close proximity to the Tainan HSR station. Diversified services such as tourist hotels, business offices, shopping and entertainment centers and office buildings will be developed to meet the needs of sightseeing, leisure, living and industry.
(3) Approximately 2.5 ha of commercial area owned and operated by the HSR will tie in with the development of the regional convention and exhibition facilities as well as the development of the industrial zones. The establishment of the Tainan Convention and Exhibition Center and related commercial service facilities will provide business exhibition, business exchange, retail shopping, dining and

Fig. 2.24 Land use layout surrounding Tainan HSR station, redrawn by the author, *original source* Map of Taiwan, Tainan Government, *Proposal of Changing the Site of Tainan HSR Station*, April, 2014

leisure and other services to meet the needs of MICE activities and industrial development businesses.

(4) Approximately 23.2 ha of residential areas are owned and managed by the HSR, of which 2.5 ha are located on the north side of the industrial zone. These can be used as demonstration areas for government to promote ecological communities. The remaining residential areas will provide local residential development needs.

Case 3: Taiwan Hsinchu HSR Station

The Hsinchu HSR station is located on the northeast side of the city of Hsinchu. It is approximately 5 km from Hsinchu Science Park and the center of Zhubei and approximately 8 km from the center of Hsinchu. Through two rail links, three highways are linked with Hsinchu and stand together with other railways. Biomedical Park, the flagship park of the knowledge economy, Hsinchu Science Park Phase III, National

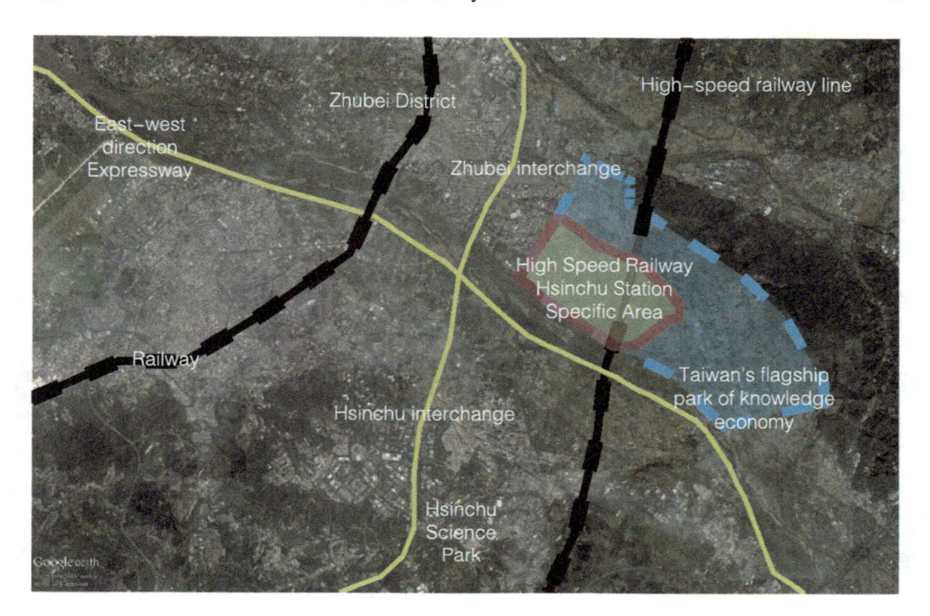

Fig. 2.25 Location of Hsinchu HSR station

Chiao Tung University, National Tsing Hua University and Taiwan University are scheduled.

The planned area of the Hsinchu HSR station is approximately 309 ha, of which 139 ha are for public facilities. The planned population is 32,000 with a residential density of 363 people per hectare. Hsinchu has a core of high-tech industries; in addition to the existing Hsinchu Science Park, ITRI and National Tsing Hua University, and National Chiao Tung University, Hsinchu Science Park Phase III, the flagship park of the knowledge economy, and a biomedical science and technology park will follow the station. Coupled with the gathering of new schools such as National Taiwan University, National Chiao Tung University, National Tsing Hua University, Taiwan University of Science and Technology and other famous tertiary institutions, Hsinchu will further enhance the competitiveness and research and development capabilities of high-tech industries, and the Hsinchu HSR station will develop a "Biomedical Technology City" with biotech, science and technology R&D capabilities. The main development strategies are as follows (Figs. 2.25 and 2.26):

(1) Cooperate with the development of Hsinchu Science City and plan the international science and technology exchange portal based on the combination of science parks, ITRI, National Tsing Hua University, National Chiao Tung University and other technology industries and human resources.
(2) Attract international high-tech industries to invest and establish factories in Taiwan. The station area will coordinate with the convenient transfer service and play the function of an overall electronic technology information center. It

Fig. 2.26 Construction status quo around Hsinchu HSR station, *source* Google Map, 2016

will introduce research, consultancy and technical service industries to promote the development of unique cultural and sightseeing resources in the district.

(3) Integrate office buildings with exhibition and presentation facilities and extremely comfortable business and leisure facilities and develop the district as a high-tech and communications business park, thus achieving the goal of the "Hsinchu Science City Development Plan."

The Hsinchu HSR site external road transport system includes the MRT network, road network and railway network (Figs. 2.27, 2.28 and 2.29):

(1) MRT network—The currently planned MRT Red Line and Blue Line of the metro area of Hsinchu will rendezvous within the designated area of the HSR and can be connected to Hsinchu station, Hsinchu Science Park, the city center and City Hall.
(2) Road network—There are road systems such as highway 1 and highway 3, east–west expressways and the 60 m park road, and the road network extends in all directions.
(3) Railway network—In the special area of the HSR, there is the Taiwan Railway Neiwan Branch station; Hsinchu station as a starting station, has a total length of 11.2 km; Hsinchu HSR station outside the mass transit rail system to facilitate the Science Park and Hsinchu Science Park Phase III. Park-area passengers take the HSR.

Summary: Cases in Taiwan

The Taiwan HSR stations have obvious location advantages; both are 5–10 km from the city center. The surrounding industrial parks promote science and education, ecology or transportation and other rich resources on a scale of 300–500 ha for the

Fig. 2.27 Planning area around the Hsinchu HSR station

Fig. 2.28 Road planning around the Hsinchu HSR station

Fig. 2.29 Land use layout surrounding Hsinchu HSR station, redrawn by the author, *original source* http://castnet.nctu.edu.tw/files/imagesdb/172/20111127094512.jpg

medium-sized' cities' HSR station specific area. With a clear development orientation relying on local industrial advantages and favorable geographical locations targeted for future development, each site has its own unique positioning. At the same time, HSR stations and the main cities have a good interactive relationship. The group of cities under the guidance of the main city relies on the industrial foundation and advantages of the main city for development. At the same time, HSR is also an important gateway for the further development of the main city. In the transportation system, a seamless connection is made between the MRT system, the Taiwan Rail Extension Line or the BRT system and the HSR ring. Together with the road network, a seamless transport system is formed. In addition, much attention has been paid to the connection with the regional central cities, other surrounding cities, and HSR as an important transportation node in the region. Image of cities are unique and distinctive, the main considerations are to combine the characteristics of transportation with HSR stations, to plan to combine local features and nearby construction, to preserve natural resources with local characteristics and to plan appropriate industrial zones and create unique urban imagery through urban design.

Cases in Japan

The Shinkansen in Japan was built four times, in 1964, 1975, 1987 and 1998, and currently forms five lines with a total distance of approximately 2000 km at a speed of approximately 320 km/h. The economic contribution of the Shinkansen station areas is concentrated in information consultation, investment consultation, advertisement,

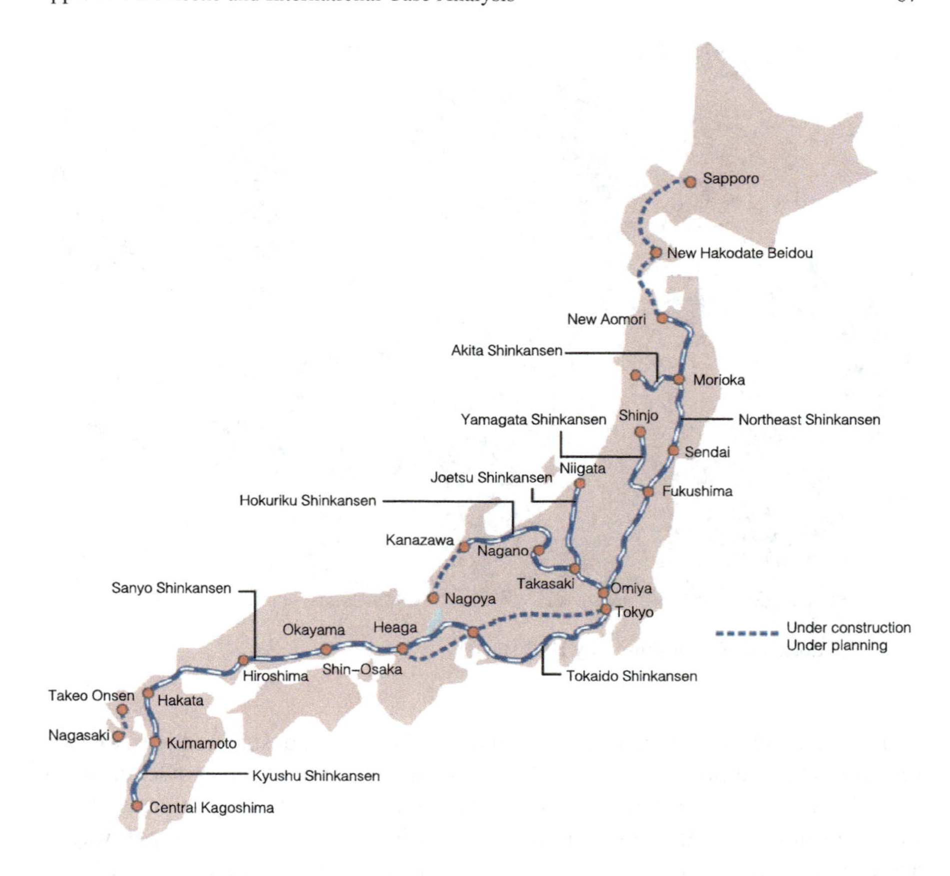

Fig. 2.30 The Shinkansen in Japan, redrawn by the author, *original source* http://en.wikipedia.org/wiki/Shinkansen

commerce, business services and real estate services. Whether in the middle of the station or the station terminal area, the catering service industry and the tourism industry have witnessed significant increases (Fig. 2.30).

Case 1: Kagoshima HSR Station

The Kagoshima HSR station is located on the west side of the city of Kagoshima, 2 km from the city center.

The development is positioned as the urban subcenter, with the core area being the ring-shaped area formed by the loop and the HSR line. Road traffic uses a ring-shaped radial structure, with the central road using the parkway form. Along the northern side of the trunk road connected with the city center, industrial development zones are

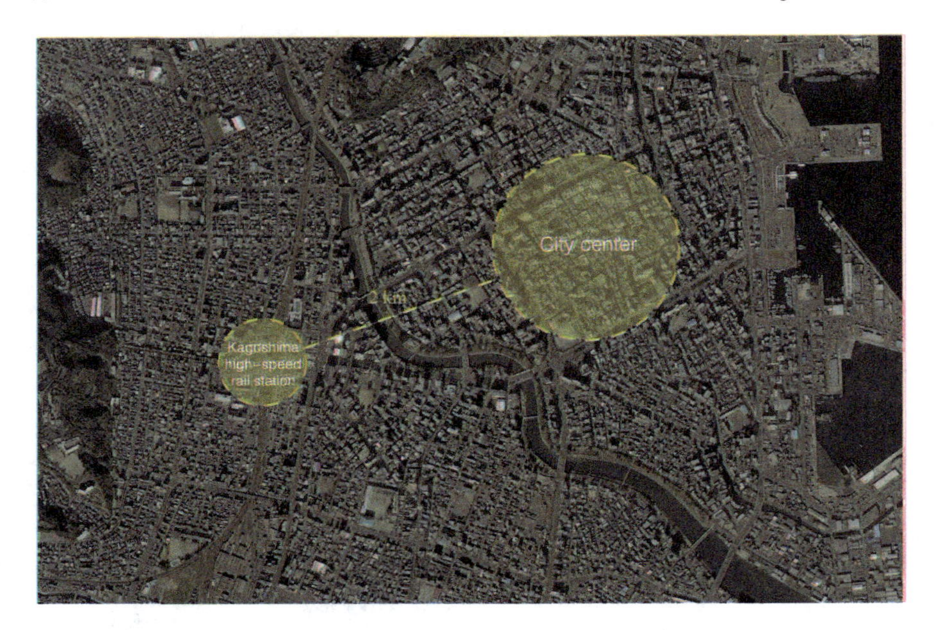

Fig. 2.31 Location of Kagoshima city HSR station

arranged on both sides which include commercial facilities, business offices, tourism services, and cultural and recreation facilities. On the south side are residential areas and related service facilities. With the station as the center, three radial traffic roads form a symmetrical structure. The central view is realized through the parkway to achieve the formation of an envisioned landscape corridor (Figs. 2.31 and 2.32).

Case 2: Kumamoto HSR Station

The Kumamoto HSR station is located in the built-up area west of the city of Kumamoto, 2 km away from the city center. As a deputy city center, the development goals are as follows (Figs. 2.33, 2.34 and 2.35):

(1) Integrate urban functions and development of commerce, business, tourism, culture, etc.;
(2) Create a convenient and friendly living environment, leading to a brand-new style of urban life;
(3) Strengthen the function of regional transport nodes, create a compact urban space and promote people's exchange. The total area around the HSR station is 358 ha, of which the key development area around the station is 63.2 ha. Road traffic planning conforms to the river to form an irregular grid. Land use features include business, tourism, culture, etc., emphasizing the integration of functions. The station area has a green square space, a station road traffic space, a river hydrophilic recreation space and living space.

Fig. 2.32 Construction status quo of the area around Kagoshima, *source* Google Map, 2016

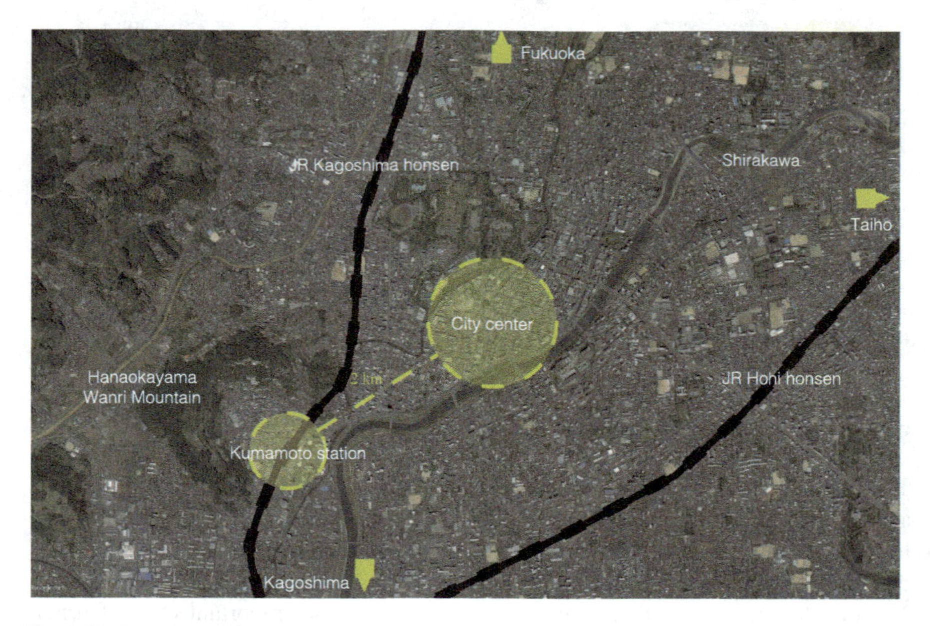

Fig. 2.33 Location of Kumamoto HSR station

Fig. 2.34 Development area around Kumamoto HSR station

Summary: Cases in Japan

The selected cases of Japan's Shinkansen are located in urban areas. Both stations are about 2 km away from the city center. They serve comprehensive functions, including commercial, office, residence, tourism, culture, etc. All are planned as the deputy center of the city, close to the city center and linked to earlier development. The development of business and trade is mature as an important core of urban development. Generally, with other railway lines, other stations, and a large number of city bus, subway and other close links, the area is the logistical distribution center of the city's flow of people. In terms of design, the area surrounding the station is modeled as a positive public space suitable for people to interact and live by integrating the rivers, roads and buildings around the station.

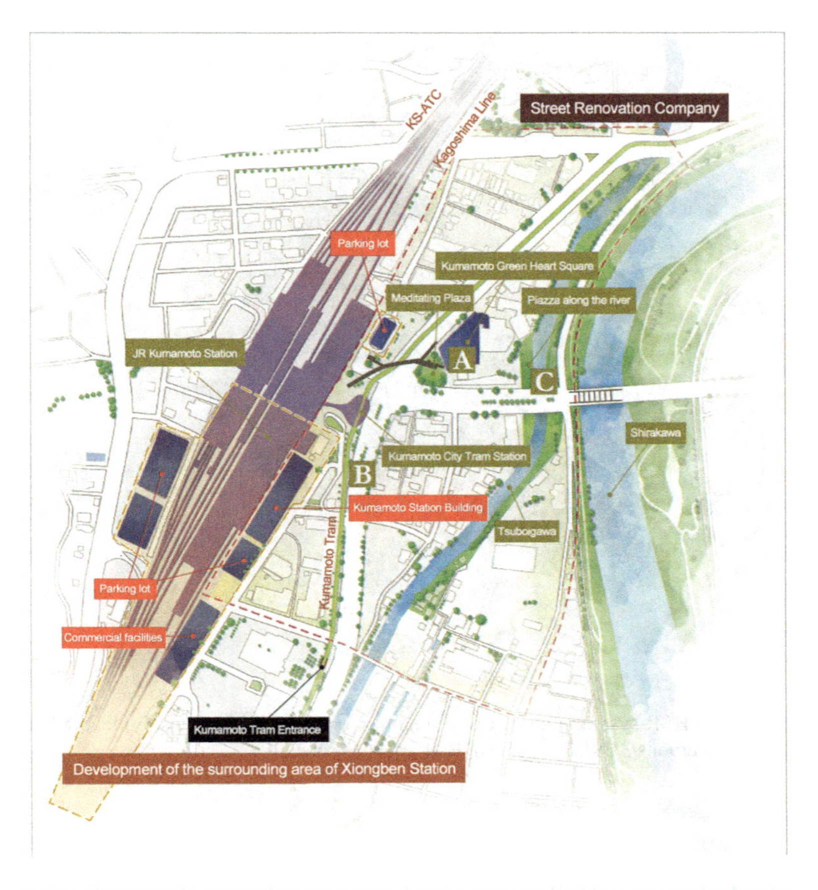

Fig. 2.35 Rendering and Site Plan of Kumamoto HSR Station, *source* http://www.jlgc.org.uk/en/news_letter/japan-study-tour-2014-to-kumamoto-and-tokyo/

Fig. 2.36 French HSR system, redrawn by the author, *original source* https://upload.wikimedia.org/wikipedia/commons/3/39/High_Speed_Rail_Map_Europe.gif

Cases in France

Currently, there are nine HSR in operation in France with a total length of 2023.6 km (Fig. 2.36).

Case 1: Lyon HSR Station

Lyon is the second-largest city in France, with a total population of 1.648 million and a population of 472,000 in the urban area; the urban area is 47.95 km^2. The demand for office space increased 43% (equivalent to an annual increase of 5.1%) between 1983 and 1990, when Train a Grande Vitesse (TGV) was completed. Some companies in Paris began to establish branches in Lyon. Company managers can travel between Paris headquarters and the Lyon branch on the same day, extending business activities in Paris to the south of France while smaller companies in Lyon are beginning to offer specialized services to many Parisian companies.

There are two train stations in Lyon: Perrache and Part-Dieu stations, both in a modern style. Perrache Station is located in the city center, 20 km away from the airport to the east of the city (Figs. 2.37, 2.38 and 2.39).

Fig. 2.37 Locations of the two HSR stations in Lyon

Pardemento Station is farther from the city center. The functions of Perrault's surrounding areas include universities, museums, churches, commercial centers, squares, cultural exposition, tourism, commerce and trade. The area of 33 ha around Pardeeg station is an "Agreement Development Zone"; its functions include administrative centers, commercial finance, offices, hotels, residences and parkland.

The opening of the TGV southeast line shortened the travel time from Lyon to Paris to 2 h for basic business travel. Part-Dieu Station is for TGV after the completion of the new line. Paldillo Station is currently the focus of most train services while Perrache Station continues to provide some regional services. The opening of the Pardement Station and the TGV has brought further development to

Fig. 2.38 Satellite photo of Perrache station, *source* Google Map, 2016

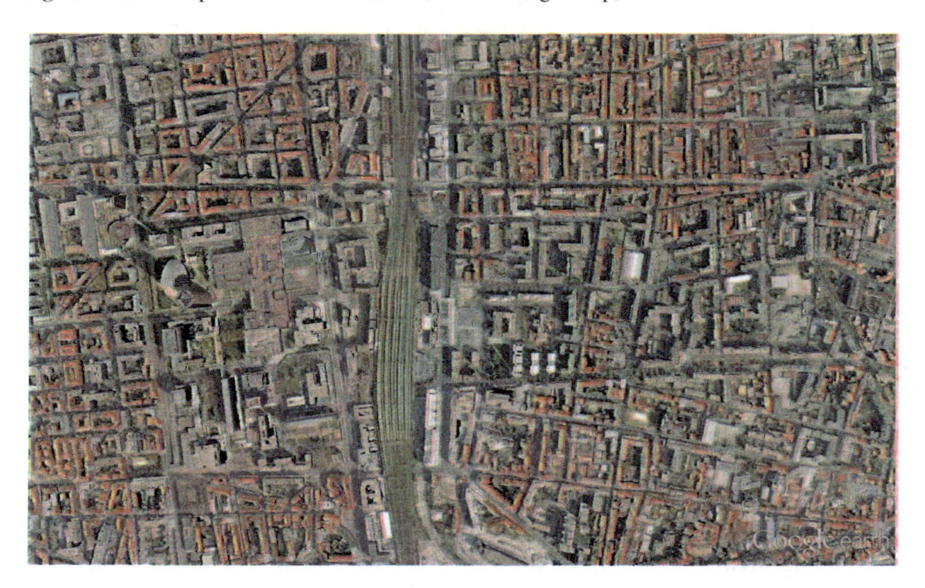

Fig. 2.39 Satellite photo of Part-Dieu station, *source* Google Map, 2016

the area around the business district. Many businesses were originally elsewhere in the city but decided to move to Paldi to be closer to the TGV in order to benefit from TGV travel. TGV also led to the development of tourism in the site area and the number of hotels where tourists stay also increased correspondingly.

The appeal of Paldiyo has continued, mainly for the following three reasons:

(1) Many years of networked development have expanded Lyon's TGV service, and currently, it is a gateway not just to Paris and the Rhone Valley, but also to other cities in France.
(2) The city's public transport system has evolved to make it easier to reach the area, and the TGV station from the satellite city that is even farther away.
(3) The business scale and commercial relocation further attracted other activities. An iconic feature has been the lack of hotels in the area. The current increase in the number of hotels indicates that TGV has led to the development of tourism in the site area, with a corresponding increase in the number of hotels.

Case 2: Lille HSR Station

Lille, a traditional industrial city in the 1970s, is located in the heart of an industrial area. Lille successfully transformed itself into a city dominated by business offices with the completion of the HSR in 1993. The TGV North Line is seen as a key to developing service industries with Lille as a node between Paris, Brussels and London. The first phase of the project includes a plan to build a "Lille Europe project," including the Convention and Exhibition Center (20,000 m^2), the European Building (25,000 m^2), the bank building (approximately 15,000 m^2), the Euralille Business Center (90,000 m^2), etc. (Fig. 2.40).

Economic growth in the area has slowed for some time due to reliance on coal, steel, cotton and other traditional products since these products have been imported

Fig. 2.40 Satellite photo of Lille HSR station, *source* Google Map, 2016

from other parts of the world where they are cheaper. Municipalities have decided to change the role of the metropolis to develop more services. The TGV North Line links Lille to Paris, Brussels and London. The TGV North Line opened in 1993 and has served the three capital cities since 1994. With the expansion of the TGV network and other high-speed lines in other countries, Lille has more opportunities for development.

After a period of development, a series of changes have taken place around the Lille site, including patterns of activity, especially land use and regeneration. The station serving TGV (Lille Europe) was built near a former military barracks section. Most of the land has been developed into commercial centers, offices, hotels and a large modern retail center. The remaining land will be used for a park. The entire site is adjacent to the old town center and has reached a certain size. The redevelopment project has led to the construction of many new office buildings, public housing and a large conference center adjacent to Lille Europe forming an important strategy for sustainable development in the region. Further development plans are under way and will be redeveloped to include a closed railway yard and other abandoned sites. Local universities were reorganized. For example, some departments were relocated to the former cotton mill in the old district to create new jobs and businesses. Old urban development benefits from redevelopment plans. At the same time, the construction of traditional universities is used by large enterprises (usually corporate headquarters). Supplementary programs for redevelopment projects have started in other parts of the metropolitan area, especially in Roubaix and Tourcoing, formerly cotton-spinning towns. Abandoned public buildings and industrial buildings have been transformed into modern offices, leisure and entertainment venues, community centers and specialty schools. For example, the Euroteleport complex in the Roubaix region is a modern retail center aimed at strengthening traditional urban centers. At the same time, residential areas in these towns have also developed improvements and new programs.

Summary: Cases in France

The arrival of HSR has enabled Lyon and Lille to flourish in different ways, especially in the service economy. These two examples illustrate the key drivers of HSR development and are equally crucial to other cities that benefit from HSR services. For example, economic and land use trends should be linked to the development of HSR. The greatest impact of HSR is on the service industry. It becomes a catalyst for sustainable development. According to the development of HSR, planning should complement existing strategies and cannot simply create activities out of thin air. HSR stations should be integrated with the city as a whole. Station site selection and function settings should be consistent with the city's development strategy. Rundown and abandoned areas (including railways and redundant industrial areas) have a chance for rejuvenation. Strong and well-oriented urban governance is an essential element, and a consistent development strategy is the key to success. Effective regional and local transport is the key to connecting various elements.

Selected examples of French HSR stations are located at the edge of the city center and integrated with other functions including administrative centers, commercial financial facilities, offices, hotels, residences and green parks. Generally, with other railway lines, train stations, city bus routes, subway and other close links, these areas become the logistical distribution center for the city's flow of people. Parkways combine landscape and transportation to shape the area around the station into a positive public space where people can interact and live.

Domestic Case Analysis
HSR Stations along the Beijing-Fuzhou HSR Line

According to the *Beijing-Fuzhou Railway (Hefei to Fuzhou Section) Pre-Feasibility Study Report*, the Beijing-Fuzhou HSR line (Hefei to Fuzhou section) starts at the north of Hefei, the capital of Anhui Province, passes through five cities in Anhui Province (Chaohu, Tongling, Wuhu, Xuancheng and Huangshan), Shangrao in Jiangxi Province and two cities in Fujian Province (Nanping and Ningde). It runs south to Fuzhou, the capital of Fujian Province. Along the line after Huangshan are Jiuhuashan, Sanqingshan, Wuyishan, other famous mountain communities, and other ecotourism destinations like Jixi, Wuyuan. The total length is approximately 808 km, of which the total length in the Anhui section is 330 km. There are 24 stations along the line (11 stations in the Anhui section). The Beijing-Fuzhou HSR line and the Beijing-Shanghai HSR line are the same lines in the Beijing-Hefei section. The section from Hefei to Fuzhou is called the Hefei-Fuzhou HSR line. There are 21 stations along the line, including one converted station (Hefei West Railway Station), 2 existing stations (Shangrao Station and Fuzhou Station under construction) and 18 new stations (Fig. 2.41).

Among them, there are two terminal stations (Hefei South Railway Station and Fuzhou Railway Station), five stations for handling some passenger train operations (Huangshan, Shangrao, North Wuyishan, Wuyishan and Nanping) and 14 general intermediate stations (West Hefei, Changlinhe, Chaohu, Wuwei, North Tongling, Nanling, Jingxian, Jingde, North Jixi, Wuyuan, Dexing, West Jian'ou and East Minqing). Tongling Station is a general intermediate station.

The selected cases of HSR stations along the Beijing-Fuzhou HSR line's southern section focused on the traffic distribution function of the HSR stations and developed the corresponding functions of commerce and trade, business offices, entertainment and living. The planned area is a minimum of approximately 1.5 km^2, with planning for 3–4 km^2 of more land use sites; the largest site is approximately 7.5 km^2, corresponding to the largest amount of estimated traffic.

According to the population size and expected flow rate of different counties and cities to determine the level of HSR stations, a station located in a prefecture-level city center area is generally 6000 m^2, and a station located in a county-level city or county is generally 3500 m^2. Special sites such as North Jixi Station (up to 12,000 m^2) or North Wuyi Station (down to 3500 m) are established separately. Tongling Station has a site construction design area of 6183 m^2, and new site size also includes Xinchaohu Station, Wuyuan Station, Jian Yang Station and North Nanping Station.

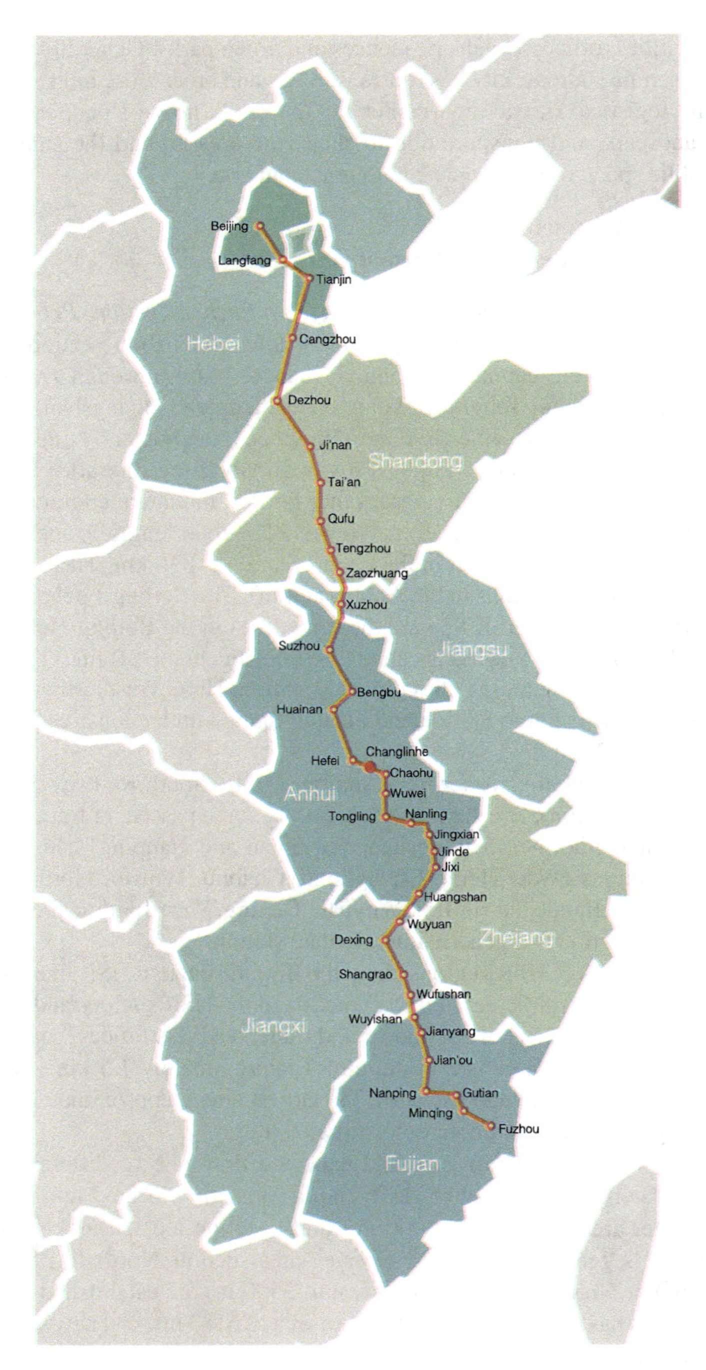

Fig. 2.41 Beijing-Fuzhou HSR line, redrawn by the author, *original source* https://www.travelchinaguide.com/china-trains/high-speed/beijing-hefei-fuzhou.htm

According to the *Beijing-Fuzhou Railway (Hefei to Fuzhou Section) Pre-Feasibility Study Report*, in 2009, the passenger flow forecast in the Fujian section from north to south of the Beijing-Fuzhou HSR line in 2015 (2015 was the completion and opening year for this line) was Wuyishan—2.5 million, Jianyang—700,000, Jianou—900,000, Nanping—4.5 million, Gutian—600,000 and Minqing—500,000.

The selected cases are Jianyang Station (East Wuyishan Station), North Wuyishan Station, Nanping Station, West Jian'ou Station, North Gutian Station and Wuyuan Station.

Case 1: Jianyang Station (East Wuyishan Station)

East Wuyishan Station is located in Yangdun Nature Village, Hengtang Village, Jiukou Town, city of Jianyang, Fujian Province. It covers an area of approximately 3.2 km^2 and was expected to have a traffic volume of 700,000 passengers per year in 2015. The site is located in the heart of the northern Fujian industrial concentration area approximately 21 km north of Wuyi Mountain, 33 km north of Wuyishan, 15 km south of Jianyang, 66 km east of Pucheng District and 63 km from Shaowu. The Jianyang urban area can be reached from Wuyishan in a scenic drive of less than 30 min. The site is next to Wuyishan International Airport.

The land area of the station is 1000 acres (67 ha), supporting a station square, bus terminal, bus hub and other facilities. The planning and construction of the new zone cover a total area of nearly 3500 acres (233 ha), supporting a number of residential and commercial facilities (Fig. 2.42).

Fig. 2.42 Pre-station planning and design of Jianyang station, *source* http://www.fujianbid.com:84/uefiles/20140904/20140904090740677З.jpg

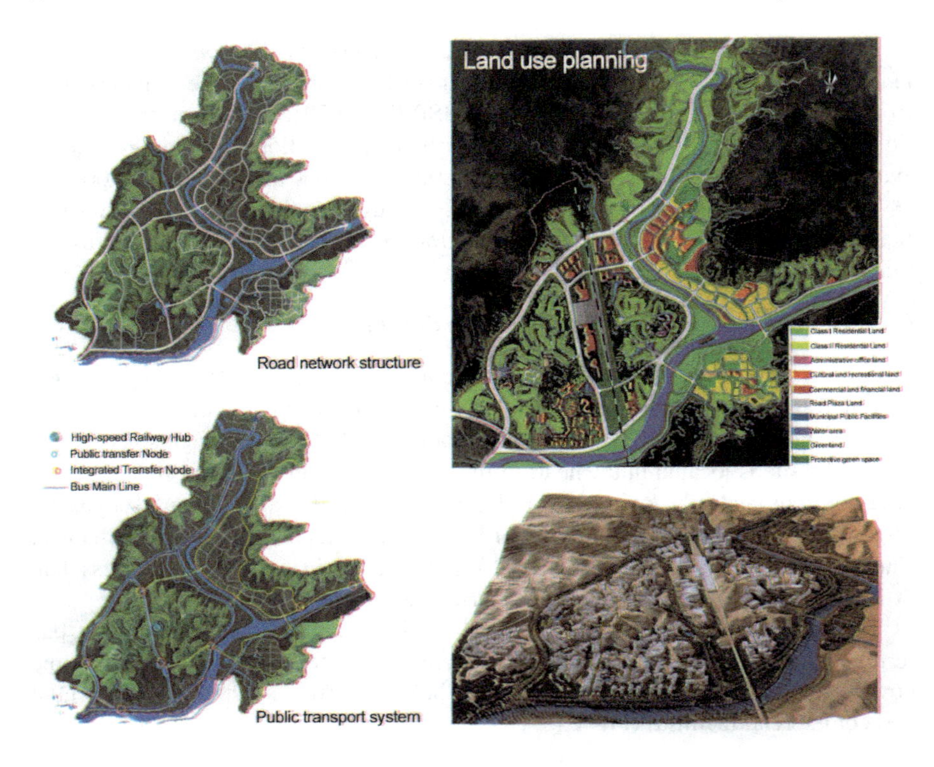

Fig. 2.43 Road network, public transportation network and Land use layout of North Wuyishan station surroundings, *source* 生态安全理念下的山地城市新区规划——武夷山市北城新区城市设计实践为例

Case 2: North Wuyishan Station

The site is located in the main urban northeast new development plots of Wuyishan. The surrounding planning area is 4 km^2, and the core area is 2 km^2. The site is approximately 3 km away from the urban area, approximately 10 km away from the airport and approximately 3 km away from the expressway entrance (Fig. 2.43).

The North Wuyishan station area provides life services and traffic distribution functions, considering commercial services, entertainment and other complex functions showing the natural characteristics of the Wuyi life services center. It was expected that in 2015, the passenger flow would be 2.5 million people per year.

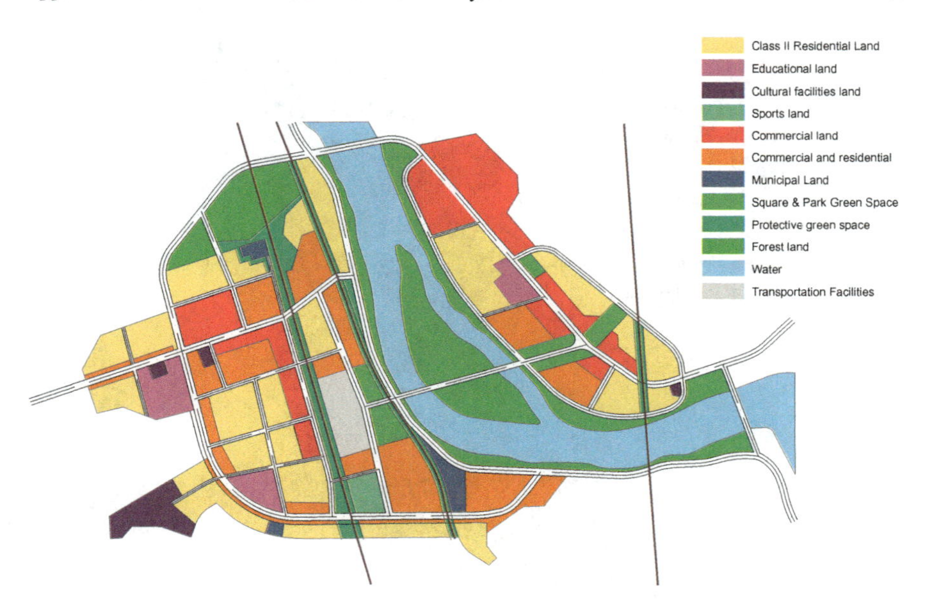

Fig. 2.44 Land use layout surrounding West Jian'ou station, *source* http://www.np.gov.cn/cms/html/npszf/2016-08-10/1145317476.html

Case 3: West Jian'ou Railway Station

The site is located on the northwest side of Jian'ou. The planned land area is 4.6 km^2. The basic functions of the planning in the area surrounding the site include transport interchange hub, catering, accommodation and the expansion of functions for business offices, business services and information services. It was expected that in 2015, the traffic volume would reach 900,000 people per year (Fig. 2.44).

Case 4: Nanping Station

A traffic flow of 4.5 million passengers per year was forecast for the site in 2015. The site development area surrounding the planning area is 7.5 km^2. The distance from the main city is approximately 11 km, and the distance from the highway entrance is 1 km. The basic positioning of the area around the site includes the important nodes of the coordinated development of the economic zone on the west side of the straits, the traffic hub of northwestern Fujian, the growth point of the east extension of Nanping, the new livable comprehensive ecological area, the integrated transportation hub of the region, the important window of the central city and the development link of Jiangnan New City. The main functions of the planning area are transportation hub, commercial services, business offices, culture and entertainment, medical education and living facilities and corresponding facilities (Fig. 2.45).

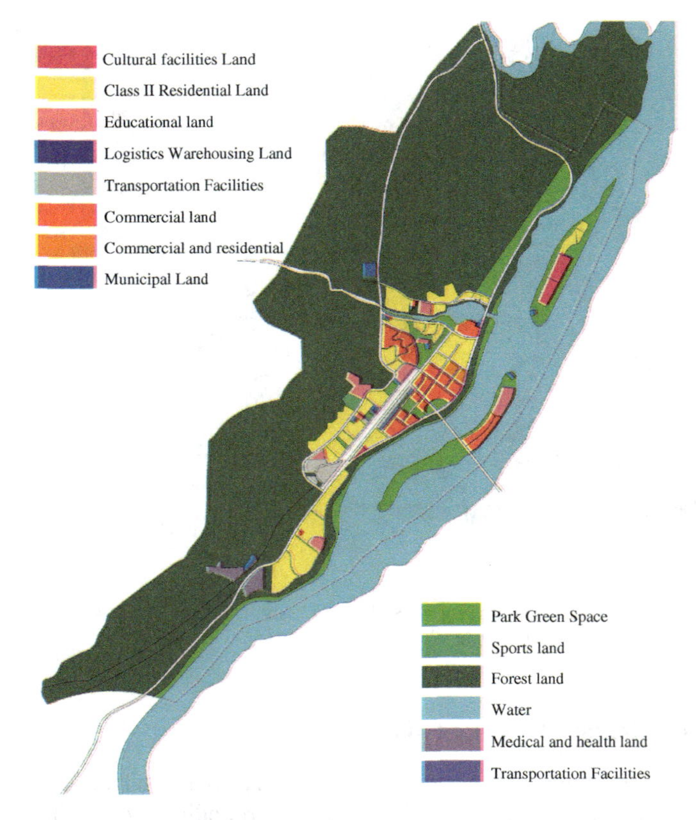

Fig. 2.45 Land use layout surrounding Nanping station, *source* http://www.npgh.gov.cn/cms/pages/340352445669240001/images/20151029042338531.jpg

Case 5: North Gutian Station

The site is approximately 16 km from the county town, and the station is a medium-sized station for HSR. The planning takes North Gutian Station as the center and the development of land north and south covers approximately 2 km, with an east–west width approximately 0.4–1.2 km and 116.5 ha of land for planning. The position plan is for a Gutian Development Zone and an economic growth point. It was expected that by 2015, the passenger traffic would be 600,000 people per year (Fig. 2.46).

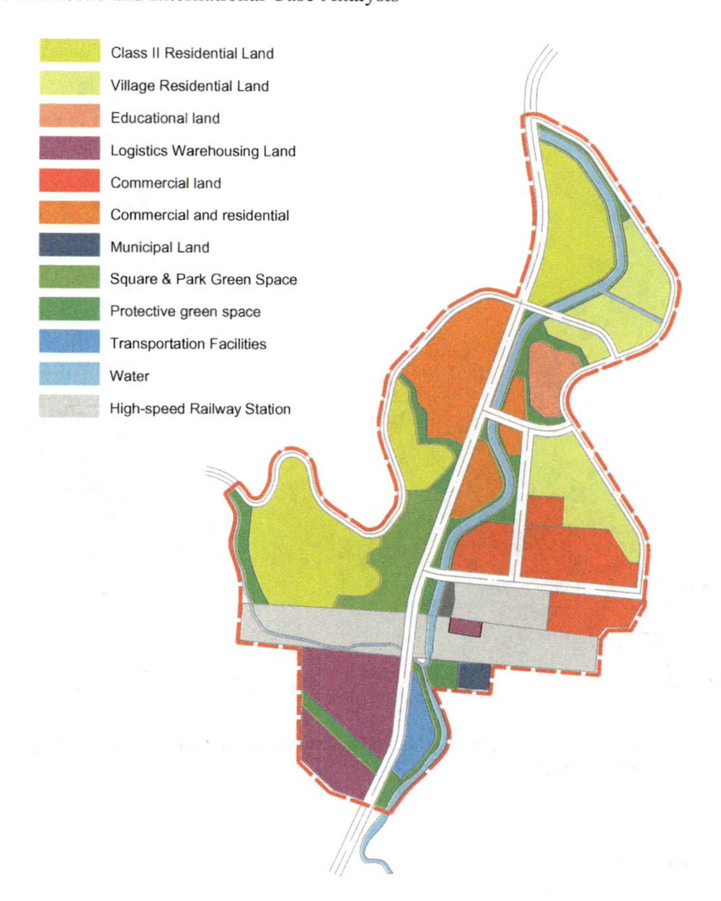

Fig. 2.46 Land use layout surrounding Gutian north station, *source* http://www.352200.com/center/uploadfile/jpg/2010-9/2010916152652960.jpg

Case 6: Wuyuan Station

The HSR station is approximately 3 km away from the county seat and is located in Meizhou Village, Chaoxi Town. The site is constructed according to the national secondary station standard, and the main building includes "2 stations and 5 lines." After the completion of the station, the area will reach 8460 m^2, with a waiting area on the fifth floor of the main building.

The station square ground-level design is 96.80 m^2 with a planning area of 6.7 ha. The auxiliary floors will contain integrated services and a large mall to be managed by an internationally renowned business operations management company. The plan is to build a well-known shopping district in northeastern Jiangxi based on tourism and the multiple advantages of a floating population at the "county-first" railway station in Fujian Province (Fig. 2.47).

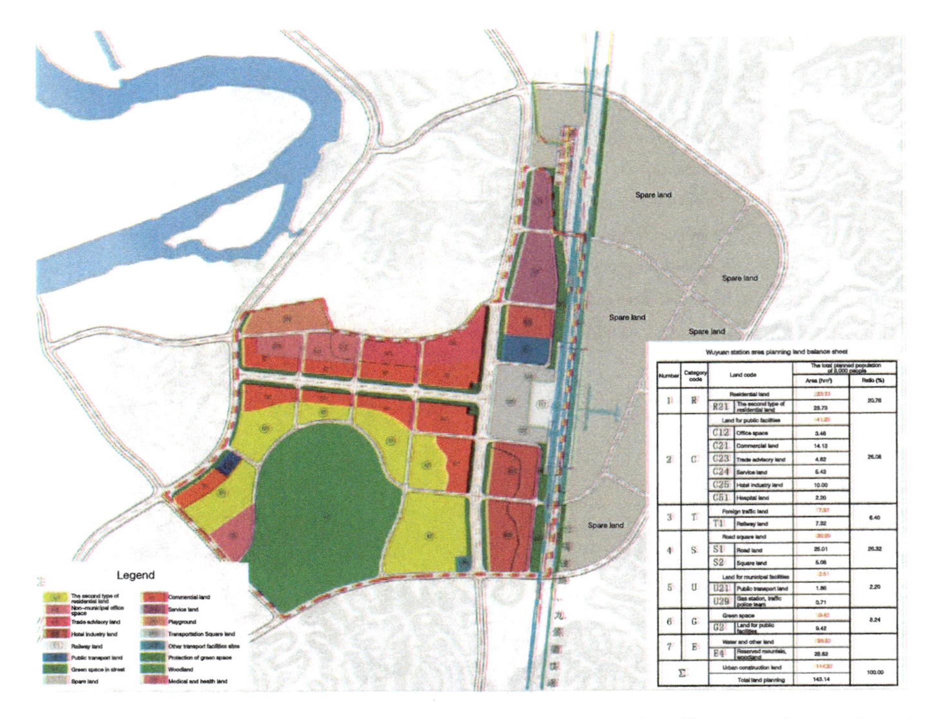

Wuyuan station area planning land balance sheet

Number	Category code	Land code		The total planned population of 8,000 people	
				Area (hm²)	Ratio (%)
1	R	Residential land		23.73	20.76
		R21	The second type of residential land	23.73	
2	C	Land for public facilities		41.25	26.08
		C12	Office space	3.46	
		C21	Commercial land	14.13	
		C23	Trade advisory land	4.82	
		C24	Service land	6.43	
		C25	Hotel industry land	10.00	
		C51	Hospital land	2.20	
3	T	Foreign traffic land		7.32	6.40
		T1	Railway land	7.32	
4	S	Road square land		30.09	26.32
		S1	Road land	25.01	
		S2	Square land	5.08	
5	U	Land for municipal facilities		2.51	2.20
		U21	Public transport land	1.86	
		U29	Gas station, traffic police team	0.71	
6	G	Green space		9.42	8.24
		G2	Land for public facilities	9.42	
7	E	Water and other land		28.82	
		E4	Reserved mountain, woodland	28.82	
Σ		Urban construction land		114.32	100.00
		Total land planning		143.14	

Fig. 2.47 Land use layout surrounding Laiyuan station, *source* http://www.srghj.gov.cn/xzgh/wyxj/ghgs/201507/2860.html

Site Planning Comparison

The selected cases of HSR stations along the Beijing-Fuzhou HSR line's southern section focused on the traffic distribution function of the HSR stations and developed the corresponding functions of commerce and trade, business offices, entertainment and living. The minimum size of land use for a station is approximately 1.5 km^2. Most of the land uses for the stations are planned for 3–4 km^2, and the largest land use area is 7.5 km^2 corresponding to the highest level of estimated traffic (Table 2.6).

Table 2.6 Comparison of new Beijing-Fuzhou HSR line stations' planning

Station	Station building area (square meters)	Planning area of station area (square kilometers)	Expected traffic (ten thousand people/year)	Location	Function
Jianyang	6000	3.2	70	15 km from downtown	Main: Center, residential, commerce and trade
Wuyishan North	3500	4	250	3 km from downtown, 3 km from highway, 10 km from airport	Main: Living, center Secondary: Business, recreation
Jianou West	4000	4.6	90	5 km from downtown	Main: Center, catering, housing Secondary: Business offices, commerce and trade, information services
Nanping	6000	7.5	450	11 km from downtown, 3 km from highway	Main: Transportation hub, commercial services, commerce and business, living Secondary: Life, culture and entertainment, health and education
Gutian North	3500	1.17	60	16 km from downtown	Main: Transportation hub, living, business services, trade and business Secondary: Health and education, support services
Wuyuan	8460	1.43	450	3 km from downtown	
Tongling North	6183	–	110	32 km from downtown	–

Source Author

References

Cervero R, Landis J (1997) Twenty years of the bay area rapid transit system: land use and development impacts. Transp Res Part A Policy Pract 31(4):309–333
Cervero R, Murakami J (2009) Rail and property development in Hong Kong: experiences and extensions. Urban Stud 46(10):2019–2043
Feng CC, Feng XB, Liu SJ (2013) Effects of high-speed rail network on the inter-provincial accessibilities in China. Prog Geogr 32(8):1187–1194
Handy S (2005) Smart growth and the transportation-land use connection: what does the research tell us?. Int Reg Sci Rev 28(2):146–167
He N, Gu BN (1998) Analysis of the role of urban rail transit in land use. Urban Mass Transit 1(4):32–36
Hess DB, Almeida TM (2007) Impact of proximity to light rail rapid transit on station-area property values in Buffalo, New York. Urban Stud 44(5):1041–1068
Immergluck D (2009) Large redevelopment initiatives, housing values and gentrification: the case of the Atlanta Beltline. Urban Stud 46:1723
Knaap GJ, Ding CR, Hopkins LD (2001) Do plans matter?: The effects of light rail plans on land values in station areas. J Plann Educ Res 21(1):32–39
Li LL, Zhang GH, Cao YL (2007) Technical method for planning and adjusting land use around rail transit stations: a case study of Suzhou. UTC J 5(1):30–36
Li LL, Zhang GH, Cao YL (2007) Approaches yo land-use planning and adjustment adjacent to rail transit stations: a case study in Suzhou. UTC J 5(1):30–36
Liu JL, Zeng XG (2004) Urban rail transit and land use integrated planning based on quantitative analysis. J China Railway Soc 26(3):13–19
Luo PF, Xu YL, Zhang NN (2004) Study on the impacts of regional accessibility of high speed rail: a case study of Nanjing to Shanghai Region. Econ Geogr 24(3):407–411
Ma AK, Cao RL, Zhang PG et al (2008) Impacts of the expansion of the railway network on reginal accessibility: a case study of the urban agglomeration along with the Jiaoji railway. J Shandong Normal University (Natural Sci) 23(2):89–93
Pol PMJ, Zhou J (2011) Governing urban developments around high-speed train stations: experiences from four European cities. Urban Plan Int 26(3):27–34
Polzin SE (1999) Transportation/Land-use relationship: public transit's impact on land use. J Urban Plann Dev 125(4):135–151
Ryan S (1999) Property values and transportation facilities: finding the transportation-land use connection. J Plann Lit 13:412–427
Shu HQ, Shi XF (2008) The effect Tokyo metropolis circle's railway system to urban spatial structure development. Urban Plann Int 23(3):105–109
Wei LH, Yan XP (2006) Driving forces and patterns of social spatial evolution in transitional urban China: a case study of Guangzhou. Geogr Geo-Inf Sci 22(1):67–72
Yang SH, Ma L, Chen S (2009) Interaction of spatial structure evolution and urban transportation. UTC J 7(5):45–48
Zhang TW (2001) The variations of urban spatial structure and its driving mechanism in 1990s in China. City Plann Rev 25(7):7–14

Chapter 3
Impact of HSR Station on the Surrounding Area

In the context of the large-scale construction of regional transportation infrastructures in China, this chapter takes the 22 cities with HSR stations located along the Beijing-Shanghai HSR line and their surrounding areas as research samples to explore the mechanism of the impact of HSR stations on the urbanization of the surrounding areas. Based on theoretical assumptions and the relevant data of those stations and the cities in which they are located from 2006 to 2012, the empirical study carried out cluster analysis and constructed multiple regression models and structural equations to further define the core elements shaping the development. In the meantime, the information collected from field surveys and plans helped us conduct a detailed analysis of the planning issues, current situations and problems of site selection, development scale, land use and connections with other transportation facilities.

On January 7, 2006, the State Council executive meeting approved the application of wheel-rail technology in the construction of the Beijing-Shanghai HSR line, the first HSR in China of a world-class standard. It provided the conditions for producing this book, and therefore, it is taken as a case here. The total length of the Beijing-Shanghai HSR line is approximately 1318 km, and it has a design speed approximately 350 km/h. The entire line with two-way tracks was newly constructed on an elevated line that parallels the pre-existing Beijing-Shanghai railway line. This HSR line passes through three municipalities (Beijing, Tianjin and Shanghai) and four provinces (Hebei, Shandong, Anhui, Jiangsu) with a total of 24 stations established in 22 cities. From project announcement to construction, the process of HSR planning exerted great impact and could be regarded as both an event and a process for a city. Consequently, using the satellite maps gathered from various sources (Google Inc., Baidu Inc. and other resources), this study traced the changes in those influenced areas between the establishment period (2006) and the operation period (2012).

This chapter focuses on the construction conditions of the 24 sites and the cities in which they are located in 2006 and 2012. Maps were drawn of construction boundaries, the road networks of surrounding areas, entrances and exits of highways, and major transportation infrastructure facilities, including airports. In addition, the development data of each city along the route were collected from the *Statistical*

L. Wang and H. Gu, *Studies on China's High-Speed Rail New Town Planning and Development*, https://doi.org/10.1007/978-981-13-6916-2_3

Yearbook of Chinese Cities and the available statistical yearbooks of other cities. The data include but are not limited to GDP, per capita GDP, number of permanent residents, industrial structure, population density, administrative area, foreign direct investment of that year, investment in fixed assets, fiscal revenue and expenditures. Together, the above information formed a GIS database of urbanization and development status in the areas surrounding high-speed rail stations, providing solid data foundations for analysis.

3.1 Theoretic Model: Potential Factors of HSR New Town Development

In this section, a cluster analysis of urbanization in the areas around high-speed rail stations was conducted to distinguish the actual development process among the station sites of the 22 cities. Based on theoretical hypotheses and samples, the study further identified explanatory variables that may affect the urbanization process and established an empirical model between urbanization and those theoretical variables.

There are three types of spatial relationship between the site selected for a rail station and a built-up urban area; namely, the former can be in the core layer, a suburb or an outer suburb of the latter. Apart from the vague characteristics of the boundary of the urban area around a rail station, it is also possible that development within a certain distance of the rail station is associated with the station site selection. Therefore, this study sets the site as the center and delimits the research area using the scope of a layer with a certain radius. According to the theoretical model of the nearby development around a transportation infrastructure, the functions in such areas are normally arranged and progress in different layers. The core layer, secondary layer and outer layer are three typical categories, and their radiuses are defined according to the time consumption of different traffic modes. Considering the wide range of groups served by HSR, the multiple forms of transportation that passengers may take to arrive at the station, and the actual construction situation around rail stations, 2, 4 and 8 km were defined, respectively, as the area of the core layer, secondary layer and outer layer (Table 3.1, Fig. 3.1). Based on GIS data, the study analyzed the size of the built-up areas together with growth and growth rate in 2006 and 2012 for those three layers and further conducted a cluster analysis.

According to the relevant theories, the following elements are determinants of how transportation facilities impact nearby urbanization: distance between HSR station and the original urban area, spatial relations with other transportation facilities, conditions of the cities in which stations are located and the current or projected passenger volume of the rail station (Hess and Almeida 2007; Immergluck 2009). The initial concept of this model is shown below.

$$U = f(D, V, G, C)$$

Table 3.1 Built-up areas layer sizes around HSR stations

City	Distance (km)	Area (ha)			Completed area accounting for the proportion of ring area		
		2006	2012	Growth rate (%)	Ratio of 2006 (%)	Ratio of 2012 (%)	Proportion increase (%)
Jinan	2	66.3	1228.9	95	5	98	93
	4	798.3	3007.0	73	21	80	59
	8	4546.7	7741.1	41	30	51	21

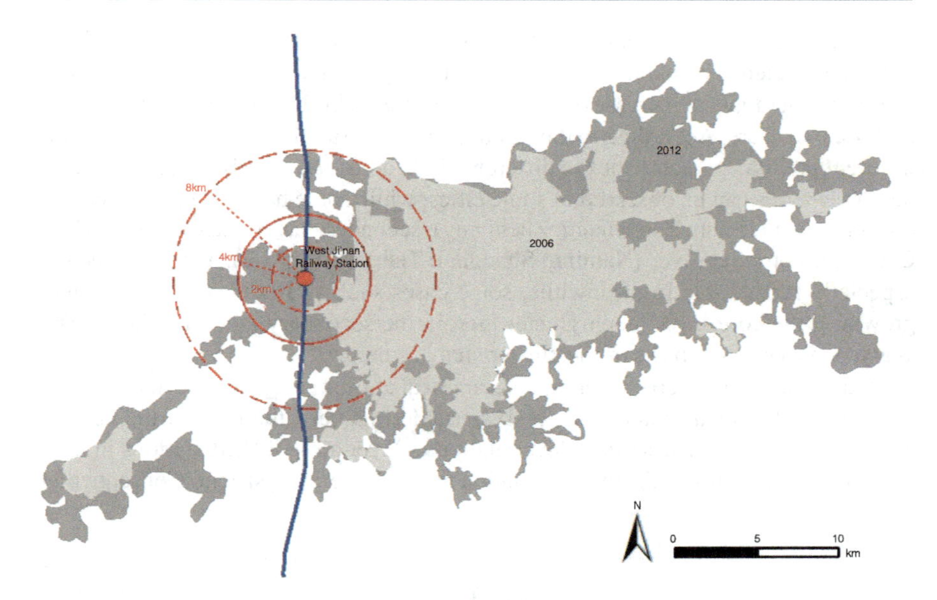

Fig. 3.1 Growing scopes of built-up areas in layers around Ji'nan HSR station (2006–2012)

The dependent variable U represents the urbanization conditions of the surrounding areas, which are signified by the growth range of the built-up area. Independent variables include D, V, G and C. Among them, D is the distance between the rail station and the original urban center, V represents the current or projected passenger volume of the rail station, G indicates the integration status of the HSR and other transportation facilities and C expresses the current development conditions of the city in which the station is located, including economic gross, industrial structure, etc.

Based on this theoretical model, an empirical model was built using data from cities along the Beijing-Shanghai HSR line. There are six dependent variables, including the ratios of the built-up area in the core layer, secondary layer and outer layer in 2006 and 2012. The independent variables include three kinds of standardized distance, from the rail station to the municipal government, the geometric city center and the nearest airport, which are calculated based on socioeconomic data in 2006 and 2010. Combining the cluster analysis, regression analysis, information gathered from site investigations and the relevant planning of the 22 cities along the line, this study extracted issues and challenges for planning and development and further proposed planning principles and paths for similar regions.

3.2 Cluster Analysis: Different Types of HSR New Towns

The sizes of the urbanized areas of the different layers in 2006 and 2012 were interpreted, extracted and calculated using satellite maps of the corresponding year. This study assumes that the increase of the urbanized area is the change in the built-up area in those two years. According to the calculated results of GIS, different cities exhibit different growth characteristics in different circle areas (layers). Balanced growth in the three layers can be observed in some cities, while some have a larger increase in the core layer (Ji'nan, Zaozhuang, Zhenjiang), secondary layer (Langfang, Xuzhou, Cangzhou) or outer layer (Nanjing, Shanghai, Tianjin) (Fig. 3.2). This increment is especially remarkable in meanwhile, some cases such as Ji'nan, which presents a growth in the core layer much greater than in the secondary and outer layer. This analysis provides the basis of the cluster analysis below.

In this study, two-step cluster and hierarchical cluster analysis were adopted, and a cluster analysis was conducted on six variables, namely, the ratio of the built-up area in the core layer, secondary layer and outer layer around HSR stations in 2006 and 2012. The results of the two calculation methods were consistent. Consequently,

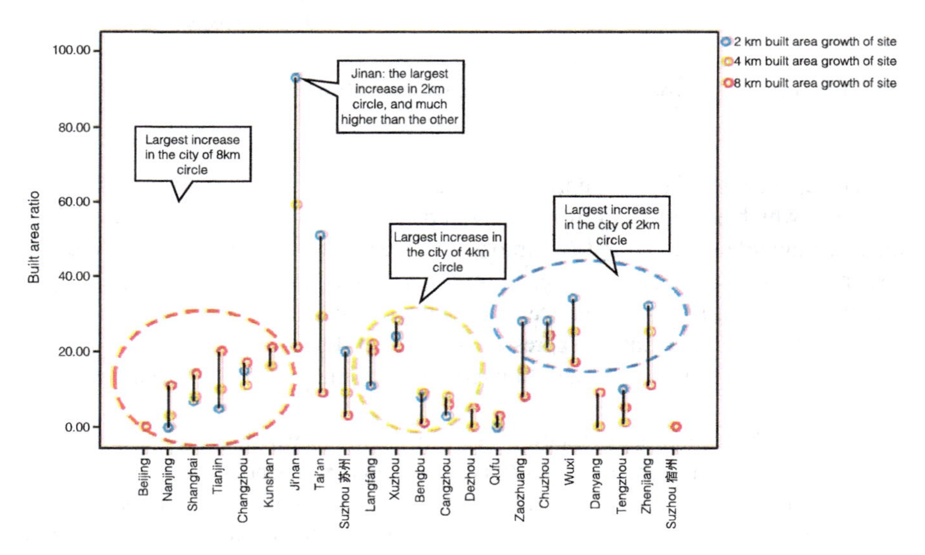

Fig. 3.2 Distribution of increase of built-up area around HSR station

the 22 cities with stations along the Beijing-Shanghai HSR line are divided into five categories (Table 3.2). Detailed definitions and analyses are shown below.

Beijing, Nanjing and Shanghai are in the first category. In 2006, the sizes of the built-up areas of those cities in all three layers, especially the core layer and secondary layer, were the largest, resulting in limited space for growth. Even though their built-up area had experienced only a slight increase by 2012, these cities still ranked first among all the sample cities. As municipalities or provincial capitals, Beijing, Nanjing and Shanghai are all megacities with a high overall urbanization rate. Among them, the HSR stations of Shanghai and Nanjing are located at the edges of existing built-up areas, and the surrounding areas have been redeveloped.

Changzhou, Kunshan and Langfang are in the second category. In 2006, the sizes of the built-up areas of those cities were the second largest. During the period from 2006 to 2012, the built-up areas experienced a steady increase at a medium speed. In Changzhou and Kunshan, the built-up areas increased the most in the secondary layer, with a radius of 4 km, while development around the rail stations was slow. The HSR stations are located at the edge of pre-existing built-up areas. Normally, such cities are midlevel cities with a high level of economic strength.

Cangzhou, Tengzhou, Qufu, Dezhou, Chuzhou, Danyang and Suzhou are in the third category. The sizes of the built-up areas in 2006 were the lowest among those of all the sample cities, and very little development was experienced in the core layer and secondary layer from 2006 to 2012. However, the development in the outer layer was faster than that in the other two layers, despite its slow speed. Such cities are typical of the cities along the Beijing-Shanghai HSR line. Normally, they have a small size and low economic strength, and the rail stations are located quite far from pre-existing built-up areas. Consequently, the urbanization around the station lacks the necessary infrastructure and fiscal investment, weakening the impetus brought by the HSR station to the overall development of the city.

Xuzhou, Bengbu, Zaozhuang, Wuxi, Zhenjiang and Tianjin are in the fourth category. The development level in all three layers was low, and the development in the core layer was the same as that in the core layer of cities in the third category. In the past six years, there were certain degrees of growth in all three layers. Although their development speeds decreased from the core layer to the outer layer, the differences are small. Among them, the development in the core layer and secondary layer far exceeded that of similar cities in the third category, and development in the outer layer surpassed that of cities in the fifth category as well. In 2006, the order of the highest urbanization level was the outer layer, secondary layer and core layer. However, this order changed to the secondary layer, core layer and outer layer, indicating planners' consideration of development surrounding the HSR stations.

Ji'nan, Tai'an and Suzhou are in the fifth category, ranked third among all the sample cities. The sizes of the built-up areas in these three cities were at a medium level in 2006, and the core layer and secondary layer experienced rapid growth during the period from 2006 to 2012. Among them, the size of the built-up area in the core layer ranked second, which is very close to that of the cities in the first category. However, development in the outer layer was stagnant and was exceeded by cities in

Table 3.2 Result 1 of cluster analysis: characteristics of increase

Cluster	City	Characteristics of 2006		Characteristics of 2012		6 years of change characteristics	
		Scale	Layer balance	Scale	Layer balance	Growth rate	Layer balance
1	Beijing, Nanjing, Shanghai	Maximum	High level of balance	Maximum	High level of balance	Slow	Balanced, no significant growth
2	Changzhou, Kunshan, Langfang	Larger	Imbalance: The level of the 8-km ring is significantly lower	Larger	Imbalance: The level of the 8-km ring is significantly lower	Middle	Balanced, have significantly increased
3	Cangzhou, Tengzhou, Qufu, Texas, Chuzhou, Danyang, Suzhou	Minimum	Low level of balance	Minimum	Low level of balance	Balanced	Balanced, no significant growth
4	Xuzhou, Bengbu, Zaozhuang, Wuxi, Zhenjiang, Tianjin	Smaller	Balanced: 2 km ring level is low	Smaller	Balanced: 2 km ring level is low	Middle	Balanced, have significantly increased
5	Jinan, Tai'an, Suzhou	Middle	Balanced: 2 km ring level is high	2 km larger, 8 km smaller	Imbalance: 2 km > 4 km > 8 km, the three layers significantly opened	Different speed	Imbalanced: 2 km layer grows rapidly, 4 km center, 8 km slow

Table 3.3 Result 2 of cluster analysis: rankings of increase

Cluster	2 km layer			4 km layer			8 km layer		
	2006	2012	Increase	2006	2012	Increase	2006	2012	Increase
1	First	First	Slow	First	First	Slow	First	First	Medium
2	Second	Third	Medium	Second	Second	Medium	Second	Second	Medium
3	Fifth	Fifth	Slow	Fifth	Fifth	Slow	Fifth	Fifth	Medium
4	Fourth	Fourth	Medium	Fourth	Fourth	Medium	Fourth	Third	Medium
5	Third	Second	Fast	Third	Third	Medium	Third	Fourth	Slow

the fourth category. This finding indicates that such cities have recently focused on development surrounding HSR stations.

In this cluster analysis (Table 3.3, Fig. 3.3), cities in the first and second categories are similar, as they all maintain a high urbanization level around HSR stations despite large differences in the size of the newly urbanized areas in the outer layer. Regarding cities in the third category, the proportion of the urbanized area around the HSR station as well as its growth rate is low. However, the construction speed is fast, and the size of the newly urbanized area is large. Finally, there are obvious differences between the layers of cities in the fifth category; namely, the core layer developed extremely rapidly, while the outer layer moved very slowly and the secondary layer maintained a medium growth rate.

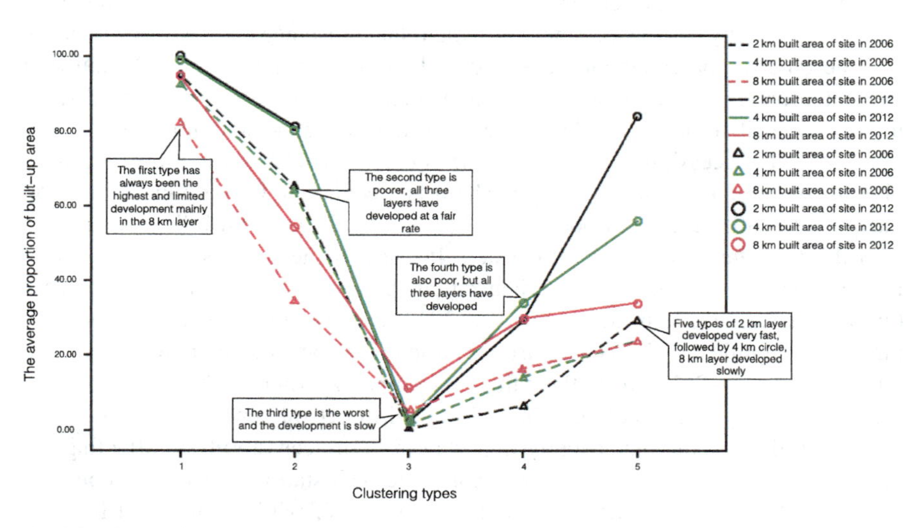

Fig. 3.3 Result of cluster analysis

3.3 Quantitative Model: Significant Factors of HSR New Town Development

This study builds a quantitative model of development factors of the areas surrounding HSR stations and the urbanization situation and adopts a structural equation model and multiple linear stepwise regression models. The two models analyze the relationship between the urban area expansion of different circles with a radius of 2, 4 and 8 km, and a series of variables. Independent variables include three distances (from the downtown, the municipal government and the nearest airport), and the quantification includes the standardization of the distance from the municipal government, the standardization of the distance from the downtown (the geometric center) and the standardization of the distance from the nearest airport. The two groups of urban social and economic data of 2006 and 2010 include GDP, per capita GDP, population density, proportion of the secondary industry, proportion of the tertiary industry, total population, administrative area, population of permanent residents, foreign direct investment (number of newly signed contracts in the current year), foreign direct investment (amount used in the current year), investment in fixed assets, fiscal revenue, fiscal expenditure and number of automobiles served. At the same time, the influences of controlling the population and economic scale on urbanization progress in the model building process are studied to avoid influences of urban scale on the development amount.

According to the data of the sample cities, the structural equation model builds two dimensions that affect the dependent variables for diverse independent variables. One is the accessibility dimension (including the related distance explanatory variable), and the other is the economic background dimension (per capita GDP, industrial structure, foreign direct investment, fiscal revenue, etc.). As shown in the results, the urbanization situation at each level in areas surrounding the station is significantly affected by the accessibility dimension. The higher the level of accessibility, the more obvious the urbanization progress is. The urban economic background dimension has positive influences, and both its degree of significance and influence are less than those of the accessibility dimension. In view of the process of urbanization, in the structural model, we take urbanization in 2006 as the intermediate process that has an influence on urbanization in 2012, and we find that the accessibility dimension significantly influences urbanization in 2006 and has an indirect influence on urbanization in 2012 due to the influence of urbanization in 2006 on that in 2012. The accessibility dimension also has additional direct influences on urbanization in 2012, reflecting the increasingly important role of accessibility in affecting the urbanization progress of the area surrounding the HSR station. The urban economic background significantly affects urbanization in 2006, while it does not have an obvious direct influence on urbanization in 2012.

Based on the clear influences of the two dimensions, the multiple linear stepwise regression model is built to further specify the influences of explanatory variables on the urbanization of the area surrounding the station. The analysis results indicate that in the layers with a radius of 2 and 4 km, the variable of distance between the station and downtown is significantly related, and the coefficient symbol is negative. This finding indicates that the farther the station is from the downtown, the less development will occur in the layers with a radius of 2 and 4 km of the area surrounding the station. Within the study scope of the layer with a radius of 8 km from the station, two variables are included in the regression model, namely, the distance between the station and downtown, and fiscal expenditure. Of those, the standardized coefficient symbol of fiscal expenditure is positive. This finding indicates that the greater the fiscal expenditure is, the higher the proportion of the built-up area in this layer will be. Regarding the standardized coefficient, the absolute value of the distance from downtown is greater than that of fiscal expenditure. Integrated with the inclusion order of variables, we can observe that the influence of the distance from downtown is greater than that of fiscal expenditure in this layer.

Two empirical models indicate results different from those of the theoretical model. Numerous explanatory variables are explained by accessibility (especially the distance from the station to downtown). This finding indicates that the location of the station has a significant influence on whether the HSR will promote the urbanization progress of the area surrounding the station. At the early stage of operation of the line, the locational condition of the station is especially more important than other variables and is an important factor that affects the development of the area surrounding the station. Fiscal expenditure exerts certain influences on the urbanization of the area surrounding the station. The economic ability of the government of the city plays a decisive role in the development situation of the area surrounding the station. This finding indicates that the development of areas surrounding traffic facilities such as HSR stations needs public investment by the government to initiate and encourage private investment.

3.4 Comparative Analysis: Planning and Reality of the Areas Surrounding HSR Stations

3.4.1 Planning Scale of the Surrounding Area of the Station

After the collection and sorting of local government Web sites, news reports and planning documents, it is found that only 15 cities have special planning for the areas surrounding the stations. The planning of some cities includes the core area and the peripheral area, while other cities have one influence scope. In addition, the planned construction scales of the areas surrounding the stations in these cities differ significantly from each other. The total peripheral area can be classified into three grades (Table 3.4).

The grade I (the planning area is more than 30 km^2) cities are Ji'nan, Dezhou, Tengzhou, Wuxi and Qufu. Among them, the city with the largest planning area is Ji'nan, with a core area of 6 km^2 and a total peripheral planning area of 55 km^2. In terms of the station type, all the stations are newly built. Among them, the stations

Table 3.4 Size of planning area around HSR station

City	Planning area (km^2)		Urban permanent population (million)
	Core area	Perimeter area	
Jinan	6	55	681
Dezhou	12.5	50	557
Tengzhou	–	48.8	160
Wuxi	2.39	45	637
Qufu	–	35	64 (Jining808)
Suzhou	–	28.9	1046
Cangzhou	–	28	713
Shanghai	–	26	2340
Changzhou	1.6	24	459
Bengbu	–	23	316
Xuzhou	–	12	858
Tianjin	–	10	1294
Zaozhuang	–	10	372
Chuzhou	–	8	394
Tai'an	–	4	549

Source Author, drawing based on the data from Internet and planning programs

of Ji'nan, Dezhou and Tengzhou are located on the edge of the built-up area, and the stations of Dezhou and Qufu are far from the built-up area of the downtown.

The grade II (the planning area is more than 20 km^2) cities are Suzhou, Cangzhou, Shanghai, Changzhou and Bengbu. Among them, the city with the largest planning area is Suzhou, whose planning area is 28.9 km^2, and the city with the smallest planning area is Bengbu, whose planning area is 23 km^2. In terms of the station type, all the stations except the Shanghai Hongqiao Station are newly built. Hongqiao Station was completely built one year earlier than the Beijing-Shanghai HSR line. The HSR stations in Suzhou, Shanghai and Changzhou are located on the edge of the built-up area. The stations in Cangzhou and Bengbu are far from the built-up area.

The grade III (the planning area is less than 20 km^2) cities are Xuzhou, Tianjin, Zaozhuang, Chuzhou and Tai'an. The city with the largest planning area is Xuzhou, with a planning area of 12 km^2, and the city with the smallest area is Tai'an, with a planning area of 4 km^2. The stations are all newly built. Excluding Chuzhou where the station is far from the built-up area, the Beijing-Shanghai HSR stations in the other four cities are located on the edge of the built-up area.

Comparison of the data of the planning scale of the area surrounding the station and the population of permanent residents in the cities shows that their relation is not obvious. Shanghai has the most permanent residents, and the planning area surrounding the station is 26 km^2, which is at the intermediate level and is even lower than that of cities with the lowest population of permanent residents, such as Qufu and Tengzhou. Of the five cities with the largest planning area surrounding the station, none is a megalopolis with a permanent resident population of more than ten million.

The area surrounding the station in Ji'nan has the largest planning area. The core area is 6 km^2, and the total planning area of the station is 55 km^2. The city with the second-largest planning area surrounding the station is Dezhou. The total planning area of the station is 50 km^2, including the core area of 12.5 km^2. The planning area of the station in Wuxi is approximately 45 km^2. The planning area surrounding the station in Xuzhou is 38 km^2, planned in two different areas, the Xuzhou HSR station area (26 km^2) and the Xuzhou HSR international business area (12 km^2). The planning areas in these four cities, with an area of more than 30 km^2, belong to grade I. The planning areas in Cangzhou (28 km^2), Shanghai (26 km^2) and Bengbu (23 km^2) belong to grade II, with an area of more than 20 km^2. The planning areas in Tianjin (the west station is approximately 10 km^2), Zaozhuang (approximately 10 km^2) and Chuzhou (8 km^2) belong to grade III, with a planning area of 10 km^2. Cities where the planning area surrounding the station is small are Tai'an, Qufu and Suzhou with planning areas of less than 5 km^2.

3.4.2 Planning Function of the Area Surrounding the Station

In terms of positioning, the comprehensive traffic hub is the basic positioning of the area surrounding the HSR station. Most cities with planning of the area surrounding the station design it as the new area or new town, which is planned as the new urban economic growth area. In terms of function, according to studies on related theories regarding the business and tourism traffic brought by the HSR, business, commerce, finance, hotels, the leisure and entertainment industry and the real estate industry are common functions that are taken into consideration in the planning of the area surrounding the station (Tables 3.5 and 3.6).

Among the fifteen cities with planning of the area surrounding the HSR station, more than half include business and commerce functions: Eight cities planned business function and nine planned commerce function, and only four have a hotel function. Cultural entertainment is also an important function of the planning of areas surrounding stations: Six cities have proposed planning with a leisure and entertainment function; some cities also have special planning for facilities such as gymnasiums, museums and libraries. The planning of areas surrounding stations has diversified living functions: affordable housing, residential and commercial buildings, large-scale residential districts, etc. Three cities have the administrative function in the planning of the area surrounding the station, and some cities city (Dezhou) has set up industrial land, hoping to develop industry in the periphery of the affected areas in accordance with the development stages and characteristics of the city.

Table 3.5 Basic planning content of Beijing-Shanghai HSR line stations

City	Station area name	Positioning	Land size	Main functions and facilities
Xuzhou	Xuzhou high-speed rail station area	An important modern integrated transport hub, an important window to showcase the modern image of Xuzhou and promote the rapid growth of the eastern region of Xuzhou, an important growth pole, deputy director of the city	26 km^2, core area of 5 km^2	Transport hub integrated area, administrative offices, commercial finance, cultural and entertainment-based integrated public service area
	High-speed rail international business district	Services in the Beijing-Shanghai high-speed rail, Beijing-Fuzhou high-speed commercial port. High-end residential and office buildings, large commercial centers in one multifunctional composite trade city, become the future of Xuzhou City, one of the core group of three cities to create a new landmark in Xuzhou, the new town	12 km^2	Corporate headquarters, executive commerce, star hotels, upscale living, modern logistics, including Ruilong Machinery Research Center, Bo Department of Science Research Building, Hao Chen Software Park, Moon Star Home City. Mini golf course, a large conference center, business center and other high-end business facilities projects
Bengbu	Bengbu South Railway Station Metro	Fully functional modern emerging city	23 km^2, core area of 8 km^2	Set of finance, science and education, trade, housing and other functions as a whole
Cangzhou	Western Metro	Services in high-end leisure, tourism, entertainment, catering, service-oriented business district	28 km^2	City gymnasium, exhibition center, square, grand theater, museum, library, two five-star hotels, international trade center, high-quality living
Dehou	High-speed rail new area	Eco-high-speed rail new district, Yi industry, livable, Yi learning, tourism, traffic business district, urban residential and integrated service area and industrial area	50 km^2, core area of 12.5 km^2	Theme parks, recreational facilities, riverside ecological community, supporting residential area

(continued)

Table 3.5 (continued)

City	Station area name	Positioning	Land size	Main functions and facilities
Tai'an	Beijing-Shanghai high-speed rail station Tai'an New District	Urban construction and public housing groups	3.7 km^2	150,000 m^2 of new passenger station square, Shin Kong Station Square, 400,000 m^2 of resettlement housing, high-rise living
Qufu	Qufu high-speed rail group	Living, commerce and trade, leisure and entertainment with high-speed rail station features, perfect, beautiful environment of the city's comprehensive development zone	Core area of 2.5 km^2	Square, living, business
Zaozhuang	Zaozhuang Metro	City living room	10 km^2	Administrative offices, culture and entertainment, commercial finance and stadiums, garden-style living groups
Chuzhou	High-speed rail station area	Chuzhou high-speed rail station in front of the district, set the traffic, commerce, cultural and entertainment and residential functions as one of the new urban areas	8 km^2	Transport, commerce, hotels and other transportation hub-based center, three living areas
Wuxi	Wuxi East Railway Station area	Wuxi emerging growth area, one of the main space carriers of modern production service functions, a new modern transportation hub with regional functions	45 km^2	Internal core area with business, living, leisure and educational functions, central supporting residential area, external green ecological area, six residential function groups

(continued)

Table 3.5 (continued)

City	Station area name	Positioning	Land size	Main functions and facilities
Suzhou	High-speed rail station area	Traffic-oriented development zone, a modern transportation hub with regional functions; an important window displaying the modern image of the north gateway of Suzhou City; the future economic vitality center of the urban area and the space carrier of modern production service functions	1.8 km^2	Business commerce, culture and entertainment

Source Author

3.4.3 Comparison Between the Planning and Development Reality of the Areas Surrounding HSR Stations

Data collected on Web sites such as local government Web sites, Wikipedia, Baidu and Google indicate that there are great differences among the land scales of the development areas along the Beijing-Shanghai HSR line. The city with the largest planning area is Ji'nan, with a core area of 6 km^2 and a total planning station area of 55 km^2. The city with the second-largest planning area of the area surrounding the station is Dezhou. The total planning area of the station is 50 km^2, including a core area of 12.5 km^2. The planning area of the station in Wuxi is approximately 45 km^2. The planning area surrounding the station in Xuzhou is 38 km^2, planned as two different areas, the Xuzhou HSR station area (26 km^2) and the Xuzhou HSR international business area (12 km^2). The planning areas in these four cities, with areas of more than 30 km^2, belong to grade I. The planning areas in Cangzhou (28 km^2), Shanghai (26 km^2) and Bengbu (23 km^2) belong to grade II, with areas of more than 20 km^2. The planning areas in Tianjin (the west station is approximately 10 km^2), Zaozhuang (approximately 10 km^2) and Chuzhou (8 km^2) belong to grade III, with a planning area of 10 km^2. The cities where the planning area surrounding the station is small are Tai'an, Qufu and Suzhou. Their planning areas are less than 5 km^2. Many cities have taken the different development contents and stages of the core area and influence area into consideration in planning. The planning and actual development situations of the five groups of cities based on cluster analysis are obtained (Fig. 3.4).

Table 3.6 Planning functions of development zones of HSR

Function	The number of configured cities	Function	The number of configured cities	Function	The number of configured cities	Function	The number of configured cities	Function	The number of configured cities
Business finance		Cultural entertainment		Living		Administration		Other industries	
Business	8	Leisure and entertainment	6	Affordable housing	3	Administration	3	Creative	1
Financial	4	Science and education	2	Commercial real estate	3	Corporate headquarters	1	Logistics	3
Financial	9	Stadium	2	Ecological settlements	2	–			
Hotel	4	Exhibition Center	1	High-end residential area	5				
–		Museum	1	–					
		Library	1						
		Theme park	1						

Source Wang et al. (2014)

Fig. 3.4 Comparison of planning construction area and actual construction area around HSR station

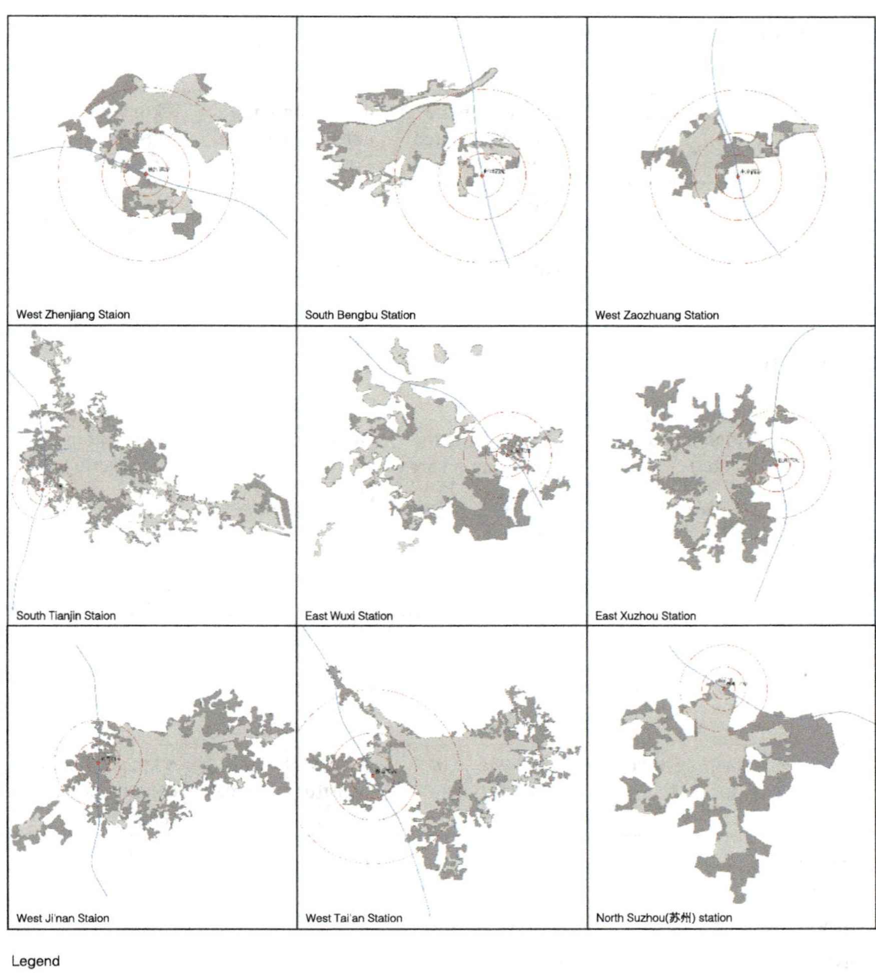

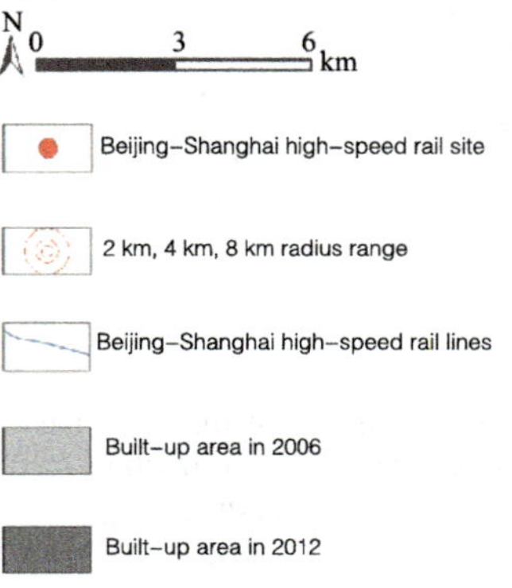

Fig. 3.4 (continued)

First group

City	Planning area (km^2)	2 km within the construction area (km^2)	4 km within the construction area (km^2)	8 km within the construction area (km^2)
Shanghai	26	1.86	6.11	39.44
Beijing	1.8	0.00	0.00	0.00
Nanjing	6	0.00	1.12	17.07

Second Group

City	Planning area (km^2)	2 km within the construction area (km^2)	4 km within the construction area (km^2)	8 km within the construction area (km^2)
Changzhou	24	1.97	3.84	26.13
Kunshan	–	2.74	5.93	32.06
Langfang	–	1.32	8.18	30.61

Third Group

City	Planning area (km^2)	2 km within the construction area (km^2)	4 km within the construction area (km^2)	8 km within the construction area (km^2)
Tengzhou	–	1.23	0.49	6.74
Dezhou	50	0.00	0.00	6.42
Chuzhou	8	3.57	8.16	35.76
Danyang	–	0.00	0.08	13.69
Cangzhou	28	0.36	3.03	8.32
Qufu	2.5	0.00	0.08	5.40
Suzhou	1.8	0.00	0.00	0.00

Fourth Group

City	Planning area (km^2)	2 km within the construction area (km^2)	4 km within the construction area (km^2)	8 km within the construction area (km^2)
Zhenjiang	–	3.95	9.50	16.81
Bengbu	23	0.99	3.28	2.86
Zaozhuang	10	3.51	5.43	11.25
Tianjin	10	1.26	7.50	59.03
Wuxi	44	4.26	9.44	26.91
Xuzhou	38	3.03	10.59	31.08

Fifth Group

City	Planning area (km^2)	2 km within the construction area (km^2)	4 km within the construction area (km^2)	8 km within the construction area (km^2)
Jinan	55	11.63	22.09	31.94
Tai'an	4	6.46	10.78	13.06
Suzhou	–	2.47	3.41	3.21

In addition to the scale, the author studies and analyzes the planning function positioning of the areas surrounding HSR stations. The basic positioning of HSR stations is as a comprehensive traffic hub. Most cities confirm the station area as a new urban economic growth area and plan it as a new area or new town with comprehensive development. In terms of function, most cities take the traffic-driving effect of HSRs, population clustering and attraction and related industrial development into consideration. According to cluster analysis, business, financial services, commerce, leisure and entertainment and real estate are the main industry types. Of those, business (eight cities), commerce (nine cities), leisure and entertainment (six cities) and real estate (13 cities) have become the main planned functions (see Table 3.6). The high-tech industry is not emphasized. Two cities have proposed science and education functions. Three cities are planning to establish administrative office buildings. One city has proposed the creative industry, and one has proposed the logistics industry. The layout of these functions takes the people-oriented characteristic of the HSR into consideration, and the functions are centered on investors, tourists and logistics personnel. One city (Dezhou) has set up industrial land, hoping to develop industry in the periphery of the affected areas in accordance with the development stages and characteristics of the city. The planning of areas surrounding different HSR stations is similar. Except for regional traffic-related functions, urban functions are set up to serve as a new area.

3.5 Key Points of Urban Development of the Areas Surrounding HSR Stations

Based on the cluster analysis, structural equation model and regression model construction, first, we define the important influence factor; then, we specifically study the areas surrounding the stations in 22 cities and specify the development details of these areas. We study and analyze the relation between location selection, scale, properties of the land and other traffic facilities and discuss problems and challenges in the development of the areas surrounding the stations, integrating the planning that we collected.

3.5.1 Site Selection

The relation between the HSR station and the local city influences the development of the area surrounding the station. Both cluster analysis and the regression model indicate the important influence of the location of the station. Hence, we analyze the relation between stations and the local cities in this section. Among the 22 sample cities, the distances between the station and the downtown (currently, the study uses the urban government as the downtown) vary significantly and have no obvious relation with the scale of the cities (that is, the distance between the station and the downtown will not become greater with an increase in the scale of the city). Stations in 11 cities are set inside or on the edge of the current built-up area, and stations in the other 11 cities are set outside the current built-up area (Wang 2011). The city with the least distance between the HSR station and the downtown is Langfang (3.3 km), and the city with the greatest distance is Suzhou (24 km).

Half of the stations of the Beijing-Shanghai HSR line are constructed outside the current built-up area, which restrains interaction between the area surrounding the station and the local city; namely, the regional traffic facilities cannot promote the development of the city, and resource integration and development of the areas surrounding the regional traffic facilities cannot be carried out easily based on the current urban facilities. For example, in Suzhou, the former Ministry of Railways selected the location by taking the speed of the railway in this area into consideration but without sufficient communication with the local planning department; as a result, the distance between the station and the downtown of Suzhou is great. Since the joining of the HSR and the ramp under the HSR is dislocated, interaction between the downtown and the HSR station is severely blocked. Currently, there is little development in the area surrounding the station. In contrast, the selection of the location of the HSR station in Shanghai was much discussed and coordinated between the Ministry of Railways and the local government, and the station was finally located in Shanghai Hongqiao as a coordinating hub of various modes of traffic, sufficiently leveraging the potential of the original facilities and promoting the development of the surrounding area. It can be seen that location selection is very important in enabling regional traffic facilities to promote the development of the local city. The active participation of local government is necessary during the selection of the location to realize the joining of the upper level and the lower level and vice versa. In addition, the development direction and existing facilities of the city should be fully considered, and a resource integration scheme should be discussed to confirm the location.

3.5.2 Planning and Development Scale of the Surrounding Area

Based on the data collected from local government Web sites and related Web sites, the study carries out a comparative analysis of the urbanization situation and the planning of the areas surrounding the HSR stations. The land scales of the development areas of Beijing-Shanghai HSR stations vary significantly. The study compares the planning and actual construction areas of the areas surrounding the stations in five groups of cities based on cluster analysis. The planning scale of the areas surrounding the stations in the first group of cities is small. Up to 2012, the areas surrounding the stations have been built. In addition, the areas surrounding the HSR stations in Shanghai and Nanjing have undergone urban expansion. The planning scale of the areas surrounding the stations in the second group of cities is small, and the completion degree of the areas surrounding the stations is high. The overall development direction of such cities does not conform the development direction of the HSR. The planning scale of the areas surrounding the stations in the third group of cities varies significantly. The completion degree of facilities in the areas surrounding these stations is low. Among these cities, the overall development direction in Chuzhou and Cangzhou conforms the development direction of the HSR, while Suzhou and other cities do not have that characteristic and even move in the opposite direction. The planning scale of the areas surrounding the HSR stations in the fourth group of cities is large, and the completion degree is moderate. Among these cities, the urban expansion of the areas surrounding the HSR stations in Zhenjiang, Zaozhuang, Xuzhou and Wuxi is obvious. The planning scale of the areas surrounding the stations in the fifth group of cities is large, and the completion degree is high. Among these cities, the development direction of Ji'nan and Tai'an conforms the development direction of the HSR. A group comparison analysis indicates that planning plays a certain guiding role, and especially when the planned development direction of the city conforms with the location of the HSR station, planning displays a positive interaction.

3.5.3 The Land Use Types of the Surrounding Areas

In planning, the basic positioning of HSR stations is as a comprehensive traffic hub. Most cities define the station as a new economic growth area and plan it as a new area or new town of comprehensive development. The planning of the areas surrounding HSR stations includes multiple functions such as business, office buildings and leisure. These plans are characterized by comprehensive development and mixed use of lands. According to a field survey, the properties of the developed land in the areas surrounding HSR stations include residential and business functions (including commerce services and business facilities).[1] In addition, there are a few public management and public service functions (including administrative offices, cultural facilities, science and education facilities and sports facilities). Green and public facility lands are set up based on the original natural conditions or functions, such as the ecological garden east of the Changzhou North Railway Station. There are a few industrial sites and logistics warehouses. This planning conforms to the characteristic of the HSR station as a facility that distributes a flow of people rather than a cargo.

[1]Classified according to the national standard *Code for Classification of Urban Land Use and Planning Standards of Development Land.*

The residential land in the areas surrounding HSR stations is grade II. It includes multilayer, medium-layer and high-rise residential facilities equipped with public facilities and traffic facilities. For example, residential projects were constructed and invested in by the Greenland Group in the areas surrounding the stations in Zhenjiang, Xuzhou and Ji'nan. There is some grade I residential lands in areas surrounding the stations in certain cities, such as the charming town of Vanke in Zhenjiang.

In taking advantage of the construction of HSRs, most cities devote their efforts to planning and constructing the business area surrounding the HSR, hoping to promote the centralized development of business service facilities in the areas surrounding HSR stations. For example, as part of the development goal of the business area in the core area of the West Railway Station in the new western urban area of Ji'nan, the Greenland Central Plaza Binfen City and Xiyuan Building have been built. High-end office buildings and experiential courtyard business buildings are currently under construction in Shiao Square in front of the Wuxi East Railway Station. Construction has begun on grade 5A office buildings and five-star hotels, which are regarded as landmark comprehensive commercial businesses and residential areas located in the headquarters base in front of the Qufu station. The development of most commercial business service facilities lags behind residential land development. The planning intention has given full play to the location value brought by the HSR, but the construction and development time are still short, and the current utilization rate of established commercial business projects is low.

The development of public management and public service land is based on the development strategy of the areas surrounding HSR stations. For example, after the completion of the Cangzhou West Railway Station, the Cangzhou municipal government built a gymnasium, international convention center, planning exhibition hall and other facilities on the edge of the built-up areas to guide the city to expand to the west. The Shandong Culture and Art Center will be established in front of the Ji'nan West Railway Station. The Xi Dong Senior High School established on the east side of the Wuxi East Railway Station led to the development and sale of many "school district housing" projects in the surrounding area. In addition, some cities, such as Zhenjiang and Tai'an, established new administrative areas around HSR stations, which stimulated the development of the areas surrounding the stations.

Based on the relevant planning statistics of each station city, 13 of the 22 station cities proposed the development of residential real estate projects, 9 cities proposed supporting commercial functions, 8 cities proposed supporting business functions and 6 cities proposed supporting leisure and entertainment functions. However, currently, the basic function of the areas surrounding the stations is residential. There is concentrated development of commercial functions in Shanghai, Nanjing and other cities, and some cities have scattered commercial facilities. The business function is still in the cultivation stage, and some cities are still constructing those facilities. The public service function is lagging. Only Cangzhou, Zhenjiang and a few other cities have established public projects such as gymnasiums and libraries. There is a gap between the actual construction and the pedestrian-oriented planning ideas, the commercial business development and the mixed development dominating.

3.5.4 Connection to Other Transportation Facilities

The connection between the HSR station and other traffic facilities determines the benefits and service level of the station. The connections between the Beijing-Shanghai HSR stations and metro stations, bus passenger terminals, bus stations and expressways are analyzed below.

Five cities along the Beijing-Shanghai HSR line, Beijing, Tianjin, Shanghai, Nanjing and Suzhou, have metro lines, all of which connect to the metro station at the HSR station, forming a combined "HSR-metro" transport hub. Four cities, Wuxi, Xuzhou, Ji'nan and Changzhou, are planning or constructing metro lines, and their HSR stations are also equipped with metro stations. We find that the combined "HSR-metro" transport hub is currently accompanied by the development of buildings in the areas surrounding the large-scale stations. This development is related to the economic strength of cities with metro lines,[2] but it also shows that the people flow brought by the two linked modes of traffic provide a foundation for comprehensive development.

Fifteen cities along the Beijing-Shanghai HSR line have bus passenger terminals in the areas surrounding the HSR stations, and 7 cities, Tianjin, Suzhou, Dezhou, Danyang, Bengbu, Zaozhuang and Suzhou, have no bus passenger terminals. The construction of the areas surrounding those 7 stations is slow. The stations in Tianjin and Suzhou are located on the edge of the built-up area, and the stations in the other 5 cities are at a certain distance from the urban area, so the construction speed is lagging. The HSR stations in all cities have bus stations that provide a link between the station and the downtown for a low volume of development and limited passenger flow.

In terms of the connection to the highway, all stations have an expressway or trunk road connecting to the downtown. The areas surrounding stations with more rapid development, such as those in Wuxi and Xuzhou, have better railway network construction, while the cities with slower development of the surrounding areas of stations, such as Dezhou and Qufu, have only a single road to the station and are separated from the nearby villages. In terms of the expressway, the connection between the ramp and the road to the station is not taken into consideration, which is the non-interconnected fly-over design. Some cities have great dislocation between the expressway ramp entrance and the HSR station due to the barriers between higher and lower levels or between different departments and regions, which further weakens the accessibility of the station, hinders the active leading role of such regional traffic facilities in urban development and reduces the efficiency of resource integration.

[2]The requirements of the state for an urban declaration of metro construction include that the urban population exceeds 3 million, GDP exceeds RMB 100 billion and the general budget of local finance exceeds RMB 10 billion. The passenger flow scale of the planned line must reach more than 30,000 people in a one-way peak hour.

3.5.5 Key Planning Issues of Areas Surrounding HSR Stations

Based on the empirical analysis of the development and planning of the areas surrounding the Beijing-Shanghai HSR stations, we find that the development of the areas surrounding the HSR stations (with the built-up area as the index) is significantly related to the relationship between the location of the HSR station and the city. The farther from the downtown the station is, the slower the development of the area surrounding it. The effect of most of the HSR stations on urban development and urbanization in the surrounding areas is not more significant than that predicted in the theory. The site selection and the integration of other traffic facilities reflect the fact that the division of administrative decisions has not led to the best integration of development resources. The weakness of local voices in the process of planning and site selection has not led to an effective local leading role in regional traffic facilities. The urban administrative level determines the level of urban stations, number of trains stops and departure times and affects the leading role of HSR in urban development. A city at a lower level also has a lower HSR service level, so it is difficult to achieve the role of HSR in promoting urban development, and a virtuous cycle of development cannot begin. At present, the development of areas surrounding HSR stations is dominated by local government investment, and in general, the cities with greater economic strength have more rapid development and more new constructions. Private investment is more focus on real estate projects. Attention should be paid to the effective mechanism of massive government investment to promote the development of the surrounding areas, thus realizing the social sharing of development benefits.

The operational time of a HSR affects the analysis results to a certain extent, namely, more development of areas surrounding the stations will occur after a longer operational time. For cities with less economic strength, additional investment or mutual economic development is needed to change the current development track and promote the development of the areas surrounding the stations. After a line is planned and before it is established and commissioned, the HSR station has a certain role in urban development, but it is not obvious. Therefore, the combination of site selection and urban planning must be strengthened, and the expected possible benefits of the station cannot be realized only through planning.

Therefore, the planning of the area surrounding the station must first connect to the overall planning of the city and then form an effective link to the urban built-up areas through effective communication, consensus building and reasonable site selection by the relevant departments at all levels so that the station has good development conditions. Second, the development planning of the station area must consider the multi-factor superposition of the scale and the functional orientation, including the relationship between the location of the station and the built-up area, the degree of convenience of the link between the HSR and other traffic facilities, the industry selection and the business land reserves of the city. The orientation of the HSR station is mostly based on the new urban growth point and must clearly define the specific orientation suitable for the development stages and characteristics of the city. Cities with economic expansion capacity and industrial development demand provide the driving force for the development of HSR station areas, and the development trend is closely related to the planning function, space creation and attraction of foreign investors and capital. The functions of the areas surrounding the stations should be dominated by business, finance and leisure and entertainment, based on the flow of people; the development of science and education and R&D can increase later in the affected peripheral area.

Finally, the planning and design of the areas surrounding the stations should be expanded based on the TOD theoretical model. The key planning points of TOD have three aspects: high-density development in the station area, mixed-use and pedestrian-friendly public spaces. The design of the HSR station area should follow these three principles. At present, due to different implementation priorities, development of the stations in most cities is dominated by traffic functions, and there are no comprehensive development modes, such as the upper cover of metro lines. In the planning and design of the areas surrounding the stations, it is necessary to promote mixed-use development of the land, enhance the interaction between various types of land, improve the vitality of the area and determine the mixed mode according to the distance from the station. At the same time, the development of the stations and the surrounding areas must break administrative barriers, realize comprehensive tridimensional development, enhance development intensity and improve the land availability factor. For the HSR system, which transports people, it is very important that the station area is designed as a pedestrian-friendly environment, and it is necessary to fully consider the degree of convenience of other traffic facilities.

HSR construction is the emphasis of public investment and regional infrastructure construction for a future period in China, which affects the development of regional and urban areas at different spatial levels. Station site selection and the planning and design of the surrounding areas should play a leading role in the planning of traffic facilities in this area. Based on the empirical study of the Beijing-Shanghai HSR line, it is beneficial for us to define the important factors that affect the development of the areas surrounding the stations, analyze the existing problems in current development, propose ideas and principles for the planning and design of regional traffic facilities in the surrounding areas and strive to promote the intensive, efficient and sustainable development of the areas around stations.

References

Hess DB, Almeida TM (2007) Impact of proximity to light rail rapid transit on station-area property values in Buffalo, New York. Urban Stud 44(5):1041–1068

Immergluck D (2009) Large redevelopment initiatives, housing values and gentrification: the case of the Atlanta Beltline. Urban Stud 46:1723

Wang L (2011) Research framework and empirical study of the impacts of HSR on urban space. Planners 27(7):13–19

Wang L, Wang C, Chen C (2014) Development and planning of the surrounding area of high-speed rail station: based on empirical study of Beijing-Shanghai line. Urban Plann Forum 4:31–37

Chapter 4
Case Analysis: Planning and Development Process of HSR Stations

The overall study of the development of the HSR new towns along the Beijing-Shanghai HSR line shows that the location of the HSR station, the economic development stage of the station-located cities themselves and financial capacity affect the development of the areas surrounding the HSR stations. To further explore the factors affecting the HSR new town and its planning, the study selected two HSR new towns with similar conditions but different development results: Wuxi and Changzhou. The study conducted a comparative analysis of their planning and development processes to determine the key factors affecting the results. This section details the planning and development process in the areas surrounding the two HSR stations in Wuxi and Changzhou and shows the changes in and causes of this process.

The comparative research methods are divided into "method of agreement" and "method of difference." The agreement method is based on similar phenomena or results, seeks important similarities in many different cases and discusses which similarities lead to similar phenomena or results. The difference method starts with different phenomena or results, finds important differences in similar cases and discusses which differences lead to different phenomena or results. The two-tailed design of the difference method deliberately selects cases at both ends (good vs. bad vs. none) and divides them into positive cases and negative cases. The positive cases prove that there are differences in the results that exist, and the negative cases prove that the lack of differences lead to lack of results and provide evidence for the positive cases (Table 4.1).

The case study of HSR site planning and development uses bilateral design in the difference method. Based on an empirical study of the development of the areas surrounding the 22 stations of the Beijing-Shanghai HSR line between 2006 and 2012, Wuxi East Station and Changzhou North Station were selected as research cases. The case selection criteria are the locations of the HSR stations and main urban areas, the stage of urban economic development and the urban financial capacity. The HSR stations in both cities are located on the edge of the built-up urban areas; the economic development includes a strong development of the secondary industry and a growth stage of the tertiary industry; being the urban financial capacity similar. Therefore, the two cases have overall similarities. There are differences in the

L. Wang and H. Gu, *Studies on China's High-Speed Rail New Town Planning and Development*, https://doi.org/10.1007/978-981-13-6916-2_4

Table 4.1 Difference comparison methods

Case	Overall similarity	Important Difference Point		Phenomenon / result
Positive case	a, b, c...	X	→	Y
Negative case	a, b, c...	X		Not Y

Redrawn by the author, Source Skocpol and Somers (1980)

development stages of the HSR new town. The basic road framework of the Wuxi HSR new town is completed, and the surrounding development has been partially completed, while the Changzhou road system is under construction, and the surrounding development volume is small. This chapter analyzes in detail the process, the main body and the key issues of the planning and development of the two cases. A case difference method is used to compare and explore the key factors affecting the implementation results of the plan. The specific basis is as follows:

(1) The cities are similar in size. For cities of different scales, the impact of HSR is very different, and there is also a great difference in the level of support that cities can provide for development around the station site. According to data from the sixth national census (2010), the permanent population of Wuxi City (excluding Jiangyin City and Yixing City) is 3.54 million, and the area of the city is 1224.85 km^2 (including 397.8 km^2 in the waters of Taihu Lake). Changzhou City (excluding Jintan City and Fuyang City) has a permanent population of 3.29 million and a total area of 1864 km^2. According to the Circular on Adjusting the Standards for the Size of Cities issued by the State Council on October 29, 2014, both cities are classified as type I large cities.[1] The similar size of the two cities makes them have similar requirements in the layout of large-scale transportation facilities and urban rail transit construction and follow similar laws and technical standards to provide the basis for the development of the areas surrounding the stations.
(2) The phase of economic development is similar (Table 4.2). According to data from the *Statistical Yearbook of Jiangsu Province* (2013), the total GDP of the two cities in 2012 was 298.656 billion yuan and 213.726 billion yuan, and the per capita GDP exceeded 10,000 US dollars; therefore, both the total and the per capita GDP of these cities belong to the primary level of the developed economy. Judging from the perspective of the composition of the tertiary industry, the two

[1]Type I large cities: Cities with a permanent population of 1 million to 5 million in urban areas are large cities, of which cities with a permanent population of 3 million to 5 million in urban areas are type I cities.

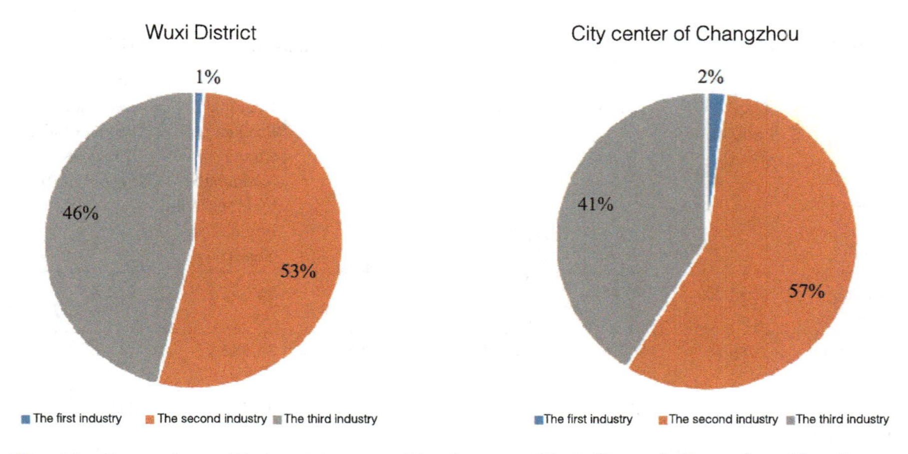

Fig. 4.1 Comparison of industrial composition between Wuxi City and Changzhou City. *Source* Author, drawing based on the data in *Statistical Yearbook*

cities share the same two-three-one structure, and their proportions are similar (Fig. 4.1), which is in line with the proportional structural characteristics that completed industrialization has in the early stages of a developed economy. From the perspective of development trends, the cities are facing a shift from production-oriented to consumption-oriented, from industrial-oriented to service-based and from labor-intensive to knowledge-intensive. Similar economic aggregates and economic development stages ensure that the impact of HSR on the cities is similar; that is, business and tourism passenger flow may promote regional development. These similarities also ensure that the city has similar positioning and appeal in the construction of the areas surrounding the station site.

(3) The spatial relationship between the site and the main urban area is similar (Fig. 4.2). The Beijing-Shanghai HSR line passes through the edge of the built-up areas of the two cities, and the station sites are close to the edge of the built-up areas of the central city. According to the research and analysis of the development of the areas surrounding the 22 HSR stations of the Beijing-Shanghai HSR line, the positional relationship between the station and the main urban area will affect the development of the areas surrounding the stations. In addition, the land ownership and use of the area surrounding the site will also affect its development. Both the Wuxi East Railway Station and the Changzhou North Railway Station are located on the edge of the central city. Before the development, the surrounding areas were collectively owned agricultural land. The conditions for development in these two areas are similar.

(4) There are differences in development. Only when different development models and development results appear in the context of similar development background, it is necessary to explore the differences between the planning and implementation processes in order to reflect the impact of the planning and

Table 4.2 Gross Domestic Product and Index of major cities in Jiangsu Province (2012)

City	Regional GDP (100 million yuan)	Primary industry	Secondary industry	Tertiary industry	Gross Regional Product per Capita (Calculated by Household Population, Yuan)	Gross Domestic Product Index (previous year = 100)
Nanjing City	4633.23	96.45	2029.97	2506.8	66,032	112.8
Wuxi urban area	2986.56	32.36	1577.4	1376.8	86,582	113
Xuzhou urban area	1779.47	52.12	967.37	759.98	57,742	116.1
Changzhou urban area	2316.26	47.64	1316.2	952.42	71,812	113.3
Suzhou urban area	3572.75	31.02	1948.71	1593.02	94,270	113
Nantong urban area	1392.81	45.94	750.27	596.59	62,132	113.3
Lianyungang urban area	437.39	28.18	221.8	187.41	42,683	111.1
Huai'an urban area	872.67	88.32	421.76	362.59	32,897	113.4
Yancheng urban area	625.76	59.88	344.17	221.71	38,785	115.2
Yangzhou urban area	989.45	21.98	562.68	404.8	71,681	114.8
Zhenjiang urban area	844.87	17.79	486.97	340.11	70,994	113.2
Taizhou urban area	566.52	13.74	346.24	206.54	67,072	114.9
Suqian urban area	374.6	55.8	192.4	137.4	26,150	114.6

Source Jiangsu Statistical Yearbook (2013)

implementation process on the development of the area surrounding the station. According to the results of cluster analysis, Wuxi belongs to the fourth type of cluster city, and Changzhou belongs to the second type of cluster city. There is a significant difference in the expansion rate of the built-up area around the two stations (Table 4.3). In addition, in the course of the investigation, it was found that the areas surrounding the Wuxi East Station and Changzhou North Station had significant differences in terms of planning scale, construction speed and scale of completion, which were conducive to distinguishing the different impacts of planning and implementation on the development of the HSR region.

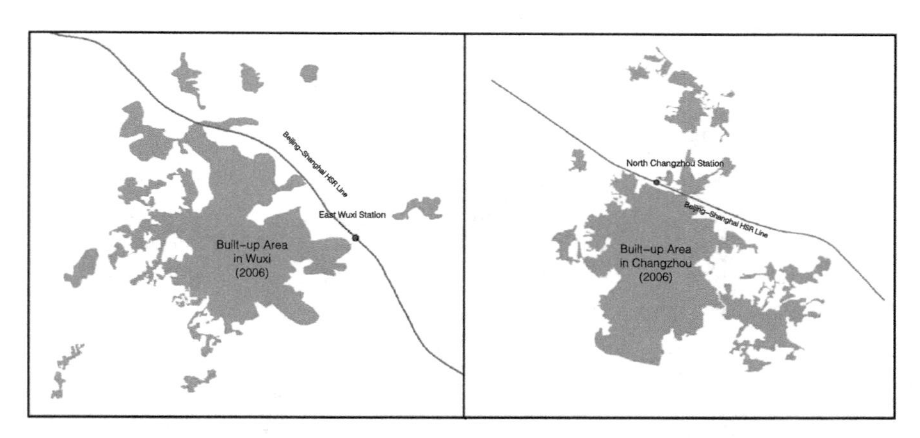

Fig. 4.2 Comparison of location relationship between HSR stations and built-up areas in Wuxi and Changzhou cities. *Source* Author, drawing based on Satellite Map

Table 4.3 Comparison of the construction of Wuxi East Station and Changzhou North Station

City	Distance (m)	Area (m^2)			Proportion of ring area		
		2006	2012	Growth rate (%)	Ratio of 2006 (%)	Ratio of 2012 (%)	Proportion increase (%)
Changzhou	2000	6,104,087	8,070,239	24	49	64	16
	4000	22,006,306	25,847,971	15%	58	69	10
	8000	61,960,730	88,091,513	30	41	58	17
Wuxi	2000	–	4,259,539	100	0	34	34
	4000	4,173,668	13,618,662	69	11	36	25
	8000	25,052,551	51,964,605	52	17	34	18

4.1 Brief History of HSR New Town of Wuxi

The Wuxi HSR new town is located east of the Wuxi built-up area and is one of the "five new towns" in Wuxi. In March 2006, with the opportunity of the Beijing-Shanghai HSR line, Wuxi built the HSR station in Xishan and started the development of the Wuxi HSR new town—Xidong new town. The planned area is 44.06 km^2, and the planned population is 320,000. In 2015, Xidong new town had a public financial budgetary revenue of 992 million yuan and the entire society's fixed asset investment of 12.57 billion yuan.

At present, Xidong new town has made breakthroughs and developments in the areas of headquarters economy, e-commerce logistics, modern commerce, the "two-vehicle" (electric vehicle, motorcycle) industry, technological innovation and eco-tourism. The headquarters building is under construction, and the three buildings of Huaxia, Tianyu and Ningtai have been completed. The Xiong Chuangrong Building has been put into use, and nine headquarters economic projects for companies such

Fig. 4.3 Wuxi HSR new town infrastructure boundary

as the Hongdou Group and Tianyu Energy have been introduced. The first phase of the Yupei e-commerce logistics park project, with a US investment of $300 million, was completed. The Cross-border E-Commerce (Wuxi) Industry Incubation Center became formally operational, the Funda Shopping Center was opened for business and the Yingyue Tiandi Business Plaza was completed. The Phoenix Mountain Ecological Park was successfully promoted, and the Mount Cuigu Automobile and Motorcycle Base in Phoenix Mountain was completed. The provincial-level Cuiping Mountain Tourist Resort was successfully approved. In 2015, the invoiced revenue of the industries in the new town was 16.23 billion yuan; the "two-vehicle" (electric vehicle and motorcycle) industry achieved invoicing of 11.3 billion yuan, an increase of 40%; and the brand concentration gradually increased. The Net new International Science and Technology Innovation Park (a total construction area of 165,000 m^2) has accelerated construction and has settled in the first central state-owned subsidiary company of Xishan City, China Railway First Municipal Rail Company.

Fig. 4.4 Satellite map of Wuxi East Station. *Source* Google Map, 2016

"Chuangronghui" has been recognized as a municipal "public space," and new businesses and new models such as Red Bean Telecom and Youcun.com have emerged. In 2015, 1750 new jobs were created.

The housing system of Xidong new town was initially constructed. In 2015, the sales of commercial space were 486,000 m^2, an increase of 8.6% over the previous year. The sales amount was 3.2 billion yuan, and the completed delivery area was about 600,000 m^2. A total of 1.95 million m^2 of the two large resettlement communities, Shanyun Jiayuan and Shuiyuan Jiayuan, was completed and delivered.

The construction of public service facilities in Xidong new town focuses on education and medical care: The investment in five schools, Xidong Senior High School, Xiehe Bilingual International School, Tianyi Experimental School, Xishan Experimental Primary School and Shuiyuan Jiayuan Kindergarten, has formed a comprehensive education system from kindergarten to high school. The Xishan People's Hospital, with a total investment of 1.08 billion yuan, a scale of 1200 beds and a design based on the standard of the top three hospitals in China, is under construction. At the same time, the ecological environment and landscape of the new town are of great importance, and the new town has accumulated 390,000 m^2 of Jiulihe Wetland Park, 240,000 m^2 of Yingyue Lake Central Park, 7.5 km of Jiulihe scenery belt, 5.8 km of Jiaoyang Road landscape belt and other large landscape projects. Along with the riverside landscape along the road, the environmental layout was constructed based on the combination of the mountain and the river, the landscape along the riverside road, and the point and line surface with the important park as the node. The public green area reached 36 square meters. A basic road pattern has been formed around the HSR stations, and the peripheral construction projects have proceeded smoothly. There is still room for development in the core area (Figs. 4.3, 4.4 and 4.5).

Fig. 4.5 Wuxi HSR new town. *Source* Author

4.2 Planning Process of HSR New Town of Wuxi

The site of the Wuxi HSR new town was established in 2006, and its planning and preparation can be divided into five stages according to the planning content: site selection period, strategic research period, comprehensive deepening period, planning period of key areas and dynamic adjustment period of plans (Fig. 4.6). The stage divisions are marked by specific decisions and planning stages.

The five phases of the planning of the Wuxi HSR new town fully reflect the comprehensiveness, hierarchy and process of HSR new town planning. Comprehensiveness is reflected in the need for different types of planning and designs in new town planning and to achieve multidimensional contents, such as functional layout, spatial structure, industrial functions, traffic organization and spatial arrangement. Hierarchy is reflected in the delineation and analysis of multilevel space from the macro- to the micro-level considering the local, regional and land levels of cities. The process is reflected in the different stages of planning and design requirements, with the previous stage of the plan providing the basis for the next stage of planning. It is very important to arrange the planning process through comprehensive and hierarchical planning.

The planning, content and characteristics of the Wuxi HSR new town are different at each stage (Table 4.4). Each phase completed the corresponding planning and preparation and supported the preparation of subsequent plans, reflecting the continuity of the plan and providing clear control and guidance for planning management and development.

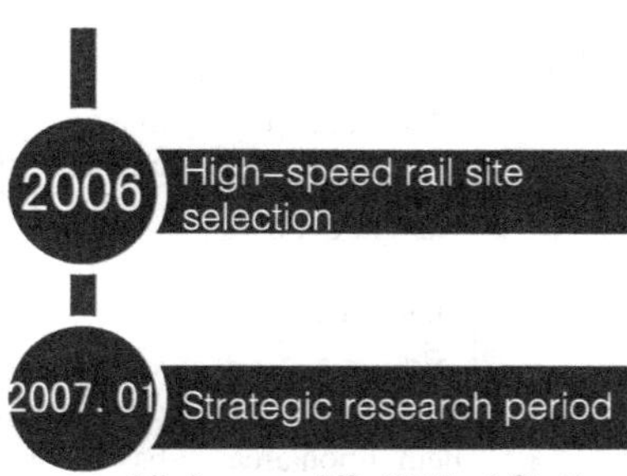

2007. 01 Strategic research period

High-speed Rail Wuxi Station International Business District Conceptual Planning
Study on the Planning of Xidong New City in Wuxi City
Wuxi High Speed Rail Station Control Detailed Planning (2008.09)

2009. 01 Full deepening

Jiuli River landscape planning
High Speed Rail Business District Industrial Development Planning
Wuxi Twelfth Five-Year Urban and Rural Planning
High-speed rail business district integrated transport planning
High-speed rail business district municipal engineering special planning
Business district road landscape planning
High-speed rail business district core area urban design
Conceptual Planning of Cuiping Mountain Forest Park
Wuxi High Speed Rail Station Controlled Detailed Planning (2010.07)

2010. 09 Key areas planning

Core District 1,2 neighborhoods urban design
Core District 4 neighborhood urban design
Xidong New City Business District open construction plan

2012. 03 Planning dynamic adjustment period

Wuxi High Speed Rail Business District Control Detailed Planning (2013.07-2014.06)

Fig. 4.6 Flowchart of planning process around Wuxi East Railway Station. *Source* Author, drawing based on research date and Wuxi government website information

Table 4.4 Summary of the planning stages of the HSR new town in Wuxi City

Stage	Time limit	Plan background	Main content	Stage characteristics
Stage one: HSR route selection and site selection	2006.12	HSR construction will start	Compare the three alternatives	–
Stage two: strategic research period	2007.01–2008.12	Nonurban areas to urban areas to change the start-up period	① Space carding, demolition area to be determined ② Clear positioning of development ③ Trunk network to be determined	Strategic, framework-based content; designed to provide the basis for follow-up work
Stage three: comprehensive deepening	2009.01–2010.08	① Scope, positioning and other strategic content to be determined ② Will soon face the construction of public investment-led period	① Space range fine-tuning ② Industrial development ③ Network system to deepen ④ Public service facilities ⑤ Regional landscape, ecological engineering	Mainly for public investment content planning and arrangements; mainly in the business area range of 45 square kilometers for the scope of work
Stage Four: Key areas Planning period	2010.09–2012.02	① Roads, municipal services, public services and other content to be determined ② Will soon face the construction period of private investment	① Combination of planning and construction ② Core design control, system of deepening	Deal with the planning conditions of private investment projects, deepen the design requirements, focus on the development of the region as the main scope of work

Source Author, drawing based on researches and planning proposal

4.2.1 Stage I: HSR Lines and Site Selection (2006.12)

In the process of selecting the site and route for Wuxi on the Beijing-Shanghai HSR line, the Ministry of Railways and the Wuxi municipal government conducted a comparison and discussion of the three options. The three schemes are parallel in Wuxi, and the main difference is the distance from the built-up areas (Fig. 4.7). The site selections are also different.

Scheme 1 was located on the original Beijing-Shanghai line, namely, the Shanghai-Nanjing intercity HSR near the original Beijing-Shanghai line and from the same period of construction, through the built-up area of the city. If this option were chosen, the optimal site selection option would undoubtedly be to renovate the original Wuxi station, expand the site scale and optimize the distribution and service functions in the areas surrounding the site. This would enable the original Wuxi station to become a common multiline station, further strengthen the status of the transportation hub and enhance the site's relation to its surrounding areas. Choosing this option would have less impact on the overall spatial structure of the city and would not involve a new linear separation. However, due to the need to cross the main city, it involved more demolition. At the same time, because the HSR lines are required overall to be straight, this solution meant additional demolition capacity might be needed in the neighboring cities of Suzhou and Changzhou. Mr. Lu, dean of the Planning and Design Institute of Wuxi City, mentioned in an interview that "at that time for HSR, in particular, the megaproject, such as the Beijing-Shanghai

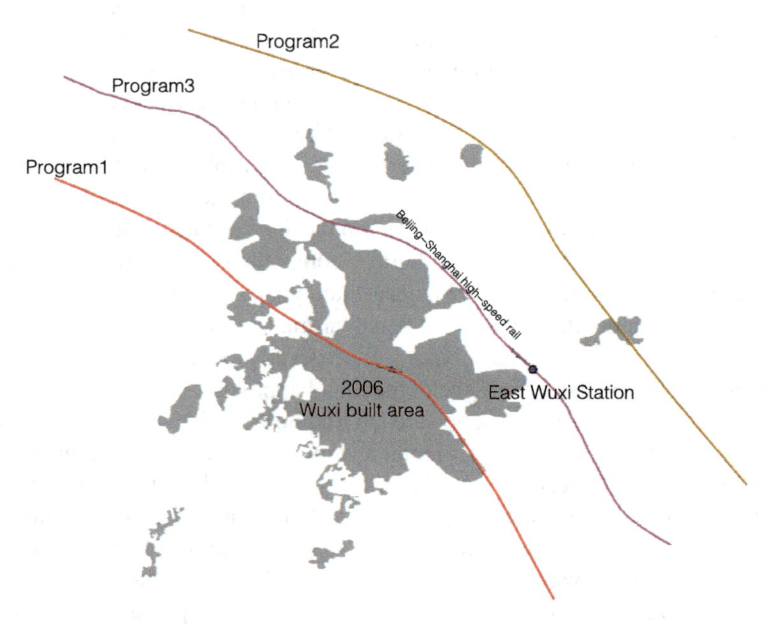

Fig. 4.7 Schematic diagram of Beijing-Shanghai HSR line selection. *Source* Author, drawing based on interviews

HSR, would have an impact on the environment and passenger flow. There is no deep research in the country, and local technical forces do not have very confident judgment. From a conservative point of view, the Beijing-Shanghai line has a greater risk in crossing the main urban area."

Scheme 2 was the Northern Line program. The overall alignment of the railway was parallel to that of Scheme 1 and the final selection plan. The overall location was the most northerly option, passing through Donggang Town under the jurisdiction of Wuxi City and Hetang Town under the jurisdiction of Jiangyin City. The area covered by this route was farmland, which could have guaranteed that the overall alignment of the line would be straight; it was also more favorable for the alignment in Suzhou Changzhou and could have reduced the demolition volume. However, the distance between the project and the edge of the built-up area of the city center at that time was more than 10 km, and the distance was too great. It would have been difficult to realize the driving role of the station in the development of the city. In addition, according to the general planning requirements of Wuxi City, the group "Donggang-Xibei (Zhangye)" belongs to the "functional subarea for the control and development of ecological industries"; the economic development is lagging, but strict pollution control is required in some parts of the area. At the same time, as the area borders Wuxi, Suzhou and Changzhou, it requires "strict control to avoid proliferation in border areas." By 2020, the group has a planned population of only 56,000. According to this scheme, the site selection could be only in the area of Donggang Town. Not only was the amount of development around the site controlled but also the station's radiation was weak and inefficient.

Scheme 3 was the final selection. The route is parallel to the Shanghai-Nanjing Railway and the Shanghai-Nanjing Intercity HSR. The basic distance between the two lines is 12 km. The line is within the scope of farmland and does not involve the large-scale demolition of built-up areas. At the same time, the line selection is closer to the main city, which ensures the radiation effect to the main city and meets the development requirements. However, to solve the site problem, the Wuxi City Planning Bureau originally proposed to locate the station in the town of Yuqiao Town, Huishan District, on the north side of the city in the direction of Jiangyin. There were three reasons: (1) It would be easy to build a comprehensive transportation hub to strengthen the connection between Wuxi and Jiangyin and promote integrated development; (2) The mature development of the township and township enterprises in the surrounding areas would provide a basis for the development of the areas surrounding the site and the potential for development and appreciation; and (3) The overall planning incorporated the construction area of Yuqiao Town into the main urban area, facilitating rapid and large-scale development around the site. However, this program was not implemented. According to the relevant personnel in the interview, "This setup plan may require a bypass line and may result in train deceleration, which is not the first choice for the Ministry of Railways. At that time, the construction of subways in Wuxi's main downtown area needed to cross the Beijing-Shanghai line and the Shanghai-Nanjing intercity HSR line, which required the approval of the Ministry of Railways. The site selection of the Beijing-Shanghai HSR station is one of many decisions made by the Wuxi municipal government and the Ministry of Railways, and this has become the 'negotiating condition' of the Ministry of Railways in the negotiations."

The final site was selected within the "Anzhen-Yanjian" group. The advantage of this site selection scheme is that Dongting Town, Fangqian Town and Meicun Town near the west side of the site were undergoing rapid development during the line selection period. The Xishan Economic and Technological Development Zone was also one of the leading industries in Wuxi City in terms of industrial development. The difficulty in the development lies in the fact that the degree of urbanization in this region is relatively low, and the development foundation is poor; in the overall planning of Wuxi City (2001–2020), it will not be incorporated into the main city until 2020. On the one hand, this difficulty shows that at the time, this area was not the major direction of the development of the city. On the other hand, the regional land use index is also relatively tight. This issue has been highlighted in the follow-up development and process.

4.2.2 Stage II: Strategic Study Period (2007.01–2008.12)

In December 2007, the Ministry of Railways approved the preliminary design of the Beijing-Shanghai HSR line, and the planning and construction of the areas surrounding the Wuxi East Station site began. According to the results of the first phase of site selection, Wuxi East Station is located at the boundary of the urban built-up area. The Xishan District, where the site is located, is an important agricultural area in Wuxi. Its "Anzhen-Eastern Pavilion" group was positioned as an "industrial area with machinery, textiles, and other industries at the core" in the "Comprehensive Development Plan for Xishan District" announced in 2006. The site is located between the township of Anzhen and the Pioneer Village Industrial Cluster. The original land surrounding it is agricultural. To clarify the development of the region after the construction of HSR stations, the Wuxi Planning Bureau began to organize related planning studies for the areas surrounding the stations.

The planning of this stage answers three questions (Table 4.5): (1) What development opportunities can the development of the Beijing-Shanghai HSR line bring, and what impact will it have on the development orientation of the district? (2) What is the scope of the impact of HSR stations, and how should the corresponding development and construction areas be defined? (3) How should spatial planning, such as how to organize road traffic and how to configure infrastructure and public service facilities, be addressed? To answer these questions, the Wuxi City Planning Bureau commissioned four companies, Atkins (UK), Easy City (Germany), World link Real Estate (China) and RTKL (US), to execute the "Conceptual Preliminary Planning of the HSR Wuxi Railway Station International Business District" (Fig. 4.8). The issues of concern for this group of planning projects are mainly the positioning of regional development strategies, major functions and industrial structures, and development

Table 4.5 Major planning achievements and contents during the strategic study period

Number	Plan name	Completion time	Planning commission department	Compiling unit	Main content	Planning range
1	*Conceptual preplanning of Wuxi HSR Station International Business District*	2007.7	Wuxi City Planning Bureau	Atkins, Erik, CenturyLink, RTKL	Strategic Positioning, Main Functions, Development Strategy, Development Timing	HSR station area (the specific scope to be clarified)
2	*Study on the Planning of Xidong New Town in Wuxi*	2008.6	Wuxi City Planning Bureau	Wuxi City Planning and Design Institute	Xidong new town, the development of positioning, spatial level	Xidong new town area of 125 km^2
3	*Wuxi HSR Station Control Detailed Planning (2008.09)*	2008.9	Wuxi City Planning Bureau	Wuxi City Planning and Design Institute	The current situation of land use, major road network planning, public service facilities preliminary planning	HSR business area of 45 km^2

Source Author, drawing based on relative planning proposals

strategies. In July 2007, a plan was formed. In conjunction with the "pre-conceptual planning," the Wuxi Planning and Design Institute completed the "Planning Study of the New District of Xidong, Wuxi" in June 2008 (Fig. 4.9). The study delineated 125 km^2 of Xidong new town and, on this basis, delineated 45.5 km^2 of the HSR business district (the center of Xidong new town) and 3.98 km^2 of the HSR core area. In addition, research on development orientation, function, industry organization and development strategy was conducted for different spatial scopes. To further control and guide the development of the space, the Wuxi City Planning Institute completed the "Regulatory Detailed Plan for Wuxi HSR Station Area (2008.09)" in September 2008. The planned HSR business district with an area of 45.5 km^2 has undergone a preliminary study and layout of the spatial structures, road traffic and public service facilities.

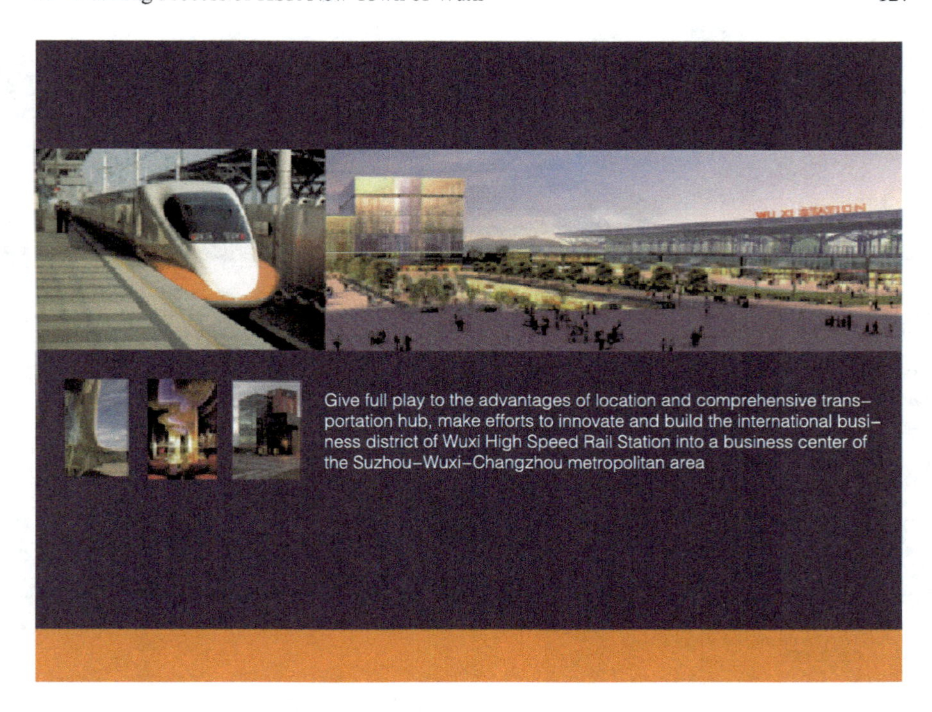

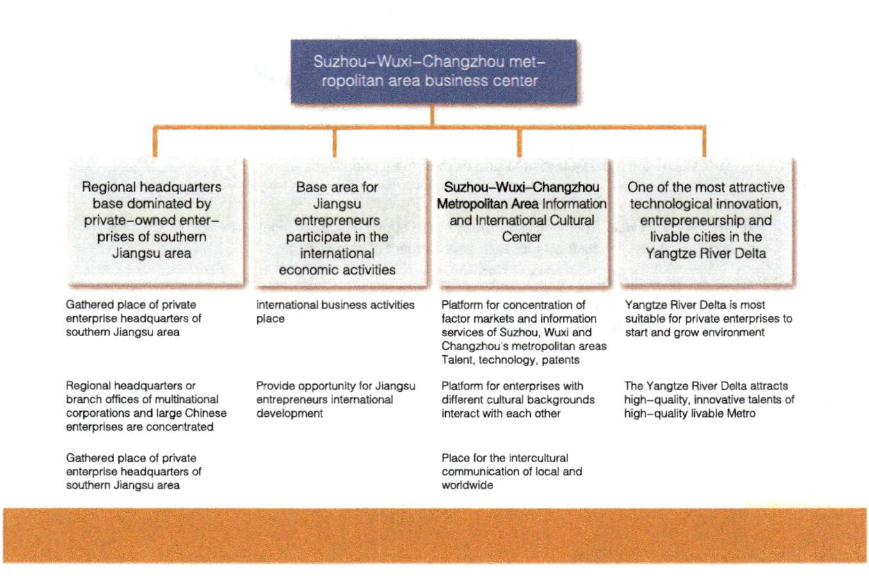

Fig. 4.8 Conceptual preplanning of the Wuxi Railway Station International Business District. *Source* Planning proposal

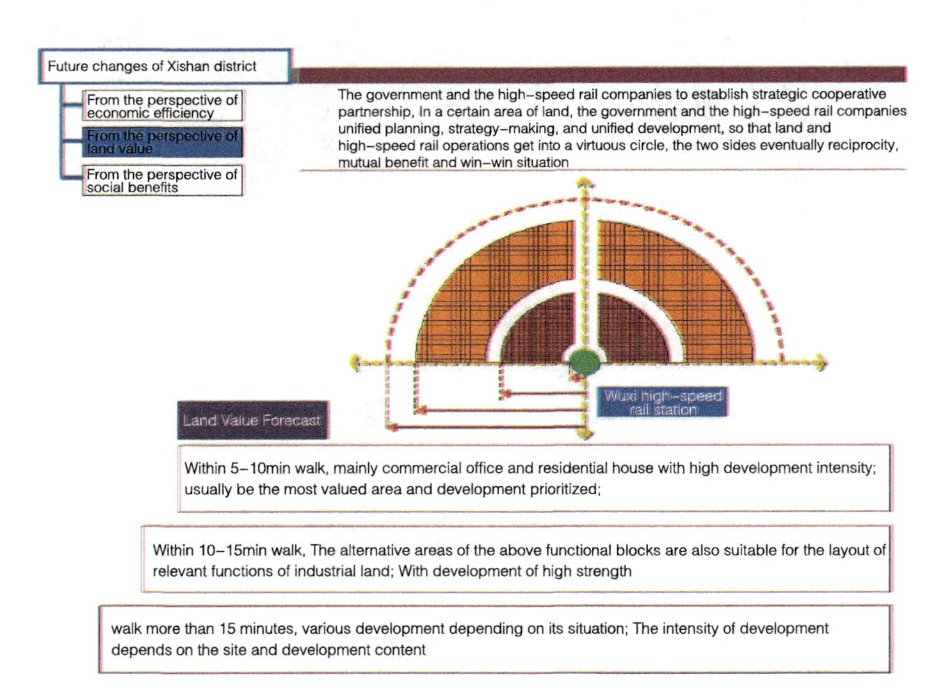

Fig. 4.8 (continued)

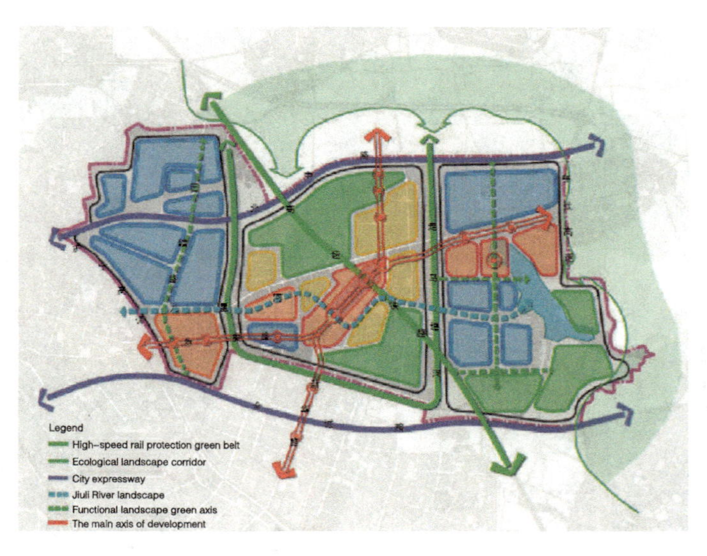

Fig. 4.9 Space development strategy of Xidong new town. *Source* Planning research of East Wuxi new town

The significance of this strategic research phase is as follows: (1) The development orientation is transformed: The former "village and township plus township enterprise gathering area" in the HSR station area is transformed into an urban functional area. (2) The spatial level is clear: The planning forms the spatial level of "Xidong New Town—Wuxi East Railway Station HSR Business District—HSR Core Area," which lays a foundation for the formulation of development policies, planning and development and management of subsequent development. (3) The scope of development and construction is clear: the demolition and land acquisition and storage as the basis for development can be carried out (Fig. 4.10). (4) The preliminary planning of important infrastructure and public service facilities is carried out, then the main road of the road network and municipal infrastructure construction can be quickly accessed.

4.2.3 Stage III: Comprehensive Deepening Period (2009.01–2010.08)

After the second stage of planning research and preparation, the HSR business district has a mature planning program and construction foundation. The construction of the spatial structures and the main parts of the main road network has begun. The management body and jurisdiction have been further defined. The demolition and land reserve work in the HSR business district is proceeding in an orderly manner. The Beijing-Shanghai HSR line has officially begun, and the HSR business district

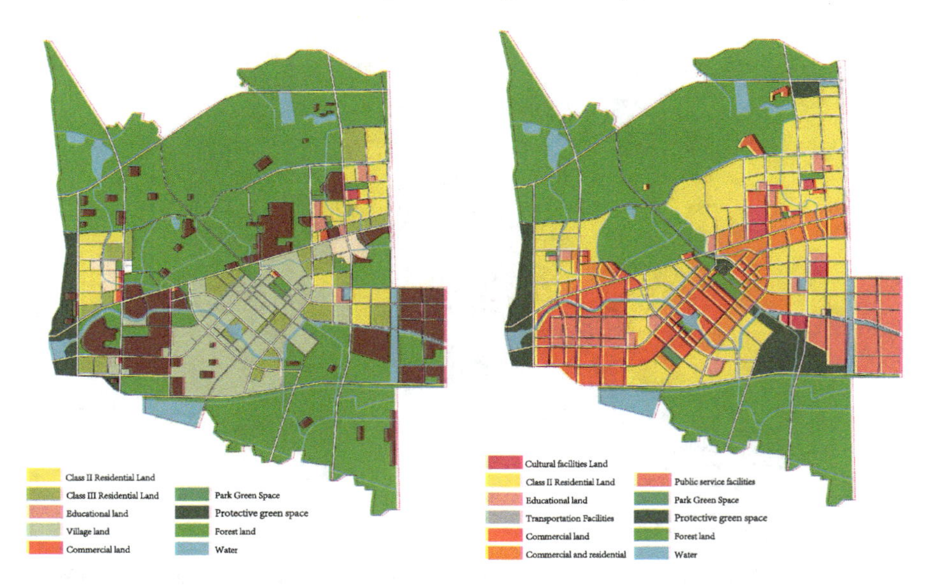

Fig. 4.10 Land use status and land use planning in the HSR business district of Wuxi East Railway Station. *Source* Regulatory plan of the area around the HSR station in Wuxi City (September 2008)

is facing specific development and construction needs. The planning documents that can be supported at this time are only the Planning Study of Xidong New Town District in Wuxi and the Regulatory Plan of the Wuxi HSR Station Area (2008.09) prepared by the Wuxi Planning Institute in 2008. It is difficult to meet the detailed standards required for project development management from the perspective of content and planning depth. Therefore, the third stage of comprehensive deepening began.

The main task of this phase was to conduct research on and planning of important special systems within the HSR business district and major projects invested in by the government to guide public investment in the next stage. The phase results include three aspects: economic industry development planning, infrastructure planning and design and key projects planning (Table 4.6).

(1) Planning for the development of the economic industry: After the establishment of the HSR Business Area Administrative Committee, Envoy Ease China and Colliers International established locations for HSR business districts, relying on the industrial planning of the original HSR stations as the basic path for industrial selection, and industrial development in the region was clarified. In October 2009, the "HSR Business District Industrial Development Plan" was completed, and the industrial space was planned and arranged. Therefore, the conceptual planning and urban design of the eastern electric-driven industrial film and the western auto-park modern commercial area were completed.
(2) Basic design and planning: The infrastructure plans completed at this stage include the "Comprehensive Transportation Plan" and "Special Plan for Munic-

Table 4.6 Major planning achievements and contents for comprehensively deepening planning

Number	Plan name	Completion time	Planning commission department	Compiling unit	Main content	Planning range
1	*Jiuli River landscape planning*	2009	HSR Business District Management Committee	Pan Asia International	Waterway adjustment, landscape node design	Project scope
2	*HSR Business District Industrial Development Planning*	2009.10		Colliers International	Industrial categories, space deployment, development timing	Business area
3	*Urban Planning for 12th Five-Year Plan of Wuxi City*	2010	Wuxi Planning and Design Bureau	Wuxi Planning and Design Bureau	Involved in this area for positioning, size, scope	Wuxi City
4	*HSR business district integrated transport planning*	2010.7	HSR Business District Management Committee	WSP	Road network refinement, road line selection and so on	Business area
5	*HSR business district municipal engineering special planning*	2010		Wuxi Planning and Design Bureau	Deepen the content of the municipal engineering system	Business area
6	*Business district road landscape planning*	2010		NITA	Road facilities, road interface, city logo	Business area

(continued)

Table 4.6 (continued)

Number	Plan name	Completion time	Planning commission department	Compiling unit	Main content	Planning range
7	*HSR business district core area urban design*	2010.7		RTKL	Core area of space model, function, development strength and so on	The core area of 2.39 square kilometers
8	*Cuipingshan Forest Park Concept Planning*	2010		–	Functional planning, site layout, means of implementation	Project scope approximately 10 square kilometers
9	*HSR business district core area underground transport planning*	2010		WSP	Underground loop project	Core area
10	*Business district water system planning*	2010.6		–	Water line, remediation strategy	Business area
11	*Wuxi HSR Station Controlled Detailed Planning (2010.07)*	2010.7		Wuxi City Planning Institute	Refinement of land use, road network adjustment and deepening, public service facilities optimization	Business area

Source Author, collection based on relative planning proposals ipal Engineering." The "Comprehensive Transportation Plan" was designed by WSP Architects. It was compiled in July 2009 and was finally approved in July 2010. Based on the basic road network framework of the "Regulatory Detailed Plan for Wuxi HSR Station Area (2008.09)," the HSR Business District Management Committee deepened the road network system within the business district. At the same time, the traffic development model of "transit priority, multimodal equilibrium, and multimodal integration" was defined. It also integrated the rail transit system, the slow-moving system and the parking system in the area, emphasizing the importance of transfers and the convenience of transportation in each development block. Under the guidance of comprehensive transportation planning, deepening the design of secondary roads and slip roads in the region laid a foundation for the refinement of the road network in the next phase. The "Special Municipal Project" includes 10 special projects, such as water supply, rainwater and pipeline integration. It was started in August 2009 and publicized at the end of November, laying a foundation for subsequent large-scale construction.

(3) Major project planning: The selection of key projects emphasizes the structural impact on important issues within the overall region and plays a significant role in embodying regional characteristics and improving urban design. The key projects carried out during this phase include the "History of the Jiuli River," the "Concept Plan for the Cuipingshan Forest Park," the "Landscape Plan for the Business District," the "Urban Area Design" and the "Core Area Underground Transportation Plan." The Jiuli River is located on the south side of the site and runs east–west through the core of the HSR business area. Cuiping Mountain is located on the northwest side of the site and constitutes an important landscape and ecological barrier together with Laoshan and Jiaoshan. The planning of Jiuli River and Cuiping Mountain is of great significance to the overall landscape structure and ecological space corridors in the area, so the execution of the planning is also an important project at this stage. Through the "Landscape Planning for the Business District," a study of the landscape features of the road systems in the business district combined with the characteristics of ecological patches and afforestation, facilities, urban signs and lighting, we hope to enhance the environmental characteristics of the business district by creating a road network system that is integrated with nature. The "core area" in the "Urban Design of the Core Area of the HSR Business District" refers to the area directly adjacent to the HSR station, which is most directly affected by the HSR station radiation and is the most densely populated portal area, with complex requirements for the rational organization of multiple transportation facilities and the creation of urban intentions. The HSR core area initially defined in the first phase of the plan was 3.98 km^2; in RTKL's design process, after the traffic flow forecast and comparison with similar cases, the area was re-range, and the area was concentrated within 2.39 km^2. The design subdivides the core area into three sections and reorganizes and designs the specific functions, development strengths, spatial patterns, landscape lines of sight and underground space of each section.

At this stage, different types of planning methodologies have been conducted with in-depth research on important issues in the development of bossiness districts nearby the HSR sites. Through the connection with each organizational unit, the Wuxi Urban Planning and Design Institute began to dynamically update the regulatory plan of the HSR business district at the beginning of 2010 and compiled the "Regulatory Plan for the Business District of Xidong New Town, Wuxi (2010.07)." The changes include narrowing the square in front of the station, canceling the extension of Xinhua Road, enlarging Yingli Lake in the Jiuli River to form a waterfront activity center and docking with plans for comprehensive transportation planning, the Jiuli River landscape, etc. These planning results need to be expressed in statutes to optimize the layout of important public service facilities such as Xidong Senior High School and Tianyi Experimental School. The multiple levels of planning and design of this stage and its achievements in the control of detailed planning provide an important foundation for follow-up construction; in 2009, the basic development and construction began.

4.2.4 *Stage IV: Planning Period for Key Areas (2010.09–2012.02)*

After the preparation of the in-depth planning study for the previous stages, the important contents of the regulatory plan for the HSR business district were determined, and the construction content to be funded by public investment was clearly defined. With the start of development and construction, the trunk road network and municipal infrastructure were completed. With the development of the comprehensive construction of public investment projects and the background of the upcoming opening of the Beijing-Shanghai HSR line, it was expected that private investment would enter the region, thus creating new requirements for planning.

The planning of this stage addressed the development and management needs in key areas. On the one hand, based on the 2010 edition of the "HSR Business District Regulatory Plan," the existing planning results were consolidated and combined with the projects that had already been built, submitted for approval and approved for construction. On the other hand, in order to ensure the construction quality of the HSR core area, further detailed design control was required for the land to be developed, and the conditions for land transfer would be included as a supplement to the contents of the regulatory plan. There were not many major plans at this stage, and the emphasis was on development management integration and detail space design (Table 4.7).

Table 4.7 Organizing planning results in key areas

Number	Plan name	Completion time	Planning commission department	Compiling unit	Main content	Planning range
1	Xidong New Town Business District Development and Construction Planning	2011	HSR Business District Management Committee	Wuxi City Planning and Design Institute	Predevelopment and construction, follow-up development scheduling	Business area
2	Core District 1, 2 neighborhood urban design	2011.4		SOM	Response to the design of land transfer control content	Core area
3	Core District 4 neighborhood urban design	2012		SOM	Design control content of dealing with the transfer of land	Core area

Source Author, drawing based on relative planning proposals

In terms of the integration of planning achievements, the "Plan for the Development and Construction of Xidong New Town Business District" was completed by the Wuxi Urban Planning Institute in February 2011. This plan is based on the 2010 regulatory plan. It reviewed roads, public service facilities, landscape facilities, etc... and clarified the land use, the construction status and planning requirements of each site.

In the urban design of core lands, the organization compiled "Urban Design of Block 1 and Block 2 in the HSR Business District" and "Urban Design of Block 4 in the HSR Business District." Blocks 1 and 2 are located in the core area of the HSR station business district and adjacent to the station's south square. They are positioned as a gateway and are identified as the most attractive areas for high-end offices and businesses. At the same time, this area also has the most complex traffic flow and the most significant impact on the functions of the business district. To guarantee the spatial quality of this critical area, relying only on controlled detailed planning that applies to the general area is insufficient. Therefore, the HSR Business District Management Committee commissioned SOM to develop a detailed urban design for the area. The contents were deepened to the section of each road, the location and scope of each open space, and the layout design of each block, including architectural design guidelines, lighting systems, marketing systems and public arts organization methods. The planning officially began in August 2010 and was completed in April 2011.

After the planning and design stage, the planning and construction of the HSR business district was effectively organized, the construction of various types of plots was clarified and a platform for detailed discussions with developers was provided for follow-up development. At the same time, the planning and management model of the HSR new town was as follows: Within the core area of the business district, control conditions for detailed land transfer were formulated on the basis of "controllable detailed planning plus urban design," and regulatory planning was used as a land transfer condition in the outlying areas.

4.2.5 Stage V: Planning Dynamic Adjustment Period (2012.03–2014.12)

Based on the planning of the previous three stages (stage II to stage IV), the planning content of the HSR business district was comprehensive and complete. The projects led by public investment were completed and put into use, and private investment projects were fully initiated. There was no increase in the content of the plan at this stage. It was intended to respond to the specific requirements of specific construction and development projects so that the regulations could be adjusted dynamically (Table 4.8).

For example, the 2013 regulatory adjustments (Fig. 4.11) included the fine-tuning of specific land area functions or the optimization of facilities within the land without large-scale structural adjustments. Therefore, when entering this stage, the technical level and related contents of the preliminary planning results were mature, and regional development had entered a period of stable planning and management.

Table 4.8 Planning dynamic adjustment period

Number	Plan name	Completion time	Planning commission department	Compiling unit	Main content	Planning range
1	Wuxi HSR Business District Control Detailed Planning (2013.07–2014.06)	2013.7–2014.6	Xidong New Town Planning Office	Wuxi City Planning Institute	Local land planning and adjustment	Business area

Source Author

(1) Adjust south side of the intersection of East Xianfeng Road and Shanhe Road, south of the intersection of Yongsheng Road and Cuishan Road and some plots along Xinhua Road Commercial Mixed Land Mixed Commercial / Residential Land to mixed commercial land;

(2) Adjust land on the south side of the intersection of East Xianfeng Road and Wenjing Road Commercial Mixed / Residential Mixed Commercial to do for commercial use / research and design sites;

(3) Adjust venues in the stadium on the south side of the intersection of Wenjing Road and Dongan Road are adjusted to the east massif at the intersection of Xinhua Road and Xishan Avenue, and the scale,to do for commercial use / research and design sites;

(4) Adjust mixed commercial and residential area on the south side of the intersection of Wenjing Road and Dongan Road to commercial land / recreation land;

(5) Adjust residential land on the east side of the intersection of South Houshan Road and Dongan Roadto do commercial / research design of land;

(6) Adding land for substations within the northwestern side of the intersection of South Runxi Road and Xishan Avenue;

(7) Adjust north of the intersection of Wanquan Road and North Xinyun Road Public facilities Land and administrative office land to the site of religious facilities;

(8) Adjust north of the intersection of Yongsheng Road and Cuishan Road Public Utilities Land / administrative office land for the administrative office space / commercial mixed use of land;

(9) According to the urban design, adjust the internal branch position on the south side of the intersection of East Xianfeng Road and Shanhe Road;

(10) Adjust land classification codes according to new standards and optimize road network;

(11) According to the actual construction needs, improve land control indicators.

Fig. 4.11 Control plan for Wuxi HSR business district June 2013 dynamic update. *Source* Official website of Wuxi Urban Planning Bureau. http://www.wxgh.gov.cn/news/viewghcg.asp?id=10240

4.3 Development Phasing and Model of HSR New Town of Wuxi

In accordance with the implementation content and the investors, the planning implementation process of the area surrounding the Wuxi HSR station could be divided into three stages (Fig. 4.12): stage I: start-up period for construction; stage II: period dominated by public investment; and stage III: period dominated by private investment. The three stages are parallel but lag behind the five stages as planned, beginning from the start-up construction in 2009.

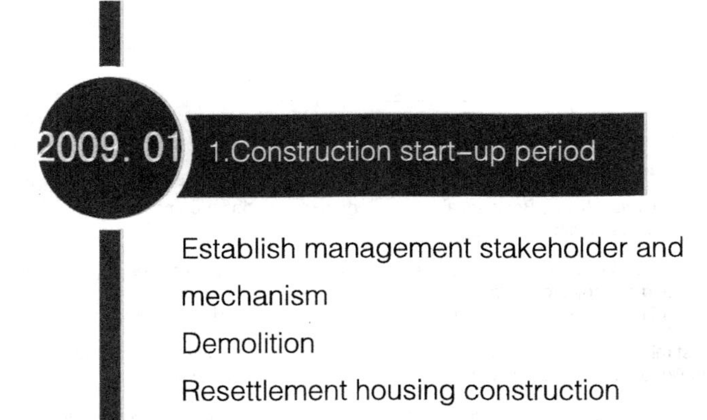

Fig. 4.12 Implementation process of the planning for the area surrounding the HSR station in Wuxi. *Source* Author, drawing based on relative news

4.3.1 *Stage I: Start-up Period for Construction (2009.01–2010.08)*

Through the site selection stage of the HSR line and its stations as well as the research stage of the development strategy as planned, the development and construction scope, as well as the development positioning of Xidong new town around the HSR station in Wuxi, are defined. Meanwhile, the main road and municipal infrastructure are preliminarily arranged. With the official launch of the construction of the Beijing-Shanghai HSR line in 2008, the Wuxi HSR business district entered the start-up period. Although the development strategy for this district was preliminarily confirmed, the management organization and management mechanism were left undecided.

From 2009 to August 2010, the implementation contents included the following four aspects: the construction of the management organization and management mechanism, the demolition and collection of land, the construction of resettlement housing and the establishment of municipal infrastructure projects. The relocation project was located to the east and west of the station (Fig. 4.13).

Fig. 4.13 Distribution map of main projects in the start-up period of construction. *Source* Author, drawing based on researches and planning proposal

1. **Construction of the management organization and mechanism**

The construction of the management organization and mechanism is the basic framework of promoting regional development. After the development strategy of Xidong new town was confirmed, the development and construction team of Xidong new town and the Management Committee of the HSR Business District were successively established. The former took charge of the development construction of the whole Xidong new town, while the latter was responsible for the HSR business district. In terms of management scope, the Management Committee of the Xishan Economy and Technology Development Zone took charge of developing and constructing the east plate and west plate of Xidong new town as a high-tech industrial park. The Management Committee of the HSR Business District specialized in developing and constructing the middle plate as the core region of Xidong new town. In terms of the management system, as the high-level officer distribution unit in the administration, the management committee concentrated on promoting the investment, development and construction of the project within the management scope. The subdistrict office took charge of social affairs management, social stability, demolition and resettlement, etc. The supportive policies included "land operation," "financial investment," "resettlement policy" and "planning and construction management," within Xidong new town. The Wuxi HSR Business District Management Committee established in Xishan District was specifically responsible for the financing, construction and investment promotion of the business district. Later, with the development demand, the subordinate unit of the management committee underwent a partial adjustment, but its main responsibilities remained. Namely, it focused on the economic construction and urban development of the HSR business district. The Construction Development Co., Ltd. of Xidong new town was jointly established by the municipal government and the Economy and Technology Development Zone with investments of 300 million and 400 million RMB, respectively. It took charge of the investment and financing activities for the construction of business districts, resource development, construction of infrastructure and other projects, and independent development and construction of other business projects.

To improve the planning approval efficiency in Xidong new town and to strengthen the planning management, the Wuxi Municipal Planning Bureau assigned an office at Xidong new town at the township level and established the subordinate General Department, Preparation Department and Licensing Department, which are responsible for the planning management within 125 km^2 of Xidong new town. So far, the HSR business district has established a special management organization for planning management, economic management, social work, investment and financing business.

2. **Demolition and the collection of land**

The demolition and the collection of land laid a foundation for development and construction work. In accordance with the *Implementation Opinion about Accelerating the Development and Construction in the HSR Business District*, the local subdistrict office of land takes charge of removing the area intended for the HSR business district.

The subdistrict office is the permanent land management organization; therefore, it is familiar with the complex social problems involved in the demolition.

The Wuxi HSR business district was predicted to require the demolition of residences over an area of 3,500,000 m^2 and nonresidential buildings over an area of 2,000,000 m^2. The three-year construction plan was predicted to require the demolition of residences over an area of 1,500,000 m^2 and nonresidential buildings over an area of 1,200,000 m^2. There were 980,000 m^2 of residences and 900,000 m^2 of nonresidential buildings in the start-up area. Up to the end of the current stage, the actual demolition area in the business district had reached 2,000,000 m^2, and the task of demolishing 3.98 km^2 in the core area stipulated in the regulatory planning in 2008 had been completely achieved. The demolition supported the subsequent development smoothly and effectively. The demolition progress depended on the following elements: (1) The subdistrict office effectively organizes and manages reasonable demolition; (2) the demolition preparation work starts in advance, including propaganda and surveys before the official demolition; and (3) relocation support is provided by the municipal government.

Wuxi has been following a comprehensive plan since 2004, and the area surrounding the HSR station is outside the urban construction area. For this reason, the land index for construction and utilization could adopt only the annual newly established construction land index of Xishan District in terms of the collection of land. Since such problems are sensitive, the personnel of the planning office in Xidong new town who participated in the interview process indicated only the existence of such an operation but refused to provide specific information.

3. **Construction of resettlement housing**

Based on the plan, resettlement housing in the HSR business district includes the Jiuli River resettlement area and the Chaqiao resettlement area. Within three years, construction of 1,600,000 m^2 of resettlement areas was commenced, and 1,200,000 m^2 of resettlement areas were completed. The resettled residents are both the original residents and the stakeholders using public service facilities in this area. Meanwhile, the usage demand for resettlement housing is also the foundation for commercial development. The resettlement housing influences the demolition work and involves social stability; therefore, the leading construction group of Xidong new town and the Management Committee of the HSR Business District had to reach a consensus to treat the construction of resettlement housing as important work at this stage.

In terms of the construction process, the construction of the resettlement housing community was included in the working plan in April 2009. The community construction drawings of Shanyun Jiayuan and Shuian Jiayuan were completed in September 2009. The planning scheme design of roads, infrastructures and other supporting facilities was completed. New resettlement housing with an area of 1,170,000 m^2 was opened at the end of 2009, and 676,000 m^2 of resettlement housing was constructed subsequently. By the end of July 2010, the main roads in Shanyun Residence and Shuian Residence had been constructed and put into use. The total quantity of resettlement housing under construction reached 2,490,000 m^2. At this stage, the total construction area of newly developed resettlement housing reached 1,814,000 m^2.

Within one year and a half, all resettlement housing of the HSR business district in the *Three-Year Work Plan* had been developed.

4. **Artery road network and water system engineering**

At this stage, the infrastructure construction was based on the *Regulatory Plan of the Wuxi HSR Station Area (2008.09)*, and the implementation contents included the artery road network and the corresponding municipal infrastructures. In accordance with the *Regulatory Plan and the Working Plan of the HSR Business District (2009)*, nine main roads were constructed, including five municipal artery roads (Dong'an Road, Xishan Avenue, Xinxi Road, Xinhua Road and Xiyu Road). All except Xiyu Road were uniformly managed and constructed by the municipal government. These main roads and the original Xidong Avenue expressway constitute the external ring-road network of the HSR business district and solve the traffic problems of travel between the business district and the main urban zone as well as the traffic flow under the business district and the radiation from each direction. Meanwhile, the HSR business district promoted the construction of Shanhe Road, Xinyun Road, Xianfeng East Road and Houshan Road in the start-up area. Those roads, together with Xinhua Road and Dong'an Road, jointly constitute the framework of the road network in the start-up area of the south square of the station. This framework lays an important foundation for the subsequent addition of public service facilities and the land transfer for initial commercial development.

The water system engineering construction includes the Jiuli River diversion project and the west extension of Furong Tang. The length of the Jiuli River diversion section is 3.4 km, and the width of the river mouth is 35 m. The engineering investment amount reached 20 million RMB. The Jiuli River diversion section is the important landscape belt in the core zone of Xidong new town. It required consideration of flood control planning, tourism channels and the urban landscape in attempting to embody natural and ecological concepts of water control. The west extension of Furong Pond starts from the old riverway of Furong Pond to the north and connects to the Jiuli River to the south. The total length is 1.2 km, and the width of the river mouth is 26.2 m. The engineering investment amount reached 6 million RMB. The riverway is designed according to the planned road network layout of the HSR station business district to improve the drainage conditions in the business core area.

The work at this stage constructed the basic framework for the development and management of the HSR business district from the perspective of planning implementation. Although the specific details set by the Management Committee of the HSR Business District were slightly adjusted in the subsequent process, the basic management frameworks, such as the division of rights and responsibility and the income distribution of the management committee and subdistrict office, were continuous, which is an important guarantee of the smooth development of the HSR business district. Meanwhile, at this stage, the demolition, resettlement and construction of the artery road network were conducted on the basis of the planning results during the research period of designating a development strategy. These construction projects depend highly on public investment by the government, and such investment improves the usage value of regional land by providing the engineering and the value foundation for the subsequent construction and land appreciation.

4.3.2 Stage II: Period Dominated by Public Investment (2010.09–2012.02)

This stage is the second stage of the development objective of Xidong new town. For the HSR business district, the stage is the official opening and operation of the Beijing-Shanghai HSR line. On the one hand, various supporting construction enabled and needed by HSR station needed to be improved; on the other hand, the HSR started to show its development and leading role. With the construction of the preliminary roads and municipal infrastructure as the framework and the stage II planning results as a basis, the following five aspects were completed in this stage: (1) Continue the final stage of the construction of road and municipal infrastructure and promote the removal of residents and the construction of resettlement housing; (2) accelerate the construction of supporting facilities at the HSR station; (3) start to build public service facilities such as schools, parks and a civic center; (4) continue to lead industry projects supported by private investors; and (5) start the addition of commercial real estate projects.

1. **Continue the final stage of construction**

In terms of the construction of roads and water system engineering, the project *Seven Roads and One River*, according to the planned and adjusted content, was comprehensively commenced in August 2010. Among the seven roads, Runxi Road is the new urban artery after the revision of the plan, Jiaoyang Middle Road is the main road that was upgraded from a branch road after the revision of the plan and the other five roads (Chongwen Road, Jin'an South Road, Chunfeng South Road, Shengyuan North Road and Hongye Road) are urban secondary artery roads. "One river" indicates a new pool project. It starts from Furong Pond to the north and ends at the Jiuli River to the south. It is a connection project for the two bodies of water in the previous phase and serves as an important component of the main river system forming the water network structure of the *Eight Vertical, Four Horizontal and One Ring* project in the business district. With the completion of the *Seven Roads and One River* project in December 2010, the *Governance Project of Five Branches and Seven Rivers* started and was completed at the end of June 2011. Until the end of this phase, namely, February 2012, the total traffic road length in the HSR business district reached 102 km, accounting for 80% of the planned mileage (excluding the highway mileage). The road system was constructed by integrating the expressway and artery roads (Fig. 4.14).

In terms of the demolition progress, at this stage, the planned public key projects and the demolition work within 10 km^2 of the start-up area were completed. The demolition work was gradually transformed into the demolition and relocation of

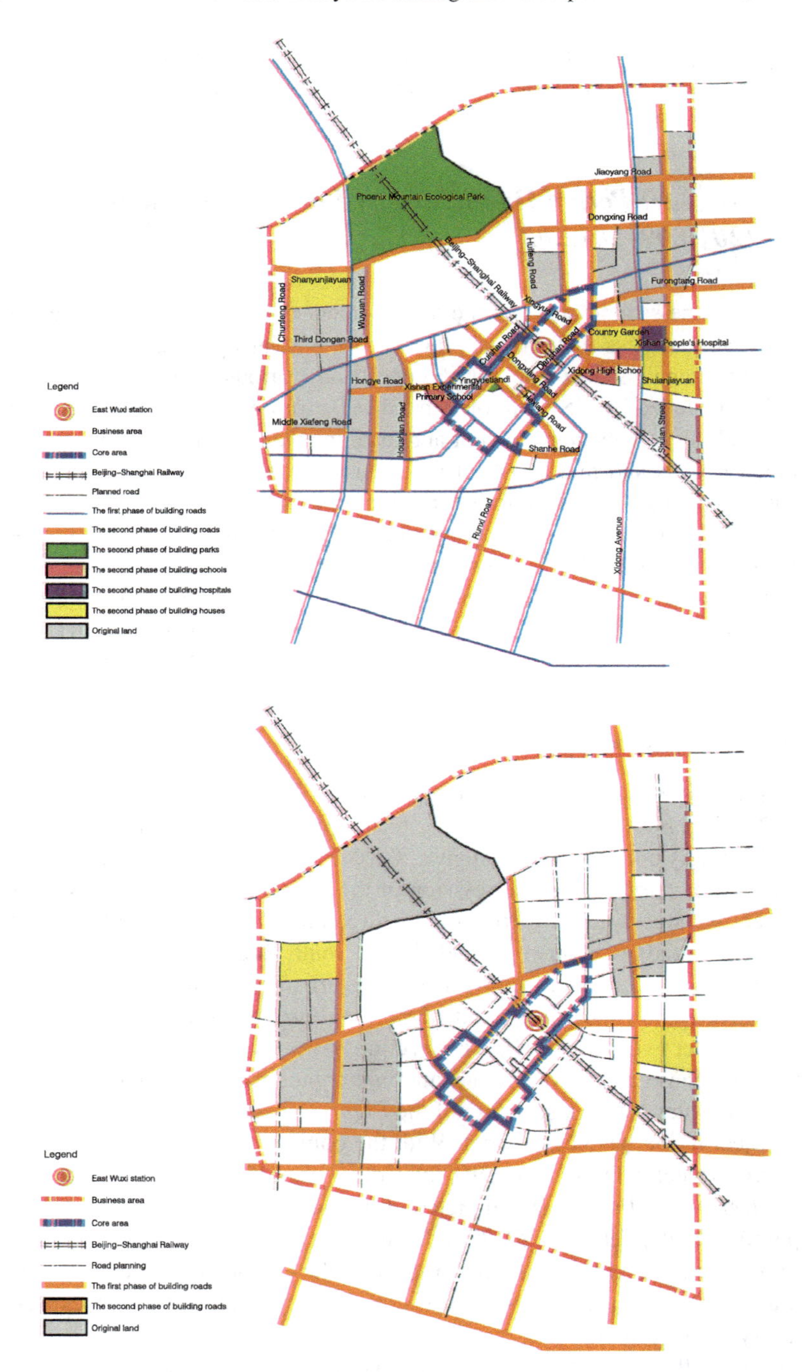

Fig. 4.14 Distribution map of the main projects in the period dominated by public investment. *Source* Author, drawing based on researches and planning proposal

blocks according to the needs for transfer and development. In terms of resettlement housing construction, Shanyun Residence A, B and C blocks and Shuian Residence A and B blocks in phase I of the resettlement housing were gradually delivered from May to August 2011. The Shuian Residence C, C+ and D blocks and the Shanyun Residence D block in phase II were officially commenced in June 2011. The total area of the four newly opened resettlement houses reached 922,400 km^2.

2. **Completion of supporting facilities at HSR station**

The Beijing-Shanghai HSR line officially opened for operation at this stage, and the station was enabled. The Wuxi East Railway Station area has two supporting facilities: An underground ring road engineered in the core zone and supporting construction engineered at the station. The underground ring road in the core zone is located to the south of Wuxi East Railway Station and below Xingwu Road, Wenjing Road, Cuishan Road and Danshan Road. It consists of the underground motor ring road of the core zone and the HSR connection channel and forms an overall layout of “one ring plus one arc.” This system has a total of 6 entrances and 4 exits. The underground garages built in the core zone of the HSR business district are connected through the underground motor system. Vehicles can also arrive directly at the Xinhua viaduct through the underground motor system, thus solving the separation problem between the office commuter traffic of the HSR business district and the arrival and departure traffic of the HSR station. The supporting project of the Beijing-Shanghai HSR line is located in the core zone of Wuxi East Railway Station New Town and is centered on the HSR station building, including comprehensive facilities such as traffic connection equipment and commercial and municipal supporting facilities in order to enable the site to achieve the regional hub function. Phase I construction included the passenger transport complex, a bus station on the east side and other equipment; phase II construction included the development of a ground plot in the south square and north square and a reserved parking garage on the east side. Both projects were commenced at the end of 2010 and were completed in September 2011, coordinating the traffic operations of the HSR.

3. **Comprehensive construction of public service facilities**

This phase focused on the construction of public service facilities. Open public service facilities include Xidong Senior High School, Wuxi Tianyi Experimental School, Shuian Jiayuan Kindergarten and Wuxi Xishan District People’s Hospital. The public service facilities constructed in this phase are located in the third layer of the HSR radiation area. The ability of the HSR to drive the development of this layer is limited, and the construction of public facilities has become an important driving force for the layer. The construction of Xidong Senior High School and Cuiping Mountain Leisure Ecological Garden as regional public service facilities in this period could effectively serve the current Anzhen town community residents and Anzhen residents moving into the town in the next year on the one hand and could attract new projects and investment on the other hand.

Merging the previous Jiangsu Province Yangjian Senior High School and Dangkou Middle School, Xidong Senior High School is the four-star middle school in Jiangsu

Province. Located northeast of Runxi Road and Wenrui Road, the new campus covers an area of 15 ha, and its investment and construction drove the development of many surrounding school district housing projects. Located west of Xidong Avenue and north of Xingyue Road, Wuxi Tianyi Experimental School has a total building area of 46,000 m^2. Each grade contains 16 classes (48 classes in total), and each class can accommodate 2400 students. The total investment in the project was RMB 150 million, and the project was completed in August 2012. Shuian Jiayuan Kindergarten is located in the resettlement area in plot B of Shuian Jiayuan; the total building area is 8776 m^2, and the total investment was RMB 26 million. It was constructed to provide supporting facilities for the Shuian Jiayuan resettlement housing community, which was to be completed in the next year. Located southwest of the HSR station business district, Xishan Wetland Park is the portal park of the business district and a subproject of the ecological rehabilitation of Taihu Lake water pollution. It covers a total area approximately 46.15 ha. The total investment in the project was RMB 150 million, and the project was completed and opened at the end of 2010.

4. **Rapid development and construction of the industrial park**

The introduction of industry and construction of the industrial park could create employment and act as an important support for farmers in the area to achieve urbanization, which encourages the population to lead the business district, thus creating demands for housing, consumption and other services and continuously promoting the development of the business district. In this phase, the construction of a good road network and municipal infrastructure in the HSR business district was conducive to attracting industry to enter. In addition, the industrial park is the beginning of investment promotion and construction work.

Industrial development is an important way to introduce the population to the HSR business district. Regional development is positioned in industrial development and transformation by utilizing the resources brought by the construction of an HSR line. Therefore, the business district plan confirms the construction of the east park and west park in the initial period of construction and optimizes the development strategy of the current west production area. On the clear basis of the preliminary plan, the importation of industry and the construction of the industrial park were strengthened in this phase. The electrically driven industrial park was always promoted. The new production base projects of Sunra and Yadea were started and constructed in 2010. Meanwhile, the Business District Management Committee started to investigate enterprises with low-end positioning, low investment intensity and low production efficiency in the district, providing a basis for updating and transforming these enterprises through the park development. As an important industry project in the HSR business district, Auto-Park Motor City aimed to build a comprehensive auto service park that integrated auto sales, service, cultural exhibition and brand promotion by introducing high-end brands. It is a pioneer in the importation of industry in the HSR business district. On April 29, 2011, during the listing-for-sale auction of the second batch of land in Wuxi, 9 plots in the Auto-Park Motor City phase I project were transferred, representing the successful promotion of industrial development in the HSR business district.

5. **Residential and business buildings invested in by private enterprises start to enter the park**

Although this stage was the development period led by public investment projects, real estate development by private enterprises was carried out successively. Due to the locational conditions of the business district being close to the HSR station, the accelerated construction of the natural landscape and public service facilities, and the low land prices, private enterprises invested in and developed residential and business projects. On November 26, 2011, sales began in the Country Garden Kaixuan Huating project, and the sales situation was good. More than 330 houses were sold in the two opening days, accounting for 75% of the total housing resources. Kaixuan Huating also became the first residential real estate project sold after the establishment of the business district. On November 10, 2011, the Kechuang Center project (Xidong Creation and Financing Building) was officially commenced and constructed. The Kechuang Center project (Xidong Creation and Financing Building) is close to the Wuxi HSR station and is located west of Xinhua Road and north of Dongxiang Road in the core area of the HSR. The total building area is 93,400 m^2, and the investment amount reached approximately RMB 500 million. This project was the first business building commenced in the HSR business district.

By surveying the specific zones of these two projects, it is not difficult to observe that the Country Garden Kaixuan Huating project is located in the inner flange of the third layer radiated by the HSR station, so it could take advantage of the relevant facilities of the HSR station and also effectively avoid being influenced by the complex flow of the station. In addition, this project is adjacent to Wuxi Tianyi Experimental School and Xidong Senior High School, under construction; Wuxi People's Hospital is also located nearby. It is clear that the previous public investment played an important pulling role in the development of this project. Located in the first layer radiated by the HSR station and close to the station, the Kechuang Center project (Xidong Creation and Financing Building) directly embodies the effect of the pulling function on business offices.

4.3.3 Stage III: Period Dominated by Private Investment (2012.03–2014.12)

With the successive completion of the projects invested in by the government in the first two stages, the conditions for the development and construction of the HSR business district gradually matured, newly established projects invested in by the government gradually decreased and private investment projects began to play a dominant role. The quantity of residential and commercial projects in the development projects increased greatly (Fig. 4.15). From the perspective of the spatial placement of the development projects, the large-scale development of the core area started in this stage (Fig. 4.16).

Fig. 4.15 Distribution map of main projects in the dominant period for private investment. *Source* Author, drawing based on researches and planning proposal

At this stage, the "three schools and three centers" invested in by the government were completed and put into use: "three schools" refer to Tianyi Middle School, Xidong Senior High School and the newly established Xiehe Bilingual School. "Three centers" refer to the district-level cultural center, civic center and medical center (Xishan District People's Hospital). New government-funded projects were ecological optimization projects, such as the Jiuli River Wetland Park and Yingyue Lake Central Park, matched with the construction of landscape belts along the Jiaoyang Road and other landscape belts intended to improve the ecological and landscape environment of the area. The construction of the resettlement housing entered the final stage. The focus of relocation shifted to the outlying areas, where the development value was lower than that of the first and second layers radiated by the HSR station. Therefore, the demolition was not widespread, as in the construction start-up phase, but was local demolition in preparation for construction projects.

Projects funded by private investment in this area are numerous. The Ever Grande Group and Long for Properties were successively located in the HSR business district. The construction area of new commercial buildings in 2012 reached 0.83 million m^2. Up to the middle of 2013, there were 1.4 million m^2 of commercial buildings under

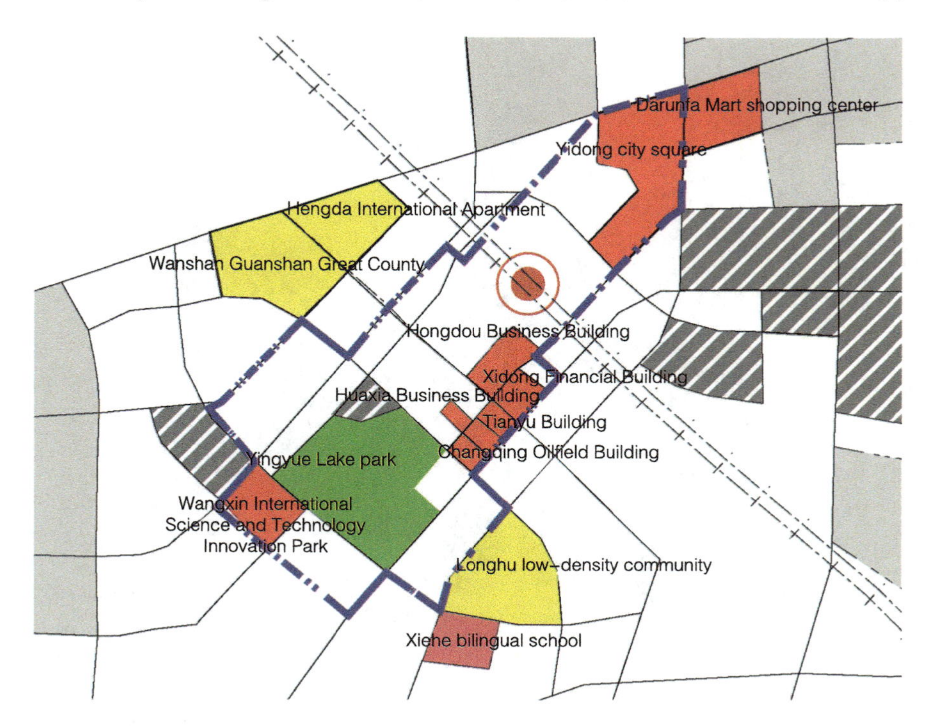

Fig. 4.16 Distribution map of main projects in the core area in the period dominated by private investment. *Source* Author, drawing based on researches and planning proposal

construction in the business district. In terms of industry, the Pioneer International Auto Plaza project was completed, and subsequent investment promotion followed. In terms of commercial buildings, 18 commercial office buildings were built from 2012 to 2013, and the area under construction reached 870,000 m^2. In terms of commercial projects, two major shopping malls, Yidong City Plaza and Runfa Shopping Center, were under construction.

4.4 Brief History of HSR New Town of Changzhou

The New Town of the HSR in Changzhou centers on Xinlong International Business City, east of Changjiang Road, south of the Shanghai–Chengdu expressway, west of Leshan Road and north of Tianhe Road, covering an area for approximately 1.6 km^2. The spatial pattern is "one axis, two centers, five squares": The one axis is the central landscape axis, the two centers are the southern "water park" and the northern "green park" and the five squares are the five major functional areas: the HSR Traffic Plaza, the Financial Welfare Plaza, the Public Service Plaza, the Science and Technology R&D Plaza and the Commercial Business Plaza. The total building area exceeds 2.5

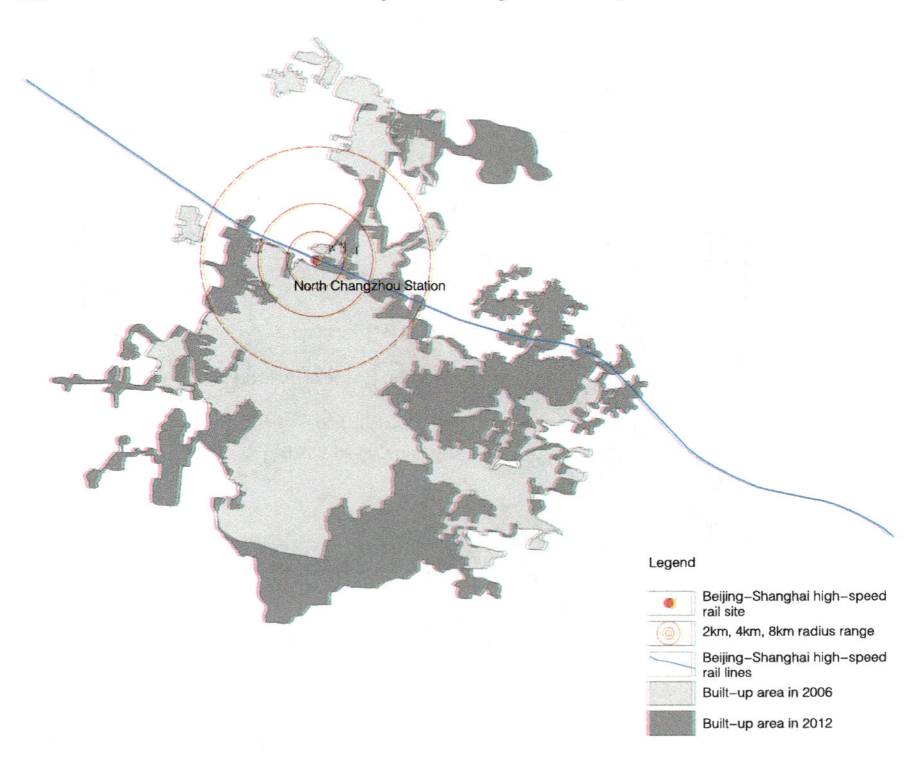

Fig. 4.17 Relationship of the Beijing-Shanghai HSR line in Changzhou, Changzhou North Railway Station and urban built-up area. *Source* Author, drawing based on Satellite Map

million m^2. The economic zone phase I project of Xinlong International Business City Headquarters started in 2014. The HSR Ecological Park and Xinlong Lake Music Park were completed as planned. The supporting road of Xinlong International Business City was publicized, and phase I of the Xinlong International Ecological Forest project was commenced. Changzhou North Railway Station is the central station of the Beijing-Shanghai HSR line. It is located at the north edge of Changzhou central urban area and is close to the Shanghai-Nanjing expressway. South of the station is the industrial land of the Changzhou high-tech district, while east of the station is the built-up residential area. The cluster analysis results show that the development of the area surrounding Changzhou North Railway Station is the second type of city, that is, the development of the area surrounding the HSR station is slow, but there is a certain level of development in the 4 and 8 km layers (Figs. 4.17, 4.18 and 4.19).

Fig. 4.18 Satellite map of Changzhou North Railway Station. *Source* Google Map 2016

Fig. 4.19 Real image of new town of HSR in Changzhou. *Source* Author

4.5 Planning Process of HSR New Town of Changzhou

According to the plans of the Changzhou Planning Component Department, the planning of the area surrounding Changzhou North Railway Station can be divided into three stages: HSR site selection period, municipal government-led planning period and district government-led planning period (Fig. 4.20).

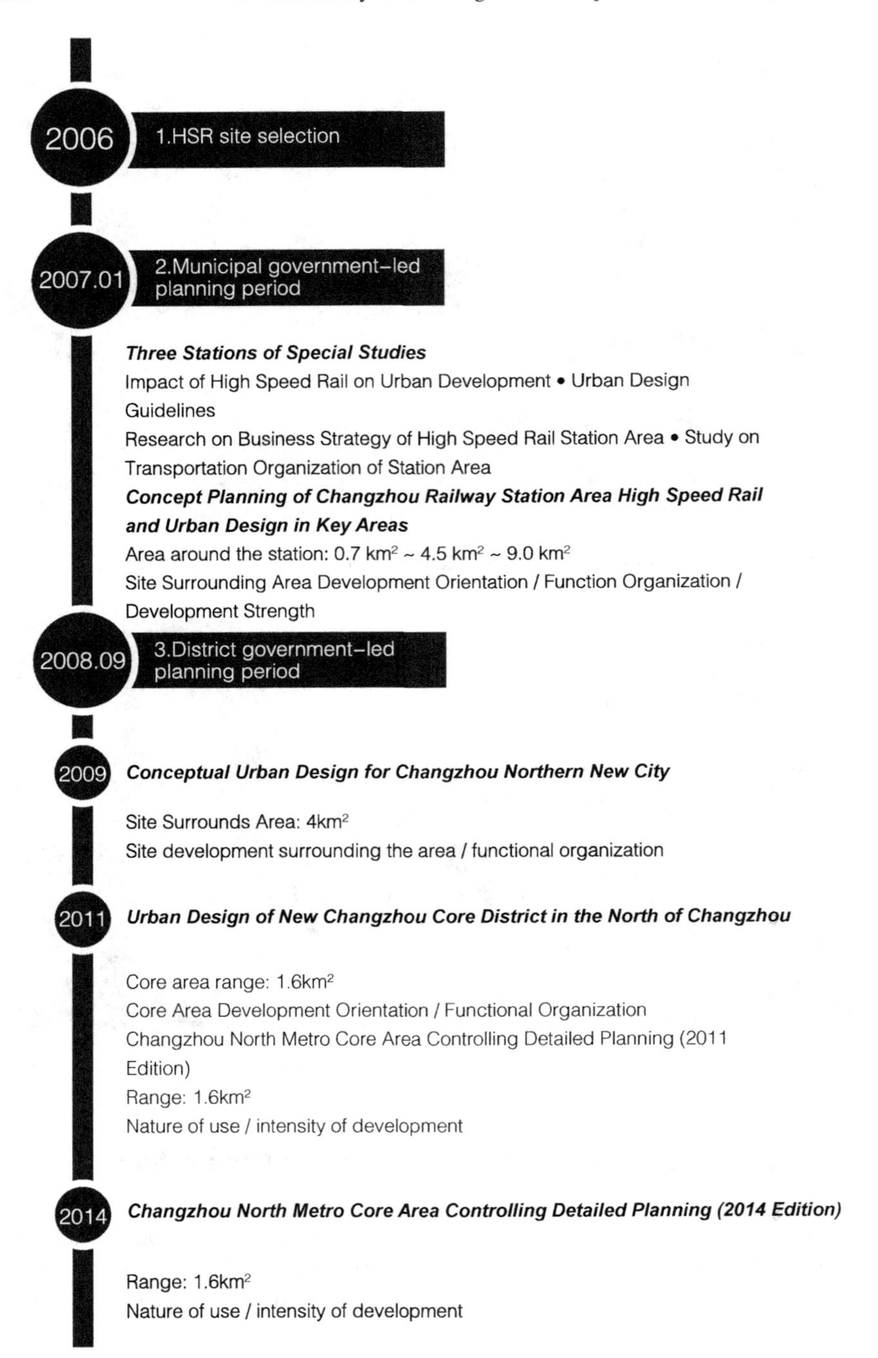

Fig. 4.20 Flow diagram of planning process for area surrounding Changzhou North Rail Station. *Source* Author, drawing based on research date and government website information

4.5.1 Stage I: Site Selection of HSR Station (2006)

Regarding the line and station site selection of the Beijing-Shanghai HSR line in Changzhou, the Ministry of Railways began to communicate with the relevant departments of the Changzhou municipal government in 2001 and reached a consensus. The designed line was to be located on the north side of the Changzhou national high-tech development zone. The site is close to the construction area of the Changzhou national high-tech development zone and is within the original Beijing-Shanghai HSR traffic corridor.

The line and site selection were based on the following reasons: ① The line was located in the original traffic corridor, and the main cross-line road had been built at the time. No cross-line development and site use efficiency problems existed, and there was no excessive impact on the overall urban spatial structure. ② The Changzhou national high-tech development zone had developed well. In 2002, the Xinbei District was established on the basis of the Changzhou national high-tech development zone. Construction began on both sides of the traffic corridor, and there was a certain foundation for the development of an area surrounding an HSR station. ③ The line selection coordinates with the surrounding city. Huang Yong, chief engineer of the Changzhou City Planning and Design Institute, stated that during the line selection of the Beijing-Shanghai HSR line, the Ministry of Railways did not provide alternative options but emphasized that surrounding cities, such as Suzhou and Wuxi, had confirmed the line and that adjustments were not feasible. Therefore, in terms of the line selection of the Beijing-Shanghai HSR line, the Ministry of Railways and Changzhou municipal government had no options for divergence.

The site selection of the Changzhou North Railway Station of the Beijing-Shanghai HSR line focused on connecting to the traffic in the main urban area and providing services for the surrounding areas. The final site is located on the edge of the central urban area of Changzhou City and the Xinlong residential area approximately 2.5 km from the administrative center of Xinbei District, 1.8 km from the crossing of the Shanghai-Nanjing expressway, 17 km from Changzhou Benniu Airport and approximately 7 km from the Luoxi expressway hub (the intersection of the Shanghai-Nanjing expressway and the Jiangdu-Yixing expressway). The south side of the site is a high-tech industrial zone, the southeast side is an administrative center and public service facilities concentration area, the east and northeast sides are high-tech zone residential areas and the north side is farmland with a small area of village construction land.

4.5.2 *Stage II: Municipal Government-Led Planning Period (2007–2008.09)*

The 2007 planning scheme of the Beijing-Shanghai HSR line and the Shanghai-Nanjing intercity HSR line had been confirmed. How would the construction of these two important railways affect the city? How should the construction of the two HSR lines be located and coordinated? How should the areas surrounding the HSR stations develop? All of these questions were important issues confronted by the Changzhou municipal government. Since 2007, the Changzhou City Planning Bureau had led the planning process, and the Changzhou City Planning and Design Institute took charge of the *Subject Study of "Three Stations"* (the three stations were Changzhou North Railway Station of the Beijing-Shanghai HSR line and Changzhou Intercity Station and the original railway station, which share one station, Table 4.9). Four subject studies were conducted to assess Changzhou North Railway Station of the Beijing-Shanghai HSR line and Changzhou Railway Station of the urban subway and national railway: ① research on the influence of the Beijing-Shanghai HSR and urban subway and national railway on the urban development of Changzhou; ② research on the business strategy in the area of the Beijing-Shanghai HSR station and the urban subway and national railway station; ③ research on the urban design

Table 4.9 Main planning achievements and contents during the municipal government-led planning process

Number	Plan name	Completion time	Planning commission department	Compiling unit	Main content	Planning range (km^2)
1	"Three Stations" special study	2008.01	Wuxi City Planning Bureau	Wuxi City Planning Institute	Site impact on the city: site development strategies around the site; traffic organization	0.7–4.5–9.0
2	Concept Plan of Changzhou Railway Station Area HSR and Urban Design in Key Areas	2008.05	Wuxi City Planning Bureau	Wuxi City Planning Institute	Development orientation, functional organization, development strength	0.7–4.5

Source Author, drawing based on relative planning proposals guidelines in the area of the Beijing-Shanghai HSR station and the urban subway and national railway station; and ④ research on the organization of traffic in the area of the Beijing-Shanghai HSR station and the urban subway and national railway station. Through the development of these subject studies, the *Conceptual Plan in the Area of Changzhou HSR Station and Urban Design in Key Areas* was formulated in May 2008 to clarify the respective influences of the three railways (HSR, urban railway and national railway). The positioning, function, form, development intensity and development strategy of the HSR station were arranged, and *the Tendering Document of Detailed Planning Scheme for the Construction of the Core Area of the Changzhou Railway Station on the Beijing-Shanghai HSR* was prepared to implement international bidding for the planning of Changzhou North Railway Station.

The planning compilation in this stage was led by the Changzhou municipal government, and the plans were compiled by the Changzhou City Planning and Design Institute as the secondary unit. Therefore, the planning contents were coordinated with the locally compiled *Comprehensive Planning for Urban Development in Changzhou City (2011–2020)*. The development plan of the area surrounding the site made a complete connection to the development orientation of the local city clusters and the plans for the surrounding commercial networks.

4.5.3 Stage III: Period Dominated by Xinbei District Government (2009–Present)

In 2009, Xinbei District organized and carried out the planning of the Northern New Town as the iconic event of this planning stage. As a national high-tech development zone, Xinbei District is at the deputy departmental level, giving it more autonomy in development. In 2009, led by the Xinbei District government, the Xinbei District Planning Bureau (Changzhou City Planning Bureau Xinbei Branch) entrusted the Urban Planning & Design Institute of Shenzhen with compiling the *Conceptual Urban Design of Changzhou Northern New Town*, where the concept for the construction of a "Northern New Town" was proposed for the first time. The planning area of the Northern New Town reached 73.23 km^2, and the whole area was within the administrative scope of Xinbei District. The current land reaches 25.67 km^2 (accounting for 35.03% of the total planned land area), while the approved land to be built reaches 16.25 km^2 (accounting for 22.19% of the total land area). The land in the planning scheme covers an area of 16.86 km^2 (accounting for 23.02% of the total land area), while the land without a planning scheme covers an area of 14.45 km^2 (accounting for 19.73% of the total land area).

Changzhou North Railway Station of the Beijing-Shanghai HSR line is located northwest of the Northern New Town, and the surrounding area includes the approved project and the unplanned land. The Northern New Town will be developed and constructed north of Changzhou and within the Xinbei District; therefore, the Beijing-Shanghai HSR line is regarded as an important construction promotion factor. The area surrounding the HSR station is positioned to be a "window of the HSR hub complex" with a land scale of approximately 4 km^2, and it is the public service center of the future expansion area of the Northern New Town. In terms of the functional structure, the HSR station partially inherited the conclusions of the analysis from the *Conceptual Plan of the Area of Changzhou HSR Station and Urban Design in Key Areas*, and Changjiang Road is used as the axis for the functional organization. In the core zone close to the HSR station, since the Xinbei District municipal government has agreed to remove the district government to the surrounding area, thus driving the development of local district-level administrative centers and business and commerce, the function in the core zone has been greatly adjusted in the plans in order to develop large-scale comprehensive business and the featured administrative offices on the north side of the station. The core zone now clearly covers an area of 1.6 km^2.

In 2011, the first round of urban design and regulatory results was formed in the core layer of the district government-led area surrounding the site. Based on the *Conceptual Urban Design of Changzhou Northern New Town*, the Xinbei District government organized an international tender of "Urban Design in Core Zone of Northern New Town in Changzhou." The core area of HSR stations is 1.6 km^2. Five enterprises, including Oubo, LAY-OUT and the Shenzhen Planning Institute, participated in the bidding. At the request of the major leaders of Xinbei District, as a reference image of Xinbei District, the HSR region should reflect the speed of development and the image of the city and be required to include large bodies of water. Therefore, all five design plans adopted the "large water plus high-level" model (Fig. 4.21). Finally, the German Oubo Company won the bid. However, in order to better reflect the government's development concept, a local design company

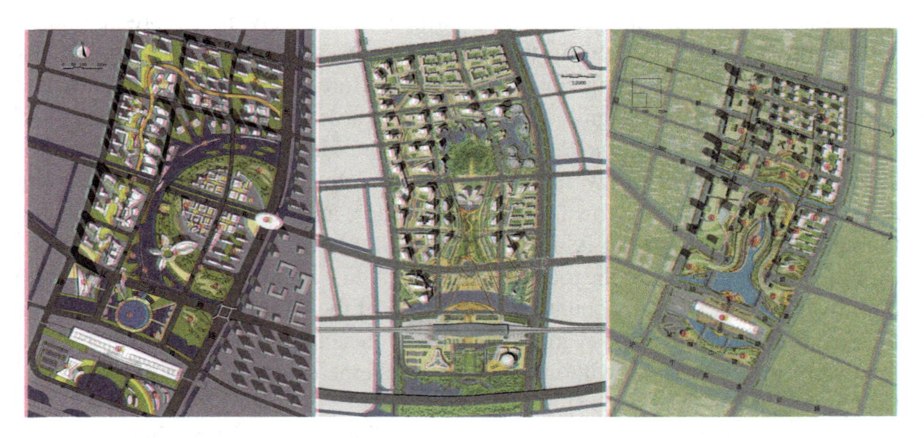

Fig. 4.21 Some of the projects in the 2011 international bidding for the urban design of the core area. *Source* Xinbei District Branch Office of Changzhou Urban Planning Bureau

in Changzhou, ICDA, eventually formed an integrated solution by combining the five schemes.

The positioning of this round of the design of the core area followed the positioning of the *Conceptual Urban Design of Changzhou Northern New Town.* However, the spatial organization unilaterally emphasized the need for the construction of large-scale bodies of water, which generated a greater conflict with the original plan and even the projects under construction. According to the urban design of the core area, in 2011, the *Regulatory Plan of Core Zone in Northern New Town of Changzhou (2011 Edition)* was completed. This round of regulatory planning had the following characteristics: ① Planning of administrative office space; the Xinbei District government intended to promote the development of surrounding commercial services and commercial finance by introducing the Xinbei District municipal government; ② the volume of the plots was higher, and the largest plot reached 9.0 FAR based on the standard of a high-density commercial business core area; and ③ set up large bodies of water before building the station, as planned (Table 4.10).

According to the information on the official Web site of the Changzhou Xinbei District, the 1.6 km^2 core area of Changzhou North Railway Station was initially known as "Xinlong International Business City." Based on contents published after 2013, the scope of "Xinlong International Business City" expanded to the east of Longjiang Road, west of the Zaojiang River, south of Provincial Highway S122 and north of the Shanghai-Chengdu expressway, with a total area of 24 km^2, and the 1.6 km^2 area of the original area surrounding the station site was confirmed as the "core zone"; meanwhile, it was also the starting area for the construction of Xinlong International Business City.

The development and construction process of the core area in 2013 was affected by the national policy, and the detailed regulatory plan was adjusted accordingly. The construction of the body of water at the site was completed in September 2013. However, in July, the State Council issued the *Notice on Party and Government Organs Stopping the Construction of New Buildings and Halls and Clearance of Office Housing*, and the plan to introduce offices for the Xinbei District government was forced to stop. It was learned from the interviews that because some original investment promotion projects were temporarily on hold due to the suspension of the municipal government's removal, in addition to the sluggishness of the real estate industry as a whole, the Xinbei District government was forced to revise the local regulations in December 2013. The *Regulatory Plan of Core Zone in Northern New Town of Changzhou (2014 Edition)* was formed (Fig. 4.22). The main modifications were as follows: Transform the original administrative office space into commercial land; according to the new classification standard for urban land, classify the business, commercial and residential comprehensive lands on original certain plots according to their main purposes, which are mainly transformed into commercial land; reserve the original two existing residential areas without demolition; and decrease the upper limit of the volume ratio for commercial plots to 3.5 from 5.

Table 4.10 Main planning achievements and contents during the district government-led planning process

Number	Plan name	Completion time	Planning commission department	Compiling unit	Main content	Planning range (km^2)
1	Conceptual urban design for Changzhou Northern New Town	2009	New North District Government, New North District Planning Branch	Shenzhen City Planning and Design Institute	Northern metro-related development content; HSR city complex positioning; function	1.6–4.0
2	Urban design of New Changzhou Core District in the North of Changzhou	2011	New North District Government, New North District Planning Branch	5 institutions such as AUBE, LAY-OUT, and Shenzhen Urban Planning and Design Institute	Function, location, space shape	1.6
3	Changzhou North Metro Core Area Controlling Detailed Planning (2011 Edition)	2011	New North District Government, New North District Planning Branch	–	The nature of the land, development intensity	1.6
4	Changzhou North Metro Core Area Controlling Detailed Planning (2014 Edition)	2014	New North District Government, New North District Planning Branch	–	Land use, development intensity adjustment	1.6

Source Author, drawing based on relative planning proposals

Fig. 4.22 Publicity and disclosure of the adjustment of the regulatory plan in 2014. *Source* Xinbei District Branch Office of Changzhou Urban Planning Bureau

4.6 Development Phasing and Model of HSR New Town of Changzhou

Overall, the development of the area surrounding Changzhou North Railway Station has been dominated by the relevant departments of Xinbei District. The main construction projects are concentrated in the 1.6 km^2 core area of Xinlong International Business City. In June 2013, the Northern Forest Park led by the forestry department was a peripheral area construction project. The construction of the area surrounding the site has been slow. The Xinbei District government expected that "in 2012, the core area plazas will be completed, and the construction of financial wealth plazas and surrounding ancillary projects will be launched. In 2013, the construction of functional areas such as science and technology research and development plazas and public service plazas will be established to ensure that the Xinlong International Business City will be fully completed in 2015." The current status of development and the established goals have great differences. Roughly, the planning process could be divided into two stages (Fig. 4.23).

4.6.1 Stage I: HSR Construction (2008.6–2011.12)

At this stage, the construction strategy for Xinlong International Business City was not confirmed. The planning led by the municipal government had weaker constraints and guidance than that of the Xinbei District government, which took charge of the development and management of the area surrounding the station. The Xinbei District government engaged in the demolition and road construction.

In terms of demolition and resettlement, the demolition due to the construction of Changzhou North Railway Station was divided into two types: The first type was the demolition due to the construction of the HSR line and station, while the other type was the demolition due to the development and construction of the station core area. The demolition and relocation required for the construction of the HSR line and station were concentrated from June to August 2008. Xinqiao Town, where the site is located, had 107 households in total, all of which were resettled in the prepared houses. The resettlement community is Xinlong, which is located in the developed area on the east side of the site. Commercial facilities, schools, hospitals and other public facilities are mature. The demolition due to the construction of the core area of the HSR started in 2010 and gradually advanced from the north side of the HSR station to Nenjiang Road (Fig. 4.24). In May 2010, the Xinbei District Land Reserve Center and the Black Peony Group concluded and signed a preliminary development contract for the HSR zone in the Northern New Town to determine a development area of approximately 17.5 km^2. The phase I development area covered an area of approximately 4.15 km^2, and the 1.6 km^2 Xinlong International Business City core zone was located within that area. The phase II development scope was the residual area of approximately 13.35 km^2. In August 2010, the demolition of

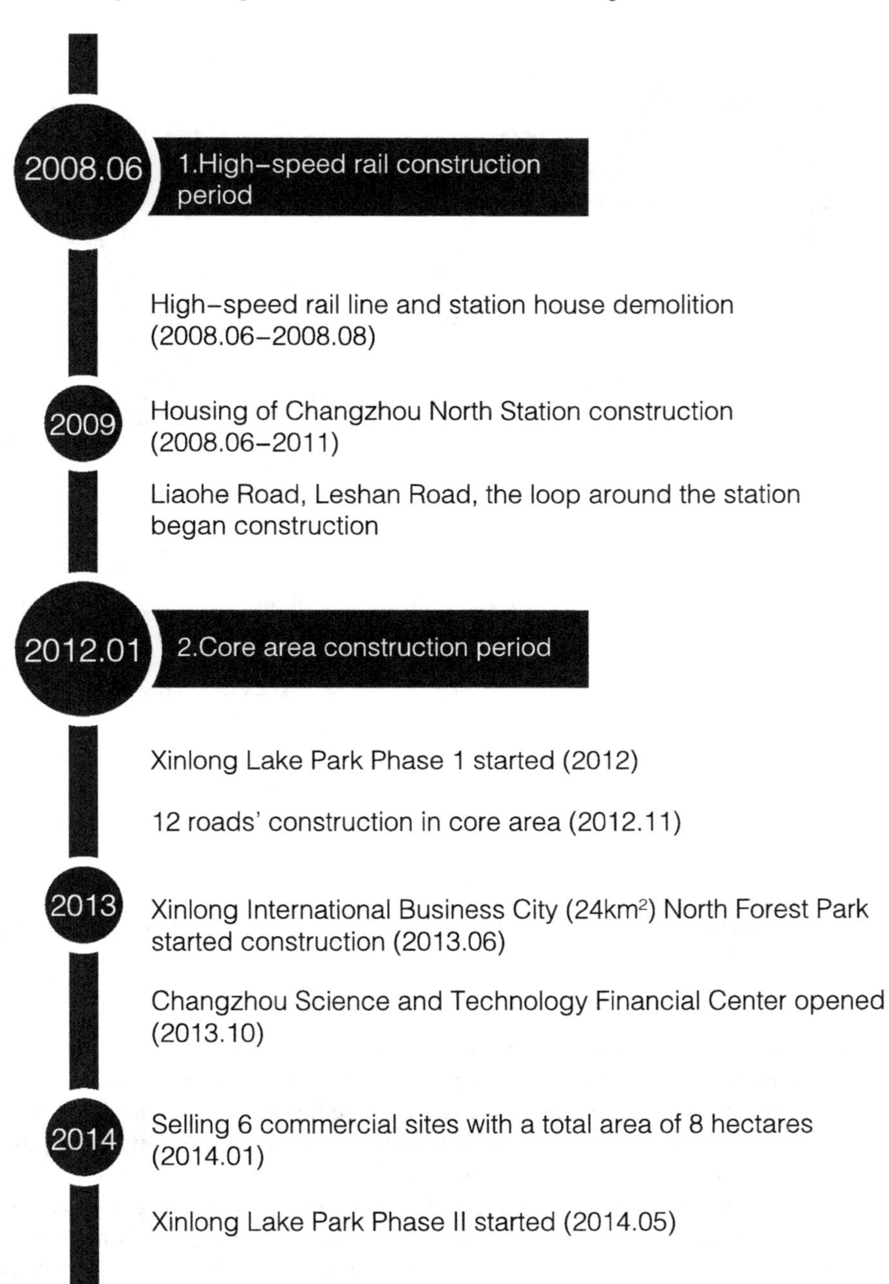

Fig. 4.23 Flow diagram of the implementation process of plans for the area surrounding the Changzhou North Rail Station. *Source* Author, drawing based on relative news

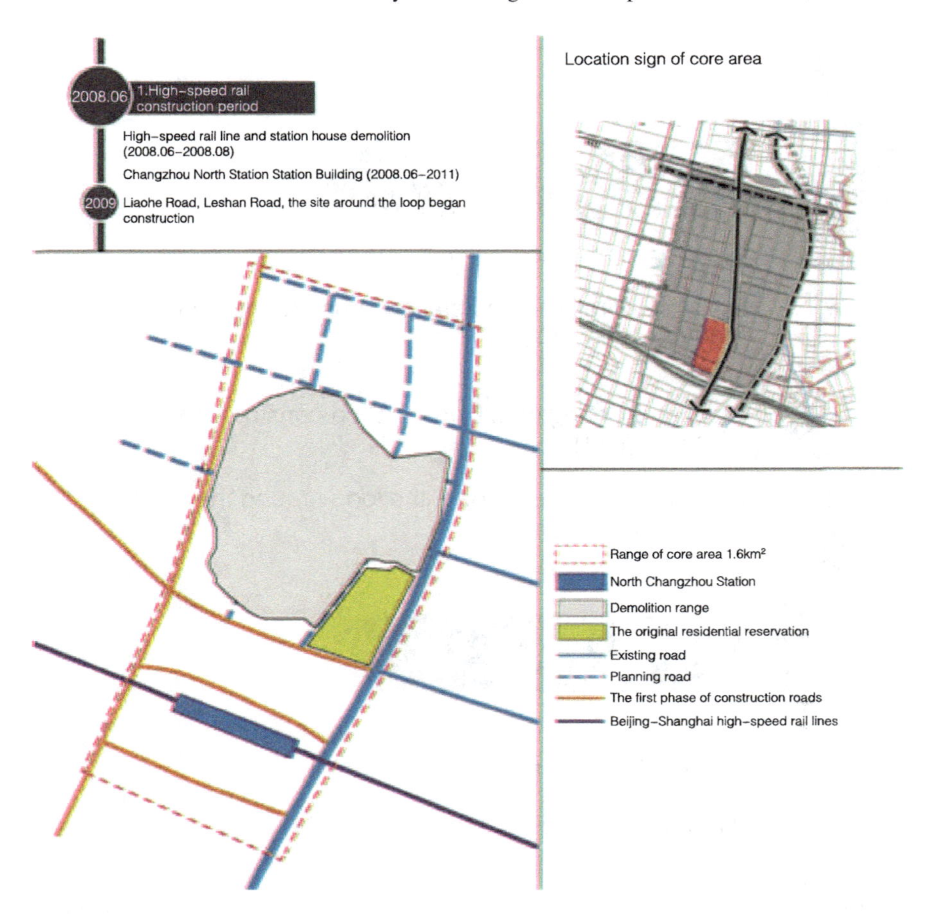

Fig. 4.24 Distribution map of projects in the stage of HSR construction. *Source* Author, drawing based on researches and planning proposal

the core zone began, and at the end of 2012, the demolition of the 1.6 km^2 area was completed. Over 1500 households were demolished. The demolition area with confirmed ownership was 313,800 m^2, and the resettlement area in the agreement was 134,800 m^2. Calculated on the basis of the area with confirmed ownership, the average unit price of house demolition was RMB 1099.8/m^2.

The roads constructed in this stage included the main roads that guarantee the connection of the HSR station and the urban region. In June 2008, construction commenced on many line sections of the Beijing-Shanghai HSR and Changzhou North Railway Station. At the beginning of 2009, construction commenced on Liaohe Road, Leshan Road and Huanzhan Road in the area surrounding the site. Leshan Road is a north–south secondary main road, which is the west boundary of the core area and connects the industrial area on the south side of the core area. Huanzhan Road is a supporting road of the area surrounding the site to meet the access demands of construction and site traffic flow. Liaohe Road is the main road in the east–west direction; it connects the site and the already built residential area on the east side of Changjiang Road and connects the main urban area to the front of the station through the intersection with Changjiang Road on the east side of the core area. The original line of Liaohe Road is located northward and far from the site. In the planning led by the municipal government, in order to meet the crowd evacuation requirement of the site, the line moved 450 m in the direction of the site. It is the only subsequent planning in the municipal government-led planning.

4.6.2 Stage II: Construction Area of Core Zone (2012–2014.6)

The construction of the core area was based on the development of public investment. Public investment projects included construction of the new Longhu Park and road network and the commencement and construction of the Changzhou financial center of science and technology invested in by the government (Fig. 4.25).

At the beginning of 2012, the construction of Longhu Park, with a lake surface width of 650 m as the main body, began. Because Liaohe Road was already under construction, it was forced to traverse downward because of the lake setting. In this way, the original rapid distribution function was weakened. The original BRT station on the road also had to be canceled, objectively resulting in construction waste. In August, the Liaohe Road opened fully to traffic. In November, the comprehensive construction of supporting roads in Xinlong International Business City began, and 12 roads, such as Xinqiao Avenue, Xinjing Avenue, Nenjiang Road and Leshan Road, were commenced. In October 2013, the first business building project in the core zone began: The Changzhou financial center of science and technology foundation, west of Leshan Road, north of Liaohe Road and east of the park to the north of the HSR station, is located in the 1.6 km^2 core zone of Xinlong International Business City and covers an area of 5 ha. The designed height of the main building is 150 m. In the same year, since the *Notice on Party and Government Organs Stopping the Construction of New Buildings and Halls and Clearance of Office Housing* was issued, the planned construction of Xinbei District government offices was suspended and the investment intended for this part of the project was also canceled accordingly.

After the publicity of the regulatory plan in January 2014, Xinlong International Business City carried out its first land bidding, auction and listing and successfully sold 6 commercial financial plots with a total area of 8 ha. In June of the same year, 5 projects on the sold land near Changjiang Road were formally commenced, and construction is expected to be completed within 15 years. In December 2014, the tendering of the e-commerce industry incubator in Xinlong International Business City began. It was expected that it would be completed in 2017. In general, the speed of construction of the area surrounding Changzhou North Railway Station has been slow, and the goal of "completely building the Xinlong International Business City by 2015" proposed by the Xinbei District government would be difficult to achieve.

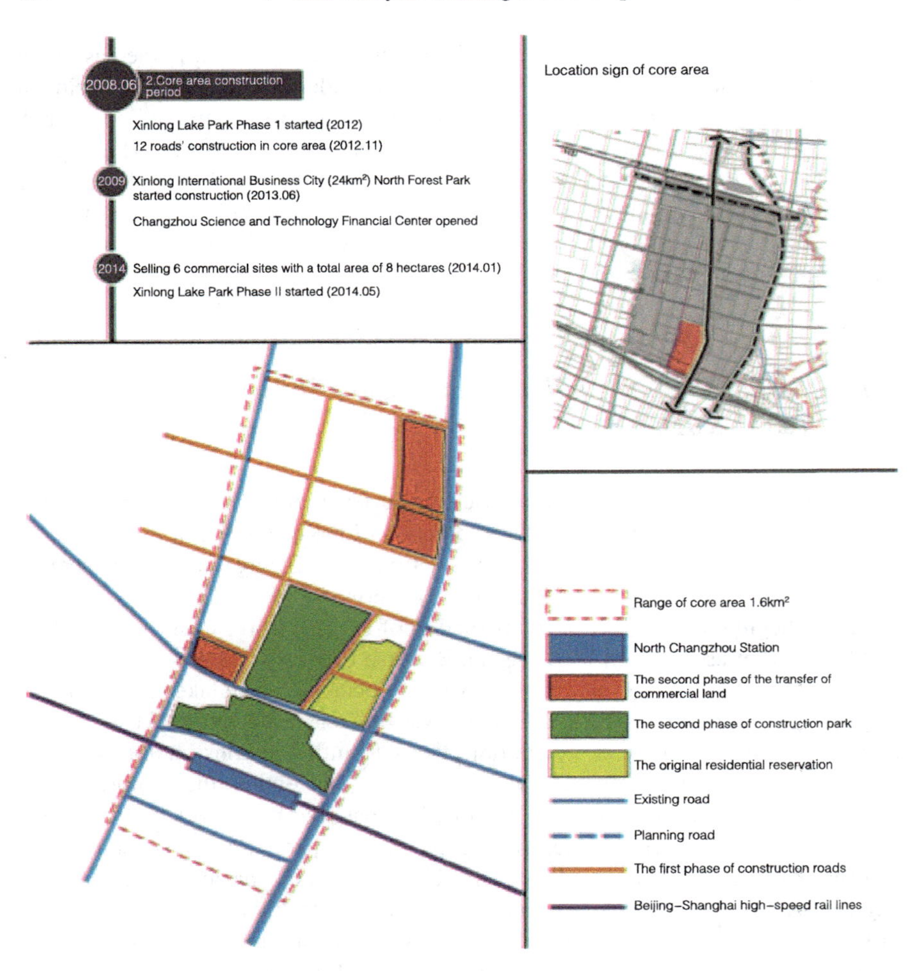

Fig. 4.25 Distribution map of projects in the stage of construction of the core area. *Source* Author, drawing based on researches and planning proposal

Reference

Skocpol T, Somers M (1980) The uses of comparative history in macrosocial inquiry. Comp Stud Soc Hist 22(02):174–197

Chapter 5
Comparative Analysis of Planning and Development of HSR New Towns

Based on the comparative analysis method, this chapter selected the new towns around the HSR in Wuxi and Hangzhou as comparative cases and analyzed their planning and development. Moreover, the research focused on the key planning content, development stakeholders and decision-making analysis of the new town around HSR stations, and a framework for comparative analysis was established based on this focus.

This study is rooted in the theory of the growth machine and the similar regime theory. The *growth machine* theory is a political and economic interpretation of urban space redevelopment proposed by John R. Logan and Harvey L. Molotch in 1987. The theory states that important attributes of space include *use value* and *exchange value*. The essence of urban redevelopment is the transformation between these two values of space. Urban space becomes a commodity, and the exchange value realized in the market is called *rent*. The exchangeable value of a plot with good accessibility, infrastructure and public service facilities is high, and the renting price is relatively high. The people who are committed to obtaining the space rent generated during the redevelopment process are called *space entrepreneurs*. Urban space entrepreneurs work together to form a *growth machine*, influence decision-making and urge cities to develop in the direction of the economic expansion and wealth accumulation that the consortium pursues, relying on its own resources. The residents' use of and dependence on the space forms the use value of the space. Space entrepreneurs in the public and private sectors and residents and community organizations that value the use value of space confront each other in the redevelopment process.

The *regime* concept was first proposed by the Fainsteins and was defined as the layer between elected government officials with power and other senior managers. The *regime* is composed of individuals in diverse government agencies, institutions and community organizations. It plays a different role at various stages of the construction period with different characteristics, including the *directive regime*, *concessionary regime* and *conserving regime*. Each role corresponds to *the period of mega projects in urban development* (mainly involving the large-scale demolition of slums, construction of large-scale public buildings or construction of commercial

L. Wang and H. Gu, *Studies on China's High-Speed Rail New Town Planning and Development*, https://doi.org/10.1007/978-981-13-6916-2_5

office buildings), *the transition period* (featuring the relative reduction of large-scale construction projects and the increased role of community organizations in the development decision-making process) and *the period of "do no harm"* (not harming the interests of the community and striving to achieve a balance between the parties in the development and construction of the project as much as possible) (Wang and Liu 2007). Clarence N. Stone further developed this concept by defining the regime as an informal but stable and continuous consortium formed between individuals in the public and private sectors to make decisions related to urban development. Elected officials, leaders of nongovernmental organizations and economic elites all participate in urban governance through their own resources. Based on the limited and decentralized authority defined by the system, the actors bridge institutional constraints and form a collaborative force that determines and implements decisions through collaboration and influences the decisions of urban development through individual collaboration across agencies.

Both the *urban regime theory* and the *growth machine theory* blur the boundaries of institutions and focus on the role of individual collaboration in the development process. They end, to a certain extent, the debate over whether urban development depends on elite rule or the decision-making of dispersed interest groups. They also neutralize the theories of economic determinism and political determinism (Lauria 1997). Based on the value judgments and driving interests of individuals, both theories treat the core decision-makers of urban development as an individual consortium with resources within and outside the system, and both interpret urban redevelopment as the process through which the consortium obtains economic value or political effect through the production or change of space. Due to the similarities between them, some theorists have classified the *growth machine theory* as a branch of the *regime theory*. However, on the one hand, the former inherits the attention of Marxism to values and reveals the purpose of space production in redevelopment. On the other hand, the *regime theory* assumes that the individual's rational choice is the premise and pays more attention to the interaction between the stakeholder of urban governance and decision-making.

In the analysis of the development planning of the new town, the two theories provide a political and economic background against which the planning takes effect from different perspectives and provide the research focus, namely the specific roles of the planning ontology, planners and managers. To better understand the decision-making rationality in the planning and development process of the new town, this chapter introduces the two theories into the planning analysis. Based on the analysis framework provided by the two theories, this chapter explores the relationship of decision-making and individual decision-making processes and decision-making content in the planning and development process. Therefore, on the basis of the above content, the comparative framework of this study compares the cases of the new towns around the HSR stations in Wuxi and Hangzhou due to their many similarities. This chapter also analyzes the reasons for changes in planning content, decision-makers and their relationships, processes, key decisions and the important factors that lead to different results.

5.1 Analysis of Key Planning of HSR New Town of Wuxi

The study found that the planning and implementation processes of the areas around the East HSR Station of Wuxi are continuous and stable, and the planning content is gradually deepening. The planning phases are converging well, without significant changes in the planning content. During the implementation process, the construction priorities and construction targets are clear at each stage. Public investment effectively stimulates private investment, and the construction goals in each phase are basically completed. The planning and implementation interact with each other. The planning content is gradually deepening along with the advancement of the implementation phase. The implementation content also achieves the goals of the previous planning phase.

5.1.1 Spatial Arrangement

According to the structural model of *three development zones* centered on public transportation stations, the overall scope of the HSR business district in Wuxi City can be divided into three layers. This scope covers the original industrial area on the west side of the station and the original center of Anzhen Town on the east side. From the first phase of the *Regulatory Plan of the Area around the HSR Station in Wuxi City (September 2008)* to the second phase of the *Regulatory Plan of Wuxi Xidong new town Business District (July 2010)*, the scope of the first and second layers around the station changed greatly. At the same time, the spatial model of the new town around the HSR station in Wuxi has changed from a more theoretical layered structure to a model of *axis + layer* that expands along the subway line. The area of the first and second layers has decreased, but the volume has increased slightly. The third layer ranging from the management scope has been stakeholdered to a small adjustment (Fig. 5.1).

On the basis of the *Research on the Planning of the Xidong new town District of Wuxi City*, the *Regulatory Plan of the Area around the HSR Station in Wuxi City (September 2008)* demarcated an area of 3.98 km^2 as the core area of the new town around the HSR station (Fig. 5.2) and planned a layered structure with the HSR station as the core. The core layer surrounds the station and extends along the arterial road, Xihu Road, to the northeast, covering an area of approximately 1.8 km with an 800 m radius. In this area, the mixed-use commercial and office land (Cb) reached 1.1 km^2, and the floor area ratio ranges from 3.0 to 6.0, with the commercial floor area ratio being slightly lower and the office floor area ratio generally higher than 5.0. The total planned station area (including the frontal square) is 0.27 km^2, with 0.15 km^2 of green land and 0.28 km^2 of road. The second layer is adjacent to the first layer and extends outward unevenly to a distance of 500–1000 m with an area of approximately 2.1 km^2. The maximum distance is 1900 m, indicating that the radius of the station is 1900 m, with an area of approximately 2.1 km^2. The land use in

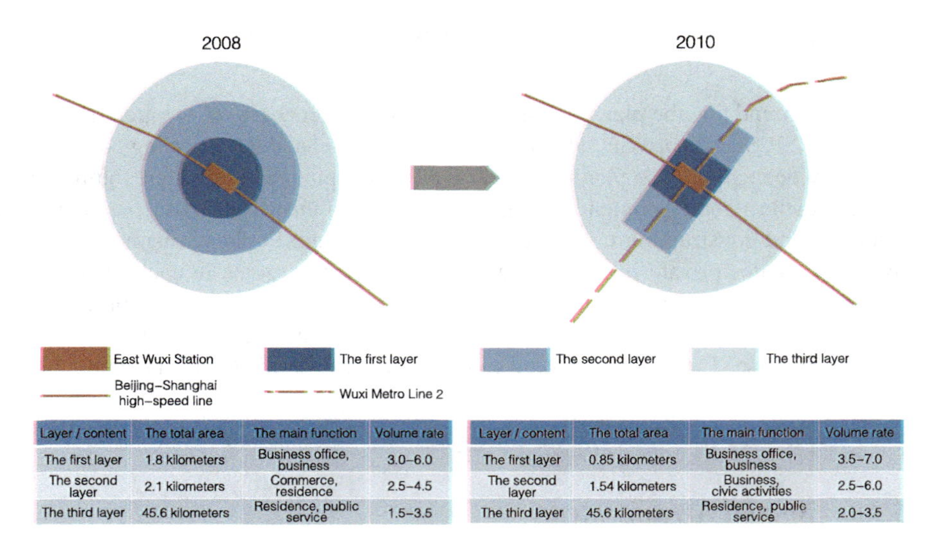

Layer / content	The total area	The main function	Volume rate
The first layer	1.8 kilometers	Business office, business	3.0–6.0
The second layer	2.1 kilometers	Commerce, residence	2.5–4.5
The third layer	45.6 kilometers	Residence, public service	1.5–3.5

Layer / content	The total area	The main function	Volume rate
The first layer	0.85 kilometers	Business office, business	3.5–7.0
The second layer	1.54 kilometers	Business, civic activities	2.5–6.0
The third layer	45.6 kilometers	Residence, public service	2.0–3.5

Fig. 5.1 Evolution pattern of land use in the area surrounding the East HSR Station of Wuxi

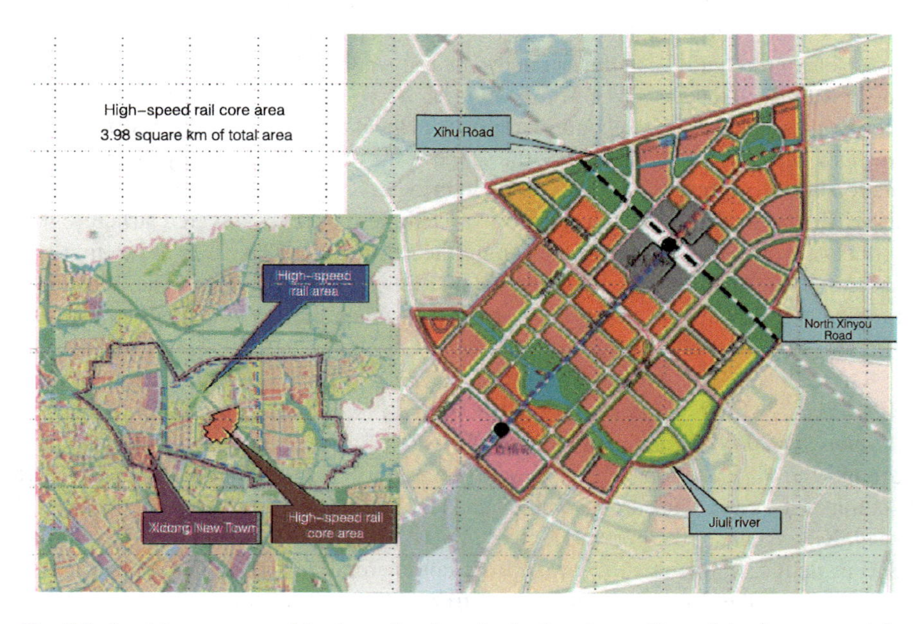

Fig. 5.2 Spatial structure and land use functions in the *Regulatory Plan of the Area around the HSR Station in Wuxi City (September 2008)*. *Source* Regulatory Plan of the Area around the HSR Station in Wuxi City (September 2008)

the second layer is mainly mixed-use commercial and residential (Cr) with an area of 0.89 km^2. At the same time, a few other land use types, including exhibition and recreation (C34, C36, with an area of 0.18 km^2), commercial and office (Cb, with an area of 0.11 km^2), parks and green space (G11, with an area of 0.26 km^2), and a green buffer (G22, with an area 0.18 km^2), are planned and designed.

Afterward, the *Regulatory Plan of Wuxi Xidong new town Business District (July 2010)* adjusted the first and second layers according to the *Urban Design of the Core Area* compiled by the RTKL company. The theoretical layer structure is weakened and converted to the belt structure forming the station-oriented development that relies on the Metro Line 2 station. The functional areas are subdivided, and different development requirements are established for each of the two parts of the second layer (Fig. 5.3).

In terms of the functional organization of land use (Fig. 5.3), the area of the first layer is 0.85 km^2 with a radiation radius of 950 m. Within the first layer, the commercial land (C2) has an area of 0.14 km^2, the commercial office area (Cb) is 0.30 km^2, and the station area (including the frontal square) is 0.13 km^2. The changes in the second layer emphasize the axial organization along Metro Line 2, located on the southwest and northwest sides of the first layer. The spatial structure is organized with the Metro Line 2 station and the Jiulihe landscape scenic node as the center, covering a total area of 1.54 km^2 with a 1900 m radius. The land use types in the second layer are mainly mixed-use commercial and office (Cb) and mixed-use commercial and residential (Cr). Both types have a high-level mix of functions.

5.1.2 *Road and Traffic*

The planning and design of the road and traffic network are mainly reflected in the two editions of the regulatory plan in 2008 and 2010. The subsequent dynamic update makes only a partial adjustment to the linearity of different roads. The goals of the road network construction in various versions of the regulatory plan are summarized in Table 5.1, and it can be seen that the road network plan has undergone the following process: *determination of the arterial road network* → *traffic research* → completion of the *road network*. There are different planning objectives for each phase of the road network, and the construction content also varies according to the actual situation.

In terms of the distance and area of the planned road network (Table 5.2), there were 19 existing and planned roads in the regulatory plan in 2008 with a distance of 88 km and 283.25 ha for the planned roads. This area accounts for 10.71% of the total urban construction land. There are 35 existing and planned roads in the regulatory plan in 2010, with a distance of 137 km (including the expressway) and 587.74 ha for the planned roads. This area accounts for 19% of the urban construction land. The road traffic plan in the regulatory plan of 2013 did not increase or decrease the number or area of the roads but makes several adjustments to the road sections with the distance and area remaining basically unchanged. The increase in road distance

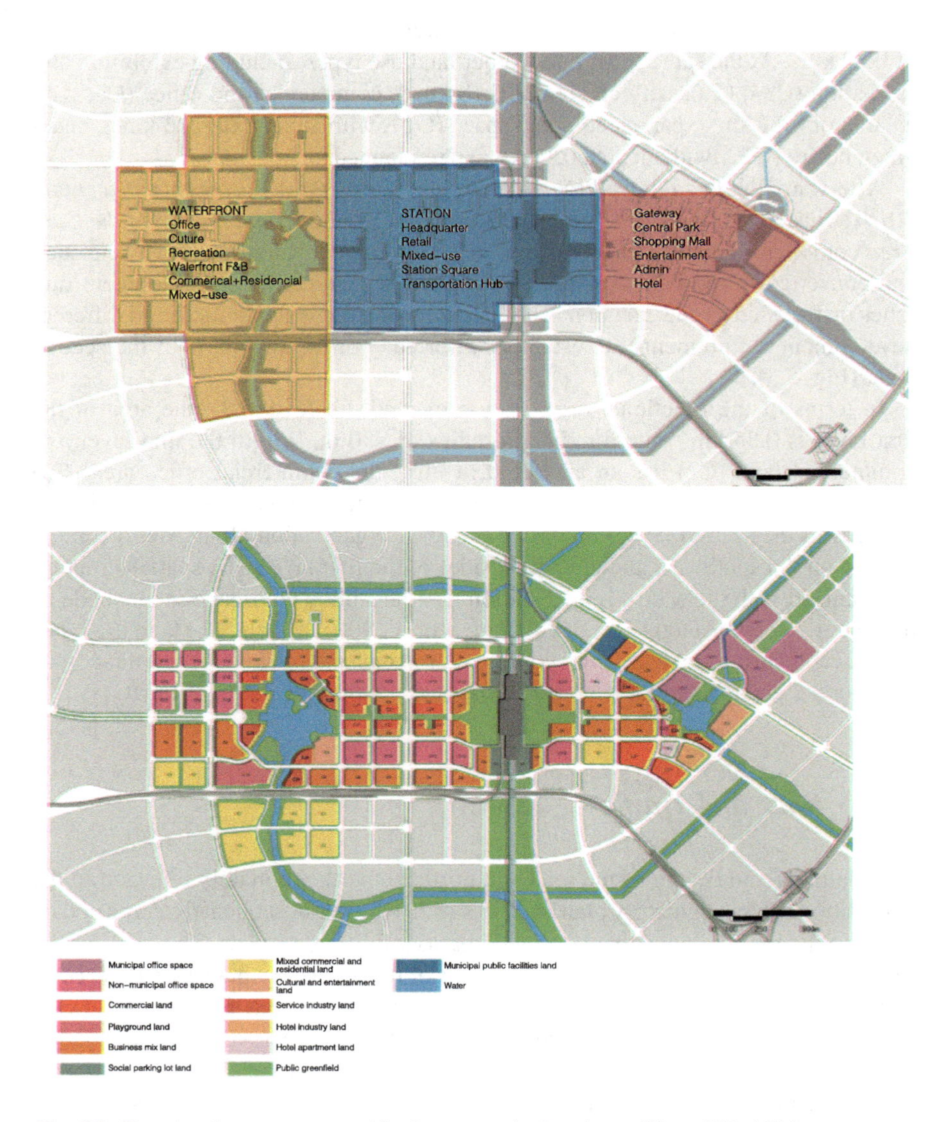

Fig. 5.3 Functional arrangement and land use types in *Regulatory Plan of Wuxi Xidong new town Business District (July 2010)*. *Source* RTKL, Urban Design of Wuxi High Speed Rail Station Core Area

Table 5.1 Planning and construction objectives of road network in different editions of the regulatory plan

Planning	Planning content
2008 version of the control rules	Road network construction objectives: a combination of radial and grid mesh road plans to optimize the road network, using the current Shanghai Road, East Road as the basis to form a radiating trunk road system with a HSR station as the center
2010 version of the control rules	Road network construction goals: based on the overall urban planning, a road network structure determined by comprehensive traffic planning and the road network density indicators of specific traffic subareas in combination with the refinement of the land use layout; the alignment of expressways, trunk roads and secondary trunk roads is optimized to deepen and improve the branch network, forming a well-structured network
2013 version of the control rules	Not involved in the construction of road network, only partial road adjustments

Source author, drawing based on regulatory plan

Table 5.2 Comparison of planning data of two types of control road network in 2008 and 2010

Grade/year	Distance (kilometers)/total distance		Road network density (kilometers/km^2)	
	2008	2010	2008	2010
Expressways	10.21 (12%)	27.53 (21%)	0.34	0.92
Arterial roads	39.22 (45%)	70.35 (54%)	1.31	2.35
Secondary trunk road	38.07 (44%)	31.8 (25%)	1.27	1.06
Total	87.5 (100%)	129.68 (100%)	2.92	4.33
Road density used in the design area does not include Roar Hill, Cui Ping Shan, Jiao Shan, Song Mountain and other mountain and ecological green areas, an area of 29.98 km^2				

Source author, drawing based on regulatory plan

and land area is mainly due to the refinement of the road network and the addition of subarterial roads and bypass roads to the plan.

In terms of changes in the grades of roads (Table 5.2) in the two editions of the regulatory plan in 2008 and 2010, the proportion of the distance of expressways increases significantly. The specific adjustments are shown below. ① Xinhua Road is extended and upgraded to an expressway, forming the *central city-HSR station area-external residential area* expressway link. ② All of Xidong Avenue is upgraded to an expressway, connecting the central city directly to the expressway in the peripheral residential area and effectively reducing the through traffic pressure in the core area. ③ A line directly connecting Xihu Road on the northwest side to the core area of the station is constructed, and the peripheral road can be quickly connected to the station without having to cross the urban road. The principle of expressway adjustment is to classify and divert traffic flow to minimize through traffic in the core area of the station and the traffic flow from and to the station.

The proportion of the distance of arterial roads also increases significantly. The specific adjustments include the following aspects. ① At the periphery of the radiating area around the station, Jiaoyang Road is constructed to connect the east and west residential areas eliminating the necessity of going through the HSR business district. ② Runxi Road (formerly Xinyou North Road) on the east side of the second layer radiating from the station is upgraded to an arterial road, and Shanhe Road on the south side is upgraded to an arterial road with the station intercepting radiating through traffic in the first and second layers.

The proportion of the distance of the subarterial roads decreases, mainly because the planned roads in the regulatory plan in 2008 focus on the first and second layers surrounding the station. Subsequently, the upgrade of many subarterial roads leads to a decrease in the total distance of the subarterial roads. In fact, in 2010, many subarterial roads are added to the third layer (i.e., the residential area) in the regulatory plan. This addition indicates that the design concept for road density and grades is evolving.

5.1.3 Public Service Facility

As a basic guarantee of regional development, public service facilities serve as important support for the vitality of the region. The *Regulatory Plan of the Area around the HSR Station (September 2008)* and the *Regulatory Plan of Wuxi Xidong new town Business District (July 2010)* are quite different in terms of the configuration of public service facilities.

The public service facilities in the *Regulatory Plan of the Area around the HSR Station in Wuxi City (September 2008)* are mainly allocated based on the population of communities in each administrative area. The community center in each residential area is configured according to the planning criteria, and no attention is paid to the content of the regional service facilities. However, the *Regulatory Plan of Wuxi Xidong new town Business District (July 2010)* divides the public service facilities into public service facilities at the regional level and at the street and community administrative level. The former serves the entire HSR business district and Xishan District, while the latter focuses on internal services for the subdistrict and community administrative units. The plan identifies regional service needs in facilities for education, medical care, culture, sports and retirement (Table 5.3).

Among them, Xidong Senior High School and Yikangyuan, a facility for elders, were projects already planned in the regulatory plan of 2008, but they experienced significant growth in the regulatory plan in 2010. Xidong Senior High School increases from 8.6 to 15.11 ha, and Yikangyuan increases from 1.83 to 4.56 ha. The People's Hospital in Xishan District is a grade-A second-class hospital that is newly added to the regulatory plan in 2010. In the allocation of public service facilities at the subdistrict and community level, the plan notes that *the allocation of public facilities should*

Table 5.3 Planning of public service facilities

Category	Content	Scale	Land area (hectares)	Position
Educational facilities	Xi Dong Senior High School	60 classes	15.11	South of Xing Yue Road, west of Jin'an Road
Medical facilities	Anzhen Hospital	300 beds	4.69	South of Dongxing Road, east of Xidong Avenue
	Xishan District People's Hospital	1000 beds	9.04	North of Xinhua Road, east of Xidong Avenue
Cultural facilities	Cultural Center	–	8.80	North of Xihu Road, west of Runxi North Road
Sports facilities	Sports Center	–	7.97	South of intersection of Dong'an Road and Wenjing Road
Pension facilities	Yikangyuan	1000 beds	4.56	South of Danshan Road, west of west Zoumatang Road
	Chaqiao Nursing Home	280 beds	1.49	South of West Street, west of Roar Hill on west side of river

Source Regulatory Plan of the Area around the HSR Station in Wuxi City (July 2010)

take the impact of factors such as construction priorities and population flow into account and is supposed to be slightly more streamlined than the previous edition of the plan. For Chaqiao Subdistrict, Anzhen Subdistrict and Shui'an Subdistrict, the plan clearly states that existing facilities at the district level should be used as much as possible with no new centralized neighborhood centers being established. In the newly built Yingyue Subdistrict, a centralized neighborhood center can be constructed.

The purpose of planning and adjusting public service facilities is to stimulate the construction of the area to be developed through the overall leading role of large-scale regional public service facilities. These public services attracted more private investment. For example, the Kaixuanhuating Project of the Country Garden Company, the earliest commercial real estate development of the region, is adjacent to Xidong Senior High School and the People's Hospital in Xishan District.

5.1.4 Analysis of the Reasons for Adjusting Planning Content

The adjustment of planning content may be due to changes in development conditions and supporting policies. These changes might reflect changes in the intent and value selection of policy-makers and might also be related to feedback regarding needs after the implementation of the plan. For the relevant adjustments that occurred during the planning and implementation of the areas surrounding the East HSR Station of Wuxi, the following reasons are the main ones according to research and analysis.

(1) The core issues to be addressed vary at different stages. Different types and stages of planning are concerned with different content. Even series of regulatory plans differ in the focus of attention and response because of differences in the preparation background and development. During the preparation of the regulatory plan in 2008, the locations of the stations along the Beijing-Shanghai HSR are determined, and the policy of using the construction of the site to drive the development of the center in Xishan District is also basically established. However, the specific construction scope, management mechanisms and other issues are not clear. The following three issues were identified by the regulatory plan in 2008. First, the scope of the HSR business district is defined; second, the main road network structure is formed through the layer structure; finally, the land to be demolished and acquired is identified. At the beginning of 2009, the government of Wuxi issued a document to clarify the position of *Xidong new town* and the *HSR business district*. Meanwhile, a leading group and a management committee are established, and their management scope and development position are clarified. These issues are based on the regulatory plan in 2008. At the same time, the primary task is the construction and demolition of major roads, which is also in line with the regulatory plan in 2008. Therefore, this edition of the regulatory plan as a basic plan for the early development period tends to focus on the structural content. It forms an important basis for the following work. By 2010, the construction of major roads and the demolition in the core area of the station are basically completed, and the construction and development of the first and second layers of the new town of the HSR station begin. Therefore, the main problems that the regulatory plan in 2010 has to address are the road network construction and land transfer issues within the region. In response to new demands, the regulatory plan in 2010 adjusted the layer structure and the road traffic network. The adjustments focus on the scope of the first and second layers. The internal road network and the types of land use are refined mainly by adjusting the layer scope and clarifying the layer functions.
(2) Differences in planning technology reserves. Through interviews with relevant personnel at the Planning and Design Institute of Wuxi City, we found that the in-depth planning study and the increase in technical reserves are important reasons for the adjustments between the regulatory plan in 2008 and the regulatory plan in 2010. During the preparation period of the regulatory plan in 2008, domestic HSR construction is still in its infancy. There are few studies on the planning

and construction of areas surrounding HSR stations, and planners can learn only from foreign research experiences and the experience of similar TOD plans for urban rail transit stations. Therefore, the layer structure that takes the station as a center is more obvious in the regulatory plan in 2008 than in the regulatory plan in 2010. In 2009, the RTKL company completed the urban design of the HSR core area. It adjusted the scope of the first and second layers by estimating the flow of the station and proposed three major functional zones for the core area (including the first and second layers), combining the landscape nodes of the Jiulihe scenic zone and the Metro Line 2 station. At the same time, on the basis of the peripheral arterial road network identified in the regulatory plan in 2008, the comprehensive traffic plan for the business district compiled in July 2009 adjusts and modifies the roads in the HSR business district through traffic forecasts. The adjustments of spatial structures and roads in the regulatory plan in 2010 reflect those changes.

In addition, the in-depth planning study brought about changes in planning concepts. In the early stage of planning, the HSR station is perceived as the main driving force. The layer structure expanding around the station is obvious, and the scope of planning becomes larger. Afterward, the planning concept is gradually transformed. It takes the HSR station and subway station as the core of the spatial organization with the supporting regional public service facilities as the main driving force. Therefore, the original layer structure is transformed into a MultiTaction, belt-type organization so that the existing equalized public service facilities are transformed into a centralized configuration that emphasizes the construction of regional public service facilities.

(3) Demands for planning control and management. In the *three-layer structure*, the first and second layers are the most significant areas radiating from the station and are the main carriers of the economic development of real estate in the area around the station. The management committee can be expected to seek high exchange values for land in this area by fully exploiting the advantage of radiating from the station, which will lead to an increase in exchange value. Therefore, this area is given special attention, and the plans are adjusted and intensified according to the development needs. After the regulatory plan in 2010 is compiled, the management committee of the HSR business district immediately commissions the US SOM company to compile the *Urban Design for Block No. 1 and No. 2 of the HSR Business District*. The planning scope of the project is the area in the first layer outside the station, and the planning goal is to further clarify the planning and design conditions for land transfer to ensure a high-quality environment in the region. It is apparent that the three layers are affected by the station to different degrees, leading to differences in the potential for incremental changes in land value. This difference in value-added potential makes the management committee of the HSR business district take different approaches to planning control and development management. For example, the third layer has large land areas and flexible land use functions, which facilitate large-scale and integrated development. The land areas of the second layer are smaller, with relatively clear and flexible (mainly on the basis

of the type of mixed land) land use functions. More resources are invested in the planning process. The first layer experiences the most planning stages and the most complex planning and design processes. The division of the blocks is specified within each project scope. The nature of the land is determined as a single use. Underground space, lighting systems and signage systems within the scope are all clearly defined in the form of design guidelines.

5.2 Analysis of Development Stakeholders and Decision-Making of HSR New Town of Wuxi

From the perspective of the organizational relationship and the important decision-makers in the planning and implementation processes, the decision-maker has a clear management scope and a right-responsibility relationship. The superior and subordinate decision-making departments have smooth communication, through which the superior decision-making department could effectively organize the subordinate decision department to promote the decision implementation, while the subordinate decision-making department could obtain the corresponding policy and economic support. In terms of major decisions, the existing "construction management team" can effectively supervise the implementation of important decisions so that the policies related to development orientation, spatial arrangement, construction sequence, etc. can maintain the necessary stability.

5.2.1 Analysis of the Decision-Maker

The development in the area around the East HSR Station of Wuxi is characterized by strong government leadership. Many institutions, such as the Ministry of Railways, municipal government, district government and management committee of the business district, influence the planning and development of the areas surrounding HSR stations. The basic relationship and the decision-making content are hierarchical in structure (Fig. 5.4).

(1) ***Ministry of Railways***

During the location selection and construction of the Beijing-Shanghai HSR, without separation between the government and enterprises, the Ministry of Railways undertakes the important functions of comprehensively managing the railway's state-owned assets, supervising the railway finances, managing and supervising the railway construction, confirming the railway investment, etc. In the construction of the Beijing-Shanghai express rail connection, on the one hand, the Ministry of Railways performs its political responsibility, namely, to promote the construction of a national HSR network. On the other hand, the construction conforms to the enterprise inter-

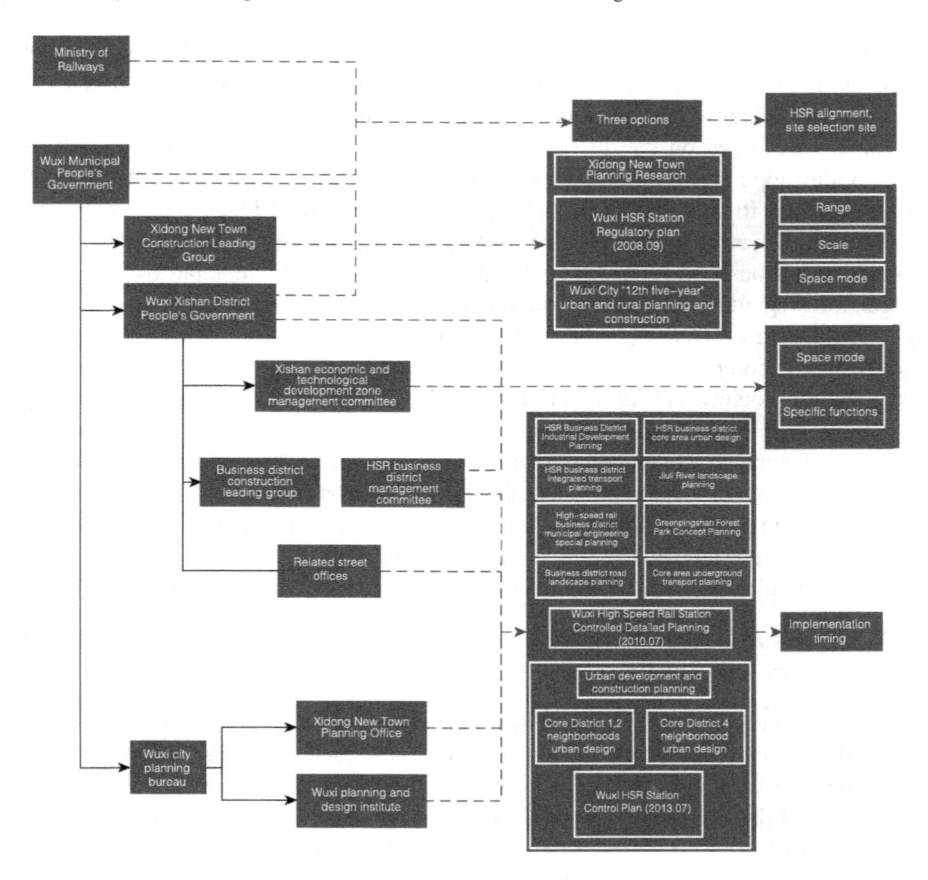

Fig. 5.4 The Decision-makers and problems of development in the area surrounding the East HSR Station of Wuxi

ests, namely, to reduce the construction costs as much as possible in the short term and to guarantee the profitability of the railway in the long term.

(2) ***Wuxi Municipal Government*** **and** ***Xishan District Government***

Since the 1980s, with the constant deepening of the reform and opening-up policy, local government forms a relatively dependent permissions and benefits relationship in the constant administrative decentralization of the central government. The direct result is that many "growth-oriented governments" form due to the "interpretation" of local government: namely, to promote urban development through city marketing, city promotion and other strategies.

As local governments at different grades, the Wuxi municipal government and the Xishan District government cannot break away from being "growth-oriented governments" in the current stage and depend on urban rapid growth to achieve economic development and political goals. The interests of the municipal government are based on the entire city, and development means the overall developmental efficiency of the entire city. The optimal allocation of development resources in each area is the approach to achieve optimal benefits, while the interests of the Xishan District government are based on the district, and the Xishan District government attracts as many development resources as possible as its approach to achieving the optimal benefits. Due to the relationship between superior and subordinate departments, the municipal government's decisions could guide and influence the district government's decisions, and the district government must obey the municipal government's decisions regarding the distribution of developmental resources.

With regard to management content, the Wuxi municipal government mainly controls the comprehensive coordination and uniform management of the politics, economics, society, etc. in the entire city; the Xishan District government mainly implements various tasks delegated by the municipal government and undertakes the social management and service provision in the administrative region. It is worth noting that, in terms of the division of authority for urban planning and construction, the "approval of urban planning," "administrative management of construction" and "management of land and resources" are uniformly managed by the municipal government, while the district government cannot manage these three important aspects.

(3) ***Xidong new town Construction Leading Group*** **and** ***HSR Business District Construction Leading Group***

The *leading group* refers to temporary agency for key works regarding which many departments conduct joint management in response to demands in the current urban management structure; its members include the main leaders and the leaders in all the relevant departments to improve management efficiency and ensure the smooth progress of key works through the regular communication and coordination of the superior department. For example, both the *Xidong new town Construction Leading Group* and the *HSR Business District Construction Leading Group* constituted such agency. In general, the leading group can be regarded as the extended special management agency for the municipal government and district government to manage the construction of *Xidong new town* and *HSR business district*. Since the leading group is not the resident office, it does not have administrative authority, but its members are responsible in each functional department and guaranteed that their decisions are implemented with administrative power.

The *Xidong new town Construction Leading Group* is created by the Wuxi municipal government after the document is approved. The executive vice mayor serves as the group leader, and the relevant department and regional leaders serve as the members. The leading group mainly harmonize the relationship *between the municipal level and the district level*. It plays a role in identifying the major problems in the development and construction processes of Xidong new town and coordinating the public resources and the semipublic resources in each department to support the development of Xidong new town. It works by regularly organizing and convening brief meetings, handling important issues in the construction and development of Xidong new town and presenting the work requirements to each department. Such an agency has the following functions: On the one hand, it can guarantee that important municipal decisions are effectively conveyed, and guarantee that the overall conception and position in the region can be achieved. On the other hand, the leading group system distributes responsibility to the specific departments so that the relevant responsible persons have stronger administrative decision-making powers and can effectively promote the work goals and implementation progress of each department as a whole.

The *HSR Business District Construction Leading Group* is created by the Xishan District government after issuing a document. The Xishan District mayor serves as the group leader, with responsible persons in all departments, committees, offices and bureaus. The leading group mainly takes charge of planning and coordinating various works within the HSR business district with district-level powers and with the management committee of the business district as its direct leadership unit. Meanwhile, this leadership group also includes subdistrict office leaders; thus, each subdistrict office owns some of the leadership. On the one hand, such a leading group can decide the position of the development of the HSR business district, the construction sequence, major projects and other key problems. In addition, these decisions not only conform to the principles confirmed by the leading group (municipal level) for the construction of Xidong new town but can also embody the district-level management intentions. On the other hand, the leading group can coordinate the economic development of the management committee of the HSR business district and the social affairs management of *each subdistrict office*.

(4) ***Management Committee of Xishan Economic and Technological Development Zone* and *Management Committee of HSR Business District***

The *management committee system* is the management mode currently implemented by most development zones in China (Zhou 2009). As a detached local government office, the management committee implements specialized management within a certain regional scope. With flexible management content, the management committee is authorized by the local government and establishes a branch according to the development demand. The management committee system has the following advantages: It can take charge of special work tasks within the specific region; purposefully establish personnel and institutions according to the development demand; and achieve high-level officer distribution in the administration. In addition, the secretary of the party working committee and the director of the management committee are the main decision-makers of the local government. The disadvantages are that the stakeholder for administrative law enforcement has more obscure qualifications, and some leaders are willing to intervene too much in regional development.

Both the *Management Committee of the Xishan Economic and Technological Development Zone* and the *Management Committee of the HSR Business District* are detached offices of the Xishan District People's Government, with the former at the deputy departmental level and the latter at the departmental level. In terms of management scope, the management committee of the Xishan Economic and Technological Development Zone manages the east area and west area of Xidong new town, with an area of 79.38 km^2, and the management committee of the HSR business district manages the middle area of Xidong new town, with an area of 45.62 km^2.

In terms of the division of functions and powers, a management committee has the following main functions and powers: ① organize and compile various plans (including economic and social development planning and various types of spatial planning) within the management scope, which is organized and implemented after the approval of the local people's government; ② review and approve the investment project (including the investment promotion, the qualification examination of the settled enterprise, etc.) within the management scope according to the regulations; ③ implement the uniform management of the land within the management scope in accordance with the planning requirements, including the level I land development of subordinate companies and the address selection suggestions and planning preliminary reviews of the settled projects; ④ plan the financial revenue and expenditure of the development area as a whole; ⑤ conduct the uniform planning and organize the construction and management of public service facilities and infrastructure facilities within the management scope; and ⑥ supervise, inspect and coordinate the relevant departments' branches in the development area. In general, the management committee takes charge of the economic development and urban construction in the management area and owns both the management function of government and the investment and construction function of enterprise.

The basic demand responsibility of the management committee lies in the sound development of the management object. Since the management objects of two separate management committees have different development orientations, the specific work of the committees is slightly different. The east area and west area of Xidong new town in the Xishan Economic and Technological Development Zone are regarded as industrial development areas, and their management committee focuses on investment promotion and economic development. The middle area of Xidong new town in the HSR business district is defined as an important urban functional area. Therefore, its management committee pays more attention to urban construction and the comprehensive development of urban functions.

(5) ***Planning Office for East HSR Station of Wuxi***

The planning office for the East HSR Station of Wuxi is the department-level detached office that was established by the Wuxi City Planning Bureau especially to coordinate the planning management of Xidong new town. The planning management of Xidong new town was originally uniformly conducted by the Xishan branch of the Wuxi City Planning Bureau. After the strategy for Xidong new town was proposed, to accelerate the project approval speed for the new town and to strengthen the management efforts, the Wuxi City Planning Bureau established the planning office for the East HSR Station of Wuxi. This agency took charge of the organization and coordination of urban planning within the new town on the east side of Wuxi, the planning permission for construction projects, the supervision of the implementation of construction engineering, etc.

5.2.2 *Important Decisions Analysis*

The important decisions in the planning and implementation process mainly included the station selection, development orientation, spatial arrangement and implementation sequence. The important decision-making logic and decision-makers appeal are analyzed by increasing the theoretical framework of spatial usage value and exchange value in machine theory.

(1) **Station location**

The Ministry of Railways and Wuxi municipal government are two actors in the decision-making process for the *route selection and station selection of the Beijing-Shanghai HSR*. The route selection and station selection for the HSR not only are an engineering technical problem but also involve political and economic problems related to the distribution and coordination of benefits. In terms of the engineering technology, the route selection and station design have special requirements regarding terrain, soil and other engineering conditions. However, the engineering requirements do not indicate the optimal or unique solutions. For example, there are three feasible schemes for the Wuxi section of the Beijing-Shanghai HSR. From the perspective of benefit, the income from the HSR construction can be generally divided into direct income and indirect income. The railway system (formerly Ministry of Railways, currently China Railway) and the local government participate directly in the benefit distribution.

In terms of the railway system, direct income is ticket prices, advertising revenue and other income brought in through transporting passengers. Direct income depends on the rapid and effective operation of the railway system. The railway must undertake to pay compensation expenses for route demolition and construction expenses for the route and stations. Therefore, the basic demand of the railway system is to keep the railway line straight to maintain the train speeds and to reduce the demolition compensation as much as possible.

For the local government, on the one hand, the HSR may bring effects over time in terms of industry, investment, etc. for the city as a whole. On the other hand, the direct development and construction around the station site could promote the urban image and bring in income from land appreciation. At the same time, since the railway generates a linear partition of the urban space, the city should undertake the resulting cross-railway development. Although most HSR tracks are overhead, they are still separated at a certain distance due to safety concerns, noise, etc. Therefore, the basic issue for the urban government is the small separation caused by the railway on the one hand and the possible development of the construction space in the area surrounding the station on the other hand to achieve the perfect urban function, and better land exchange value and usage value at the same time.

From the perspective of the space value, the railway system pursues the usage value of the space occupied by the tracks and the station while the urban government pursues the usage value in the area surrounding the station to achieve the urban

Table 5.4 Analysis of the interest claims of Ministry of Railways and local government

Main stakeholder	Ministry of railways (Railway corporation)	Local government
Seeking the main benefits	Direct benefits	Indirect benefits
The main source	Railway use function	Land appreciation around the site
Space value	The use of lines and sites	The exchange value of land around the site
Best benefit	Straight lines, low value of the original space	Site near the city, significant value-added space

function and obtain the exchange value from the land development (Table 5.4). Such a perspective is used to evaluate the three schemes for the East HSR Station of Wuxi.

Scheme I: Beijing-Shanghai Corridor Plan. The railway track occupies the original urban space, so the railway system needs to undertake a higher *exchange value* to achieve the railway usage value. For Wuxi City, the Beijing-Shanghai HSR parallels the original Shanghai-Nanjing railway line so that the enterprise, passenger flow, information, etc. in Wuxi, Shanghai, Beijing, etc. could improve the previous contact speed and efficiency. However, such a scheme would cause a large amount of demolition because the railway passes through the built-up area, the cost of achieving the exchange value would be high, and the space appreciation potential in the surrounding area is relatively limited. Therefore, such a scheme does not conform to the actual benefit of both parties.

Scheme II: Northern Line Scheme. The railway mainly occupies farmland, so the railway system needs to pay the lowest exchange value. However, for Wuxi City, the usage value of the station declines relative to the too-great distance from the main urban area. Meanwhile, since the development surrounding the station is restricted by environmental requirements, it is difficult to attempt to drive development and growth through the surrounding development. In addition, it is difficult to obtain the exchange value for the land transactions. Therefore, such a scheme is the optimal selection of the railway system defined by the Wuxi municipal government.

Scheme III: Construction Scheme. Although the railway detours slightly, it mainly passes through farmland. The station is close to the main urban area, where certain development foundations could support development on a larger scale to achieve the exchange value of the land surrounding the station.

(2) **Development orientation**

The development orientation of the HSR business district is jointly formulated by the municipal government and the district government; the *Xidong new town Construction Leading Group* is an important conduit for the communication of decisions. The positioning of the HSR business district is finally defined as a *transfer center with a regional transportation junction, a gathering area for the high-end service industry and the subcenter on the east side of Wuxi*. In general, the Wuxi municipal government and Xishan District government reach a higher level of consensus in terms of the development orientation in the area surrounding the East HSR Station of Wuxi because such an orientation could meet the core demands of both parties on the one hand and such a consensus benefits from the guarantee function of the management system and mechanism on the other hand.

From the perspective of the Wuxi municipal government, the optimal station selection scheme is to establish the station in Huishan District, which is preferred by the Ministry of Railways. First, it is difficult to promote the integrated development of Wuxi Jiangyin by utilizing the radiation effect of the HSR station; that is, it is difficult to achieve the idea of positioning the development in the region surrounding the HSR station on the regional level. Second, on the urban level, *the commercial center in the old city* and the *Taihu new town commercial center* in Wuxi City have good development foundations and possess more accumulative capital input, so the commercial business development in the area surrounding the HSR station does not have a great advantage when competing with them. The overall spatial strategy of building *three areas* (central old urban area, Taihu new urban area and landscape tourism and leisure area) centered on Taihu Lake is promoted continuously and consistently, but the East HSR Station of the Wuxi area of the Beijing-Shanghai HSR is outside that scope, so huge economic, social and administrative costs would undoubtedly be encountered if the development strategy were greatly adjusted toward the east. Meanwhile, the railway contact between Wuxi and all cities in the Yangtze River Delta depends more on the Shanghai-Nanjing intercity HSR (more shop flight shifts); therefore, the status of the East HSR Station of Wuxi of the Beijing-Shanghai HSR is relatively lower, and the station has a relatively limited pulling function for the surrounding region. Therefore, the East HSR Station of the Wuxi area of the Beijing-Shanghai HSR is not suitable for being constructed as the urban gateway or city-level center. Finally, at the regional (district) level, the original center of Xishan District is located southwest of the Beijing-Shanghai HSR and has been incorporated into the central urban area as planned, while the manufacturing industry is mainly located on the northwest side of Xishan District; thus, the original district center radiates and drives the development area more weakly. In addition, the original industry in Xishan District is mainly textile machinery and other traditional manufacturing industries, with high pollution and low added value. Therefore, building a new district-level center by virtue of the construction opportunity of a HSR station could certainly improve the urbanization rate, enhance the construction requirements and promote overall economic development. In addition, the construction of a new center could drive the upgrading and transformation of the original industry, strengthen the radiation of urban functions and expand the spatial range to achieve the development of a spanning traffic line. Based on the above issues, in terms of the development orientation of the area surrounding the East HSR Station of Wuxi (Xidong new town area) of the Beijing-Shanghai HSR, the Wuxi municipal government emphasizes its transportation junction position in *The Twelfth Five-Year Plan for Construction and Development*, namely, as the *regional transportation junction of the Yangtze River Delta* and the *strategic center of connecting the Beijing-Shanghai radiation* and also defines its function as a district-level center and driver of industry, namely the *Xishan economic, political and economic cultural center* and *gathering area of high-end industry*.

From the perspective of the district government, Xidong new town originally belonged to the *Anzhen-Dongting* area. In accordance with the 2006 *Comprehensive Development Area Plan in Xishan*, Xidong new town is defined as the industrial area that centers on machinery, textile, etc. as the overall development orientation, and the land planned for use in the HSR business district is currently the residential and agricultural land of villages and towns. For the development and construction of Xidong new town and the HSR business district brought by the construction of the HSR station, from the perspective of space production, first, the space usage value changes mainly from the original agricultural land to residential, commercial and industrial land, and the traditional industrial land with lower added value changes to modern industrial land with high added value and high technical content. The former change in usage value substantially creates the demand space between production and consumption through urbanization, while the latter change in usage value could effectively promote the land production rate. Both changes in value jointly promote the continuous income-generating ability of the land. Second, in the exchange process of the space usage value, "the district government could obtain its exchange value in a shorter period through the land development, and such income could be invested in urban public services to eliminate the barrier of the large-scale public facility construction restricting the urban development" (Zhao 2014). Therefore, the development orientation of the area surrounding the Wuxi railway station of the Beijing-Shanghai HSR, whether on the regional level, urban level or district level, presents development opportunities for the district government. The differences mainly lie in a higher return in economic development and government performance if a high level of orientation can be achieved; meanwhile, a high level of orientation also means the input of more resources (capital, policy, manpower, etc.) and unavoidable homogeneous competition. For the area surrounding the East HSR Station of Wuxi, the development resources depend on the input of the municipal government to a great extent, for instance, in conducting the uniform management and uniform construction of the main roads, providing the registered capital for Xidong new town, providing policy support for the land indicators, supporting the land expropriation policy, etc. If built as the city-level center, the new town must compete with the original city center. However, obviously, the existing business center and commercial center have a very strong first-mover advantage. In addition, the level of administrative power must inevitably be included in the district government's consideration of the development orientation of the area surrounding the station. Therefore, remaining consistent with the municipal government in terms of the development orientation problem is a dominant strategy that the municipal government practices to best utilize municipal resources and achieve its own interests.

(3) **Spatial arrangement**

The spatial arrangement in the area surrounding the East HSR Station of Wuxi is jointly decided by the Xishan District people's government, the management committee of the HSR business district and the management committee of the Xishan

Fig. 5.5 The division of layers of the areas surrounding the East HSR Station of Wuxi

Economic and Technological Development Zone. The Xishan District people's government controls the spatial arrangement as a whole to confirm the overall function and development scale of the area surrounding the station. The management committee of the HSR business district designates the specific layer structure and confirms the land nature, land mode, development intensity, etc. of each layer by researching the organizational planning. The management committee of the Xishan Economic and Technological Development Zone proposes collaborative functional arrangements of the business district by combining them with its own development demands, such as facility sharing in the third layer and the uniform construction of settlement buildings.

In terms of the layer component and the dominant function, three layers are planned for the area surrounding the station (Fig. 5.5). The first layer is the 0.85 km^2 scope centered on the station core and focuses on commercial and business offices, with a defined land nature and single land usage as a whole. Because this layer is set up with the HSR station and the subway station, with very convenient traffic options, it could support business purposes with a higher space usage value. It acts as the district center that the district government hopes to build. The second layer is the 1.54 km scope on the northeast and southwest sides. It emphasizes mixed uses of the land and focuses on commercial, residential and mixed commercial and office uses. In addition, it establishes citizen activities that could attract the flow of people when combined with the landscape node. The third layer is within the HSR business district, with a building area of 31 km^2, and is mainly used for residences and public

Table 5.5 Comparison chart of land use modes in each layer of the area surrounding the East HSR Station of Wuxi

Range/content		Area (km^2)	Functional content	Lot size (ha)	Volume rate
Core area	The first layer	0.85	Business offices, transportation	0.5–1.5	3.5–7.0
	The second layer	1.54	Commercial, public services	2.5–4.0	2.5–6.0
Peripheral area	The third layer	45 (Construction area of 31 km^2)	Commercial, public services	5.0–16.0	2.0–3.5

Source author, drawing based on relative data

services. It could serve the eastern and western industrial sectors of Xidong new town and provide residential space (Table 5.5).

In terms of the land mode (Fig. 5.6), within the same area scope, the first layer has the greatest quantity of divided plots, the block street has a small dimension, and the minimum area for a single plot generally does not exceed 1.0 ha. The second layer has the second-greatest quantity of divided plots, and the area of a single plot is generally between 2.5 and 3.5 ha. The plots in the third layer are roughly divided, the block street has a large dimension, and the maximum area of a single plot is 16 ha. Such a land mode is generated because the development of business offices in the first and second layers in the development process with the same land area could bring a higher exchange value, but at the same time, stricter design controls are needed to guarantee good space quality. In contrast, for the peripheral layer, the overall development is focused on residences, and the greater dimensions and rough division are beneficial for the free organization of roads in the community under the current conditions. Influenced by the concept of transit-oriented development (TOD), the small block street land mode is adopted around the station to give priority to pedestrians.

In terms of development intensity, the land development strength in the first layer is generally greater, and the floor area ratio is above 3.5. Most of the floor area ratios in the second layer are between 2.5 and 3.5, and a partial plot ratio can reach 6.0, according to the location and functional development demand. The ratio in the third layer is generally lower, between 2.0 and 2.5. The floor area ratio is confirmed in three aspects: ① Improve the ratio to achieve the best possible exchange value of land and future usage value; ② the ratio is directly related to the plot function, and the plots used for business and commercial purposes could support a higher ratio; and ③ it is necessary to balance the exchange value and the costs created by a high ratio. An excessive ratio needs more cost input for infrastructure support. The exchange value of the development process is certainly not optimal and may cause a decline in the

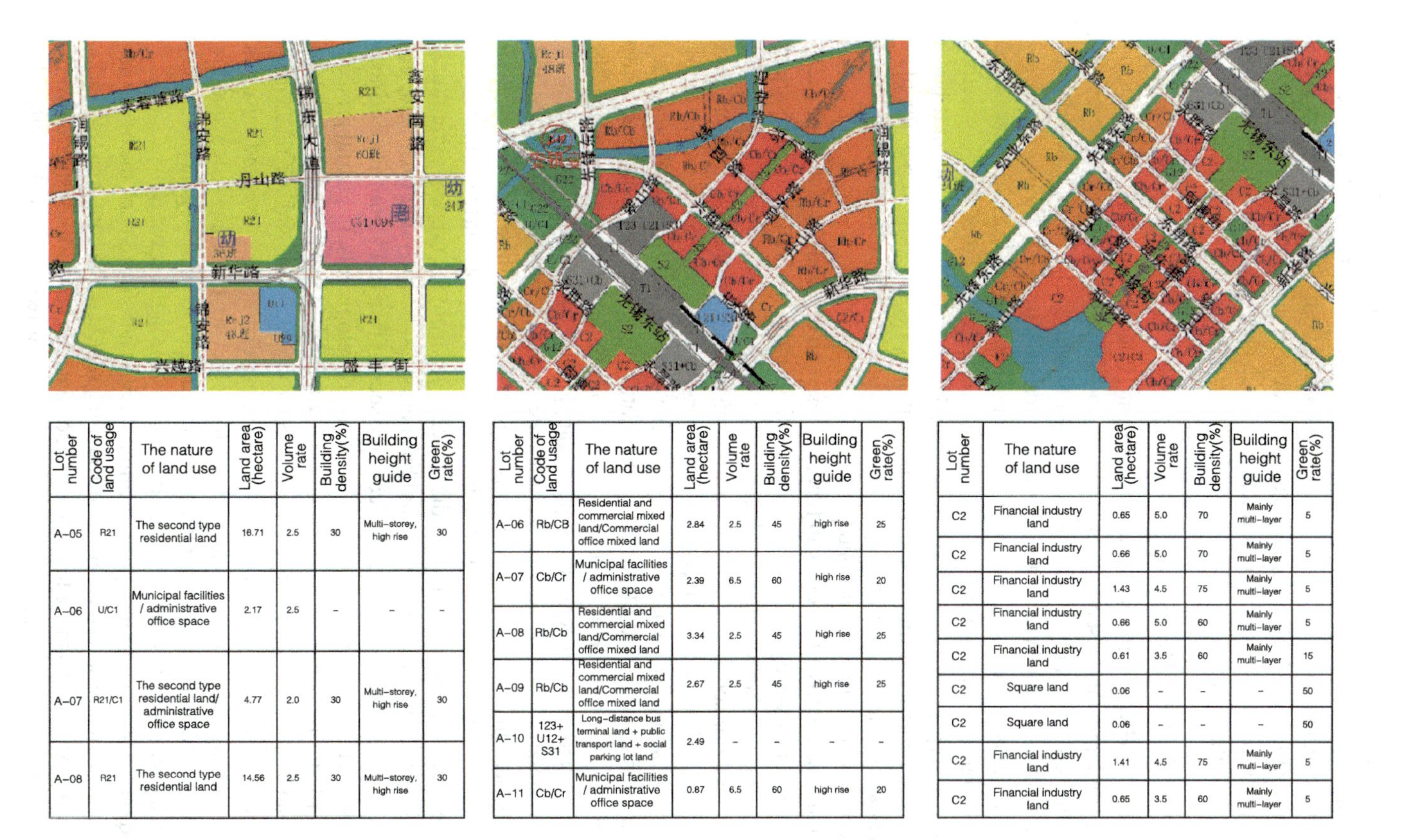

Lot number	Code of land usage	The nature of land use	Land area (hectare)	Volume rate	Building density(%)	Building height guide	Green rate(%)
A–05	R21	The second type residential land	16.71	2.5	30	Multi–storey, high rise	30
A–06	U/C1	Municipal facilities / administrative office space	2.17	2.5	–	–	–
A–07	R21/C1	The second type residential land/ administrative office space	4.77	2.0	30	Multi–storey, high rise	30
A–08	R21	The second type residential land	14.56	2.5	30	Multi–storey, high rise	30

Lot number	Code of land usage	The nature of land use	Land area (hectare)	Volume rate	Building density(%)	Building height guide	Green rate(%)
A–06	Rb/CB	Residential and commercial mixed land/Commercial office mixed land	2.84	2.5	45	high rise	25
A–07	Cb/Cr	Municipal facilities / administrative office space	2.39	6.5	60	high rise	20
A–08	Rb/Cb	Residential and commercial mixed land/Commercial office mixed land	3.34	2.5	45	high rise	25
A–09	Rb/Cb	Residential and commercial mixed land/Commercial office mixed land	2.67	2.5	45	high rise	25
A–10	123+ U12+ S31	Long–distance bus terminal land + public transport land + social parking lot land	2.49	–	–	–	–
A–11	Cb/Cr	Municipal facilities / administrative office space	0.87	6.5	60	high rise	20

Lot number	The nature of land use	Land area (hectare)	Volume rate	Building density(%)	Building height guide	Green rate(%)
C2	Financial industry land	0.65	5.0	70	Mainly multi–layer	5
C2	Financial industry land	0.66	5.0	70	Mainly multi–layer	5
C2	Financial industry land	1.43	4.5	75	Mainly multi–layer	5
C2	Financial industry land	0.66	5.0	60	Mainly multi–layer	5
C2	Financial industry land	0.61	3.5	60	Mainly multi–layer	15
C2	Square land	0.06	–	–	–	50
C2	Square land	0.06	–	–	–	50
C2	Financial industry land	1.41	4.5	75	Mainly multi–layer	5
C2	Financial industry land	0.65	3.5	60	Mainly multi–layer	5

Fig. 5.6 Comparison of the land use mode in each layer of the area surrounding the East HSR Station of Wuxi. *Source* Regulatory Plan of the Area around the HSR Station in Wuxi City (July 2010)

space quality; causing the exchange value could not be achieved. Therefore, the first layer and the second layer, which focus on public investment and take commercial and business functions as the core functions, have a high development intensity, while the third layer, which takes peripheral residences and public services as the core functions, has low development intensity, which also conforms to TOD's basic assumption of the ratio.

(4) **Implementation sequence**

In terms of the HSR business district of the East HSR Station of Wuxi, the planning and implementation sequence are led by the management committee of the HSR business district and the Xidong new town Planning Office, and the relevant subdistrict office and the Wuxi City Design and Research Institute provide the corresponding administrative and technical support. In general, the construction sequence of the HSR business district has the following two features: From the perspective of investment type, it is of the development mode, with public investment driving private investment; from the perspective of spatial scale, the time series for plan implementation transitions gradually from the surrounding region (the third layer) to the core zone (the first and second layers).

From the perspective of investment type, the research reviews the plan implementation project at each stage (Table 5.6). The findings indicate that public investment experiences the gradual change of *plane* $\rightarrow$ *line* $\rightarrow$ *point*. From the initial demolition and land collection and storage to the construction of the road network and infrastructure and specific projects, the project, proportion and involvement scope for public investment decrease gradually. During the "start-up period of construction" and "dominant period of public investment," the construction projects focus on demolition and resettlement, road network construction and municipal facility construction, and their content is basically completed. During the medium term of the second stage, the public service facilities are comprehensively constructed and completely brought into use in the preliminary period of the third phase. Since the preliminary construction of these public investment projects provides the basic condition, the usage value of the business district changes from the original agricultural land to urban residential, commercial and business uses. For private investment, the preferred investment is in industrial projects, the second is in residential and business building projects, and the last is in business project developments. Industrial projects have fewer external dependencies and are likely to undergo independent development. Meanwhile, private investment could create a usage demand for residential and business buildings, further cultivating a commercial atmosphere and business activity. The usage demand is the foundation for achieving the usage value, and the space usage value can be created only when the usage demand is generated, thus gaining the exchange value of the space in the transaction process.

Based on the analysis of the space location of construction projects, the construction projects gradually spread from the surrounding region to the core area (Fig. 5.7). The construction projects in the first stage include the road network construction covering the business district scope and the construction of two large-scale settlement

Table 5.6 Timeline of the construction period of different projects

Construction content / Time		2009	2010	2011	2012	2013	2014
Mainly public investment	Demolition	Core area demolition		Gradual periphery demolition			
	Housing construction	Resettlement housing I					
	Road network construction	The main road network construction					
	Municipal facilities built	Municipal infrastructure / water system renovation					
	Public facilities		Schools, hospitals, parks, civic centers				
Mainly private investment	Industrial projects		Start to enter		Full operation		
	Residential real estate					Full operation	
	Commercial real estate						
	Business real estate						
			Shades of color mean the amount of construction: light color, fewer construction projects				

Source author, drawing based on relative data

communities. The construction emphasis in the second stage still focuses on the surrounding region, but the projects gradually become closer to the core area scope, and the Xishan Experimental Elementary School, Xidong Senior High School and Country Garden Kiting Huating Community are located outside the core zone. The construction projects in the third stage not only involve continuous construction in the surrounding region but also cover the core zone. From the analysis perspective of space value, the three layers are different in content for usage value transfer (Table 5.7). The usage value from inside to outside transfers from agricultural use to business offices, business and citizen activities, daily living and public services. From the perspective of the change in the usage value, the residential and public service purposes of the third layer are the foundation for the two layers in the core zone to achieve their usage value; therefore, the third layer is preferable for construction in terms of the construction sequence.

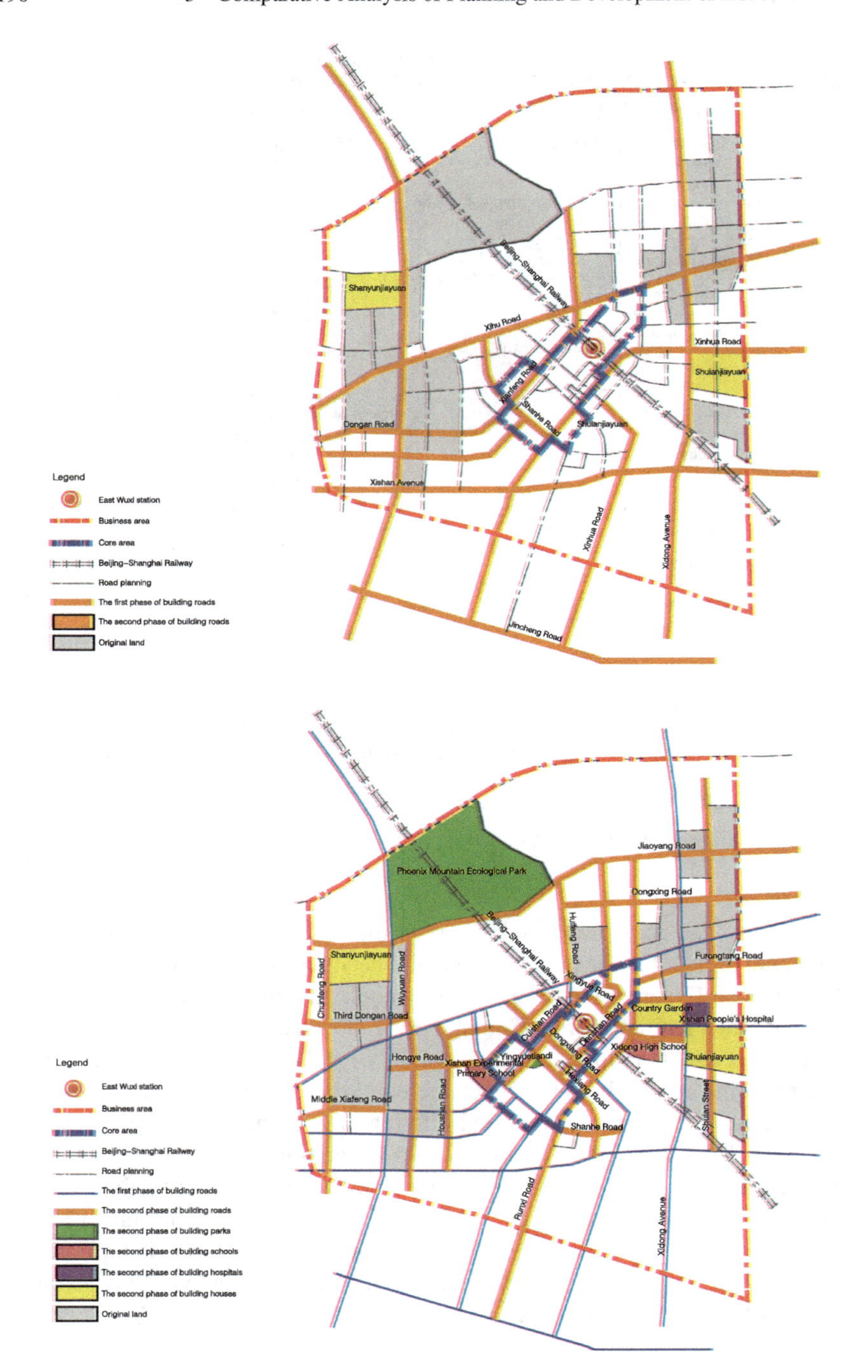

Fig. 5.7 Distribution of construction projects in each stage

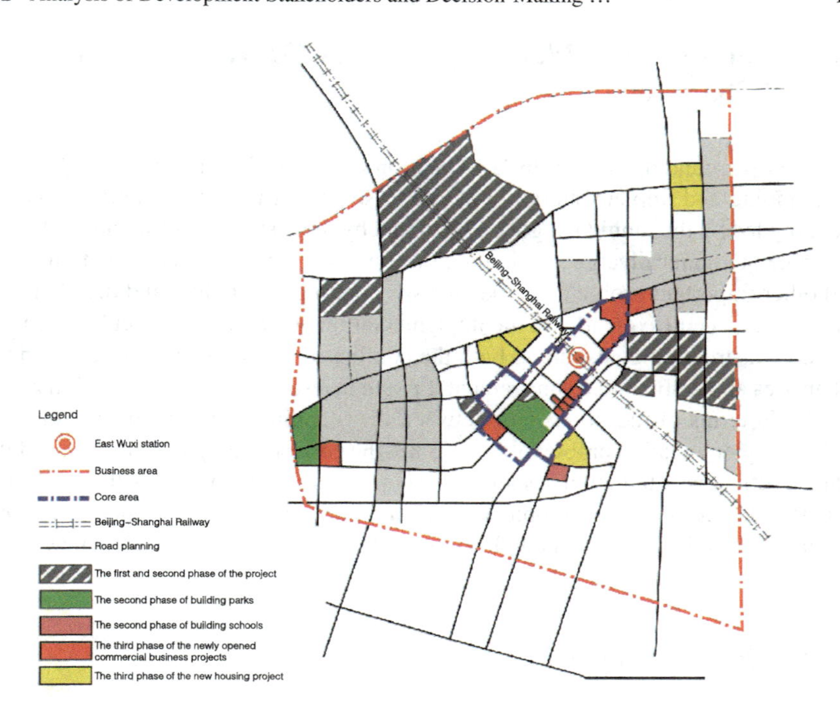

Fig. 5.7 (continued)

Table 5.7 Comparison of the spatial value of each layer

Range/content		Use value conversion	Use value conversion conditions	Space exchange value
Core area	The first layer	Farmland → business offices, transportation	HSR station, subway station drive; living and public services package	High
	The second layer	Farmland → business, commercial and residential, public services	Subway station, landscape center drive; commercial and residential and public service facilities	Middle
Peripheral area	The third layer	Farmland → living, public services	Public facilities	Low

5.3 Analysis of Key Planning Content of HSR New Town of Changzhou

The area surrounding the North HSR Station of Changzhou has obvious *faults* in the planning and implementation process. It is difficult to coordinate the stages of planning led by the municipal government and by the district government. Both parties generate major divergences in the development orientation, construction scope and other important problems in the area surrounding the station, and the plan contents are adjusted frequently. In the implementation process, it is difficult to achieve phased construction goals due to both the conflict of decisions on two levels in the urban area and ambiguous management. Private investment around the station grows slowly. In terms of the interaction between the planning and implementation, it is mainly implementation problems that prompt the corresponding modifications of the plan, so efficient interaction does not develop. This section analyzes the confirmation and changes of the development orientation, development intensity, functional organization and other important planning content of the Changzhou HSR new town.

5.3.1 Development Orientation

The development orientation in the area surrounding the North HSR Station of Changzhou has obvious differences at the planning stages led by the municipal government and by the district government. The HSR station's leading role in the surrounding area shall be considered for the development orientation of the area surrounding the station, and the HSR shall also be placed within a certain spatial scope for overall consideration to analyze the development resources of the area surrounding the station and coordinate the overall development demand of the area. The planning led by both the municipal government and the district government fully considers the leading role of the HSR station. The starting potential of the Changzhou HSR station is analyzed on the basis of the development construction case in the area surrounding the HSR station locally and abroad. Since both the municipal government and the district government handle the relationship between the HSR and the current plan and facilities on different spatial levels and on the basis of different survey backgrounds, the development orientation has great differences. The plan led by the municipal government complies with the general planning logic based on the research, is organized according to the basic mode of a central urban grouping, and aims to perfect the function of the current cities and each urban area by virtue of the opportunity for HSR construction. In contrast, the plan led by the Xinbei District government complies with the development path of the new town in the north, takes the HSR station as its developmental center, and fully relies on the driving effect of HSR on urban development to form new function nodes, thus driving urbanization and land development in the surrounding area.

At the planning stage led by the municipal government, the Municipal Planning Institute delimits the research level as *central urban area-Xinlong group area surrounding the station*. For the central urban area, the North HSR Station of Changzhou is located on Changjiang Road (a main route for urban development), Changzhou City, and the construction of the HSR station promotes the northward movement of urban construction activity and then stimulates development north of the city so that the urban form of Changzhou gradually transitions to *multicenter* from *a single center*. However, the North HSR Station of Changzhou is some distance from the city center, which centers on the intercity Changzhou Station and the Xinbei District government, and is near the developing urban commerce subcenter. Although Xinbei District would establish the commercial business function in combination with the station, it is not the key to the regional development. In the interview, we learned that "The HSR station is used as the passenger terminal, and the urban commercial center area will still be located in the main city area, which depends on the south and north avenues of Yanling Road and other commercial streets due to the geography, traffic, human activities and other factors." In contrast, for the *Xinlong group* as the transportation junction, the HSR station shall mutually connect to the group T-shaped partition level of the administrative trade finance axis in the original plan and the public facilities route to jointly build a subcenter north of the city. The development of the area surrounding the HSR station does not bring direct change to the current administrative and business route planning.

As specified in the *Regional Concept Plan of the HSR Changzhou Station and Urban Design in the Key Region* led by the municipal government, the HSR area is a portal instead of a *heart*. The HSR area is a gate leading to the central urban area, and its advantages can be utilized to develop industry on the basis of guaranteeing smooth passage through the gate.

Based on such issues, the area surrounding the HSR station in this urban design is defined as ① an emerging growth area of Changzhou; ② a new modern transportation junction with a regional function; and ③ the center with the most vitality north of the city and a main spatial carrier with modern service functions.

Beginning at the urban level, the *emerging growth area of Changzhou* drives the northern expansion of the overall urban space of Changzhou by utilizing the HSR. The *new modern transportation junction with a regional function* emphasizes both the HSR station and the surrounding area, focusing on the collection and distribution of passengers. According to the overall planning description of Changzhou City, *north of the city* refers to Xinlong and Xingang with relatively lagging development in Changzhou City, which differs from the Xinbei District, or the *new town in the north*. The *modern service function*, with broad content, mainly refers to the retail, catering, leisure and entertainment, etc. that are close to the station and meet the demands of passengers and the surrounding residents in combination with the specific plan content as well as the business offices radiating from the station, but with a greater distance. The orientation of the business offices is mainly influenced by the high-tech industrial zone, the industrial area surrounding the port, and the planned east–west administrative trade finance axis. Based on this orientation, the *Stakeholder Study of HSR* delimits three basic spatial levels to further define the scope of the area surrounding the station: ① within 0.7 km^2 in the area surrounding the station square, including the public activity area of the passenger flow distribution plaza in front of the station; ② within 4.5 km^2 in the peripheral area of the station, including the business offices along Changjiang Road (a main route for urban development) and the residential area and science park driven by the road; and ③ a new area of approximately 9 km^2, including the urbanized area due to the northern expansion of the city.

At the planning stage led by the district-level government, the function of the area surrounding the HSR station changes several times. The research is based on the administrative scope of Xinbei District, and finally, the spatial level of the *new town in the north—Xinlong International Business City-Business City core area* is formed. In 2009, the Xinbei District government leads and compiles the *Conceptual Urban Design of the New town in the North*. The *new town in the north* is located in the southeast part of Xinbei District, and the east side and the south side coincide with the administrative boundaries. The scope delimitation of the *new town in the north* fundamentally aims to integrate the developmental resources of Xinbei District and promote urbanization construction in Xinbei District. Therefore, in the functional composition of *one center, two routes, five blocks and five urban complexes* in the *Conceptual Urban Design of the New town in the North*, the HSR station belongs to the Xinlong urban function block. The planning sets a development goal of a *HSR hub stakeholder—window on the city* that centers on the station, with a scale of 4 km^2, and is defined as follows: *the area surrounding the HSR station is the core traffic element and the new town in the north is the future public service center that will be developed into the future block core of the new town in the north* (Fig. 5.8). The commercial recreational area, administrative office area, mixed functional area and public service area are set within the scope. The areas close to the HSR station are the administrative office area and a large-scale commercial complex, and the Xinbei District government intentionally moves the district government to this area at this time to drive the surrounding development. The functional status of the HSR station is improved in the plan led by this level of government.

After the Beijing-Shanghai HSR opens to traffic in 2011, the scope within 1.6 km^2 surrounding the station is defined as the *Changzhou station center of the Beijing-Shanghai HSR in the new town in the north of Changzhou*, and 5 companies are organized for the urban design. In the request for proposals, this area is defined as the *business district integrating international functions and a new center highlighting an ecological image* (Fig. 5.8). Both the orientation and the specific functional content differ from the statement in the relevant document regarding the new town in the north. In September 2011, the Xinbei District government issues a report on its official Web site that the area within 1.6 km^2 around the station is named *Xinlong International Business City*. The report indicates that "the government will strive to basically construct the core zone square until 2012 in accordance with the objective orientation of 'international first class and domestic leading,' will start the construction of the financial square and surrounding supporting projects, and will start the construction of the R&D square of science and technology, the public services square and other functional areas in 2013 to ensure the comprehensive construction of Xinlong International Business City in 2015."

In news published on the official Web site of the national high-tech district urban construction network of Changzhou in September 2013, the scope of Xinlong

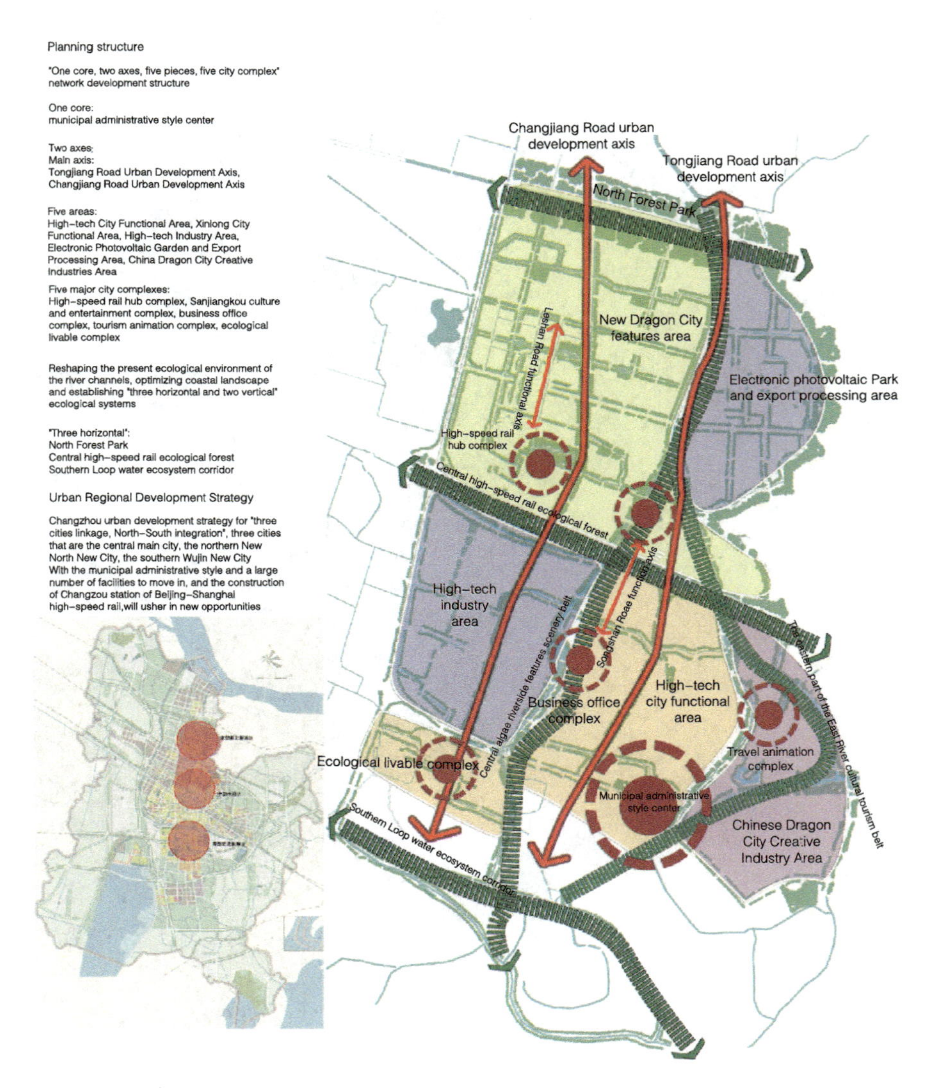

Fig. 5.8 Relevant content of the area surrounding the HSR station in the *Conceptual Design of the New town in the North*. *Source* Xinbei District Branch Office of Changzhou Urban Planning Bureau, *Conceptual Design of the New town in the North*

Fig. 5.8 (continued)

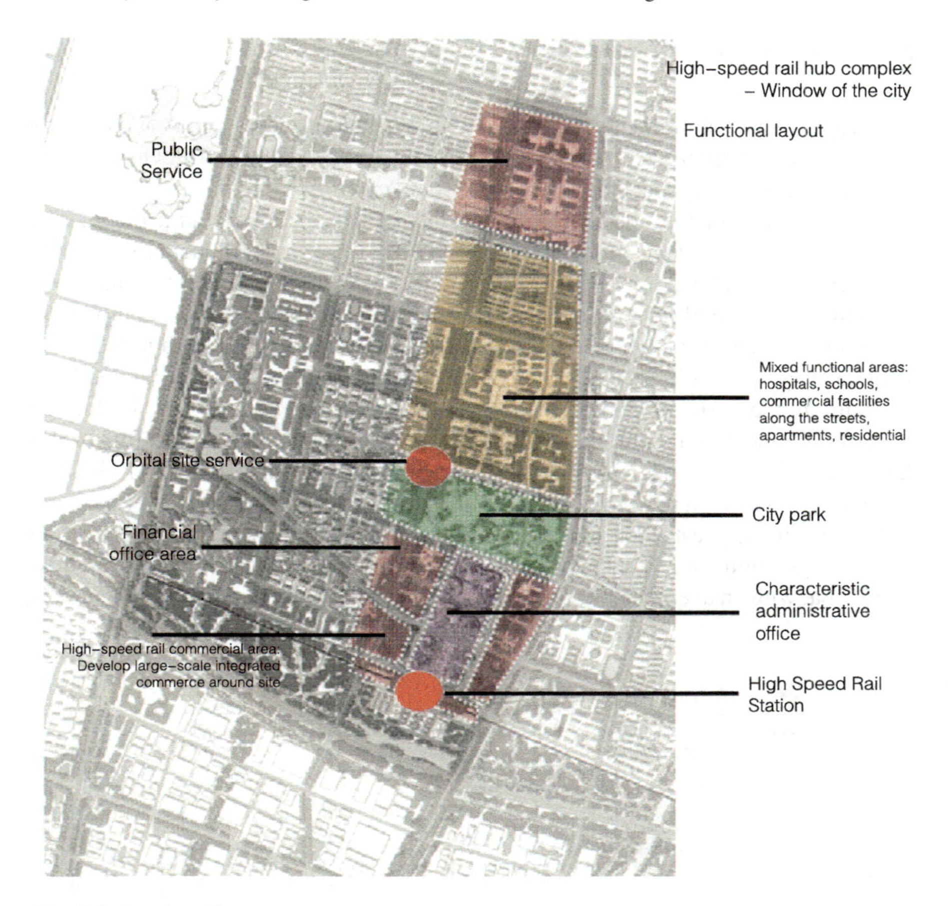

Fig. 5.8 (continued)

International Business City is designated as follows after adjustment: Xinlong International Business City is located east of Longjiang Road, west of the Zaojiang River, south of provincial highway S122, and north of the Shanghai-Chengdu highway, and the total planning area is 24 km^2.[1] The original area within 1.6 km^2 is delimited as the *core area of Xinlong International Business City*, and its development orientation is not adjusted in principle.

[1] The New North District government of Changzhou City issued relevant opinions on the construction of Xinlong International Business City. http://www.czndcjj.gov/newscenter/2277.htm.

5.3.2 Development Intensity

According to the requirements for the plan implementation, adjustments of the development intensity mainly center on the core zone. The development intensity in the central zone is mainly specified in the 2008 *Study on Three Stations*, the 2011 design and control plan of the relevant cities and the 2014 control plan adjustment. The overall development intensity in the core zone undergoes an initial increase and then a decrease, and the maximum change amplitude reaches 100% (Figs. 5.9 and 5.10).

In the core zone of 0.7 km^2 delimited in the *Study on Three Stations* compiled in 2008, in addition to the planned traffic station land (S) and park green space (G1), the total area of other lands reaches 59.33 ha, and the average floor area ratio is 1.98. In addition, the business land (C1) has the highest floor area ratio (3.5). Within the area of 1.6 km^2 delimited in the 2011 control plan, in addition to the traffic station land and park green space, the total area of development plots reaches 76.14 ha, and the average floor area ratio reaches 4.54. In addition, the mixed land use of business offices has the highest floor area ratio of 9.0. After the dynamic adjustment of the control regulations in 2014, the total area of the development land changes slightly, to 75.7 ha, with an average floor area ratio of 4.31. The floor area ratio in each plot is

(1) Project name: "Changzhou Northern New Town of central area of Beijing Shanghai HSR Changzhou Station Concept Urban Design and Site Plan "

(2) Planning Goal: establishing international functional business area, Eco-based new center

(3) Planning Area: (see Annex 6)

The research area covers: Shanghai-Chengdu Expressway to the south, Nenjiang Road to the north, Changjiang Road to the east, Xinlong Second Road to the west, and the area around the HSR station covering an area of nearly 3 square kilometers (partially extended to the east side of the Changjiang Road). Combined with the concept of overall urban design of 18 square kilometers, further study determines the functional blocks, spatial structures and urban intentions in this area, and coordinates the land use layout to provide the master planning support for the detailed design of 1.6 square kilometers.

Detailed design area: Shanghai-Chengdu Expressway to the south, to the north of Tianhe Street, west of Leshan Road, east of the Yangtze River Road, an area of about 1.6 square kilometers.

(4) Proposal of Beijing-Shanghai HSR surrounding area functional layout (see Annex 7)

Lot 1: High Speed Rail Highway Terminal;
Lot 2: High Speed Rail Station South Square;
Lot 3: Temporary as HSR green area, plan to build business hotels;
Lot 4: High Speed Rail Station House;
Lot 5: Super high-rise hotel or low-rise multi-storey public building (symmetrical with Metro 7);
Lot 6: High Speed Rail Station North Plaza, the main surface of the landscape;
Lot 7: Technology and Culture Exhibition Center;
Lot 9: District-level administrative center and the public square, government square;
Lot 8, 10, 11, 13, 16, 17, 18: Commercial and business office;
Lot 14, 15: urban green space;

The 1,2,4 plots for the construction projects have been started, in principle, the design needs to be retained, no functional adjustments. On the 12th part of the block is built in high-rise residential area, the remaining part of the planning and construction of business office.

Fig. 5.9 The description of the positioning and functional designation of the area surrounding the HSR station in international bidding documents. *Source* Xinbei District Branch Office of Changzhou Urban Planning Bureau, bidding documents

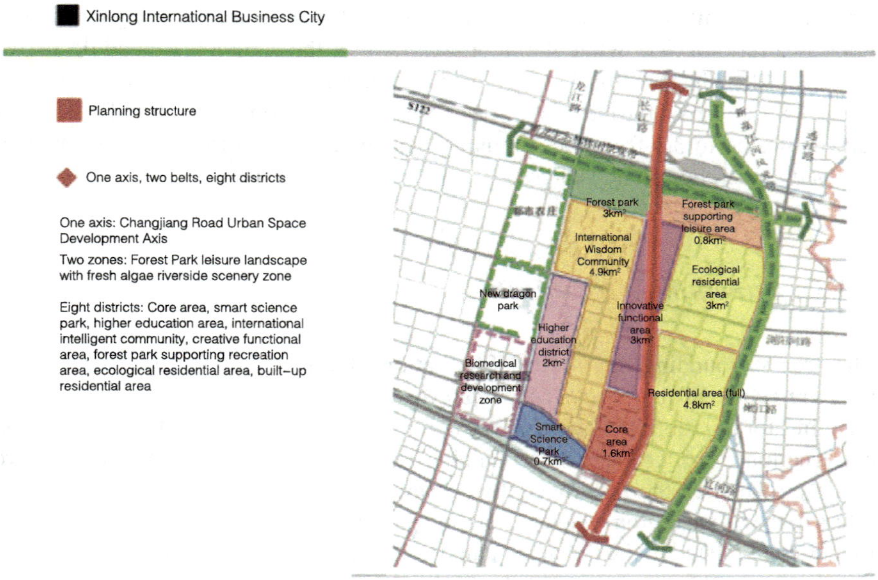

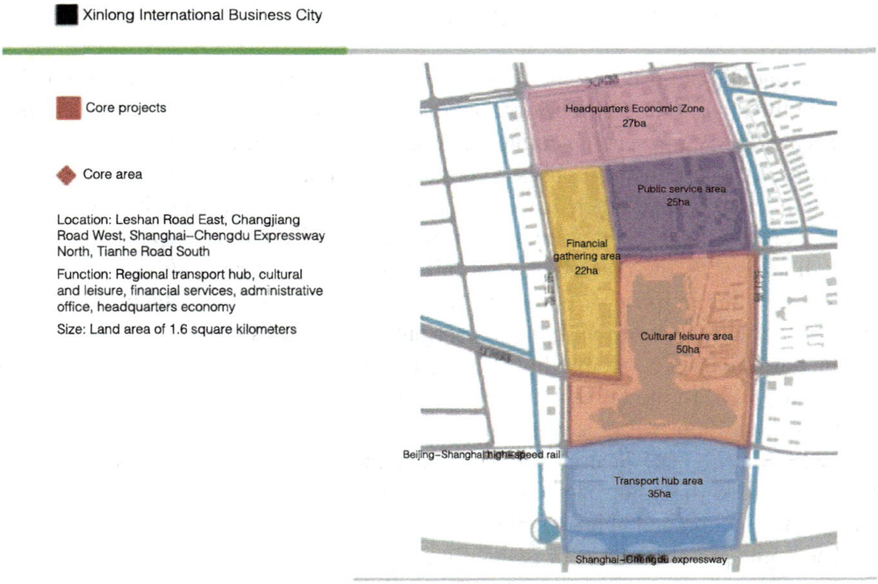

Xinlong International Business City

Planning structure

One axis, two belts, eight districts

One axis: Changjiang Road Urban Space Development Axis

Two zones: Forest Park leisure landscape with fresh algae riverside scenery zone

Eight districts: Core area, smart science park, higher education area, international intelligent community, creative functional area, forest park supporting recreation area, ecological residential area, built-up residential area

Forest park 3km²
Forest park supporting leisure area 0.8km²
International Wisdom Community 4.9km²
Ecological residential area 3km²
New dragon park
Innovative functional area 3km²
Higher education district 2km²
Biomedical research and development zone
Residential area (full) 4.8km²
Smart Science Park 0.7km²
Core area 1.6km²

Xinlong International Business City

Core projects

Core area

Location: Leshan Road East, Changjiang Road West, Shanghai–Chengdu Expressway North, Tianhe Road South

Function: Regional transport hub, cultural and leisure, financial services, administrative office, headquarters economy

Size: Land area of 1.6 square kilometers

Headquarters Economic Zone 27ba
Public service area 25ha
Financial gathering area 22ha
Cultural leisure area 50ha
Beijing–Shanghai high-speed rail
Transport hub area 35ha
Shanghai–Chengdu expressway

Fig. 5.10 Spatial arrangement and scope adjustment of the core area of Xinlong International Business City. *Source* Xinbei District Branch Office of Changzhou Urban Planning Bureau, *Brief of Xinlong International Business City*

classified and unified according to the function, with business land having the highest floor area ratio of 5–6, while the floor area ratio of the commercial land decreases to 1.5–2. After the preliminary adjustment of the control plan in 2014, the floor area ratio of 5 transferred business lands decreases, with the floor area ratio of 4 business lands adjusted to 3.5 from the originally planned 5.0. Therefore, the development intensity undergoes great changes.

5.3.3 Functional Organization

The area surrounding the North HSR Station of Changzhou can be divided into the *core zone* and the *peripheral region* for analysis. From the perspective of the planning formulation, the *core zone* is the direct radiation zone of the HSR station construction considered by each planning institute, and its scale and function are directly related to the passenger flow of the HSR station. The *peripheral region* refers to the urban region whose development is driven by the HSR, and its scale and function are related to the financial situation and the development demand in Xinbei District. From the perspective of the plan implementation, in several planning and implementation conditions compiled by Changzhou, the *core zone* is the highest-priority development zone, while the *peripheral region* is a later expansion zone (Table 5.8).

The function of the core zone changes in the four plans. The commercial function is set in the plan of the four periods, but the specific functions vary. In the *Research of Three Stations*, the specific commercial function is "a comprehensive service area of business hotel industry focuses on hotels, restaurants, leisure and entertainment, shopping, catering and other services, and mainly develops into a hub service commercial area by centering on the Changzhou Station hub area." In the *Plan of the New town in the North*, the commercial organization form is *large-scale comprehensive business* for the purpose of commercial recreation. The service object is not only limited to the transportation junction user but also expands to leisure and shopping for the surrounding residents. In the control rules of 2011, the commercial land is mainly adjacent to the HSR station and is located south of Liaohe Road, which is basically consistent with the relevant plan content of the new town in the north. In the 2014 control plan, the core zone is divided into two parts: one part is adjacent to the station, and the other part is located east and north of a large body of water in front of the station and approximately 500 m from the station.

The commercial function was mainly proposed in the 2009 *Conceptual Urban Design of the New town in the North*. Later, in the 2011 design and control plan of the relevant cities and the 2014 control plan update, the commercial function is the main function in the core zone but is to be changed according to the terms of the specific positioning and development intensity. The administrative offices' function is not mentioned in the *Study on Three Stations* led by the municipal government in 2008, so the middle part of the Xinlong group and the north side of the station are set up on the basis of the original plan. In the *Conceptual Urban Design of the New town in the*

Table 5.8 Comparison of the planning functions for the North HSR Station of Changzhou

Time	Core area			Peripheral area	
2008	Three stations study	0.7 km^2	Commercial, convention and exhibition area, theme park	4.5 km^2	Residential area, business offices area, cultural and entertainment area, science and technology park
2009	Northern metro	1.6 km^2	Business, administrative offices, business offices	4 km^2	Public services area, residential area, cultural park area, integrated area
2011	Control regulations 1	1.6 km^2	Artificial lake park, commercial, financial, headquarters business, public services area	Not involved	Not involved
2014	Control regulations 2	1.6 km^2	Artificial lake park, business, commerce	24 km^2	Science and technology innovation area, residential area

North led by Xinbei District in 2009, since the Xinbei District government proposes to establish the district government in the core zone, the administrative offices' function is included and is defined in the urban design and control plan in 2011. However, the general office of the CPC Central Committee and the general office of the State Council print and issue the *Notice of Party and Government Organization about Stopping the Establishment of New Building Halls and Cleaning Up Office Housing* in July 2013, which directly affects the removal plan of the Xinbei District government, making this move difficult to implement; thus, the administrative offices land is adjusted as business and commercial offices land in the control plan update for the core zone in 2014.

The great artificial lake park in front of the station is located north of Liaohe Road, south of Xinqiao Street, east of Renhe Road, and west of Chongyi Road. The total planned land is 17.35 ha. People can watch the water in the park, and a musical fountain is included in the design. This artificial lake garden is included in the design requirements in the 2011 international request for proposals for urban design. The Changzhou Zhusen Company integrates the urban design scheme on the basis of the opinions of the Xinbei District government. This artificial lake park is basically shaped and is confirmed in the 2011 regulatory planning. The Black Peony Group, with shareholders that include the management committee of the Changzhou

V .Land use function and scale

1. Land use function:
A comprehensive area with living, commercial, office, business commerce, culture and entertainment, transportation, emerging industries, trading submit is one of the integrated area.

Commercial: stores, supermarkets, fashion boutiques and specialty stores street, convenience facilities (financial institutions, administrative services, etc.), restaurants, hotels, etc.

Trade and Business: Office buildings for station-related administration, highly informative industry, circulation enterprise.

Cultural and entertainment: entertainment facilities for passengers, surrounding residents and corporate staff.

Trading Submit: local industry & business, high-tech industry.

2. Scale (0.5–0.6 square km)
Housing area of Changzhou HSR station is about 0.06 square km, of which 0.035 square km of HSR station, 0.02 square km of bus station and 0.005 square km of rail transit station. Business and Finance construction area of about 0.2 square km, hospitality and office construction area of about 0.15 square km, education and research construction area of about 0.08 square km, cultural and entertainment construction area of about 0.01– 0.02 square km, Trading and Expo construction area of about 0.1 square km.

Fig. 5.11 The description of functional designation of core area in the international bidding documents. *Source* Xinbei District Branch Office of Changzhou Urban Planning Bureau, bidding documents

National Economic and Technological Development Zone, invests in and constructs the lake in 2012.

In the plans led by the municipal government and the district government, the peripheral regions lag behind the core zone in terms of confirmed development. There is no large-scale advance in actual construction. For the peripheral regions influenced by the HSR station, the Changzhou Planning Institute first delimited 4.5 km^2 in 2008 according to the passenger capacity of HSR and the area affected by similar stations. The *Conceptual Urban Design of the New town in the North* basically continues the direction of the *Study on Three Stations* and delimits roughly 4 km^2. The original scope of *Xinlong International Business City* proposed in 2011 includes only a 1.6 km^2 core zone. It is readjusted as 24 km^2 in 2013 due to the development construction and management demand. The surrounding region has a more obscure description of the planning function and is delimited with a large-scale functional area. The special functions and operation types of the plots are temporarily not specified (Fig. 5.11).

5.3.4 Summary of Reason for Content Adjustment

(1) **Interest demand difference of decision-makers**

During the planning period led by the municipal government and the district government, the area surrounding the North HSR Station of Changzhou changes greatly in the special scale, development orientation, development intensity, etc.; the root cause is the difference in the interest demands of the city-level and district-level decision-makers. For the municipal government, the development of the region surrounding the HSR station is to be included in the overall urban development and will initially conform to the overall urban benefit. For example, although the area surrounding the HSR station could develop a commercial function, the plan temporarily does not include a large-scale commercial function in order to avoid homogeneous competition in consideration of the overall commercial real estate demand in the whole city and the construction of current commercial centers. For the district government, the administrative scope gives priority to the benefit. If the development opportunity of a business function caused by the construction of a HSR station could bring a benefit to the district government in land transfer fees, tax revenues, etc., the district government will conduct large-scale investment around the station to attract commercial businesses.

(2) **Development policy influence**

External policy, especially the integral mandatory instruction issued by China, will directly influence local specific decisions. Once the *Notice of Party and Government Organization about Stopping the Establishment of New Building Halls and Cleaning up Office Housing* is issued, the original plan that the Xinbei District government will move to the HSR station is suspended. Later, the investment projects that rely on the relocation of the district government is withdrawn. The development in the core zone of Xinlong International Business City is stagnant. Thus, the plan of building a *municipal business center* in the area surrounding the HSR station is difficult to achieve. The original plan must be adjusted and amended accordingly.

(3) **Individual role of leader**

In the interview process, the relevant planners indicate that some adjustments of the plan and construction in the area surrounding the North HSR Station of Changzhou emerge from the leaders' vision. For example, the shape of the North HSR Station of Changzhou is similar to that of a fish. The body of water is planned to correspond to the station, and the fountain and water surface are designed to promote the regional image and attract the residents for activity at the same time. Therefore, a large-scale water park is planned close to the north side of the station, but as a result, the original express road is to be built underground, increasing the construction cost. The construction of the Xinlong Lake Park characterized by the large body of water and fountain is rapidly implemented, but due to a lack of popularity, the service efficiency is lower, while the operation and maintenance costs are higher (Fig. 5.12).

6. Urban public space design

Identify the urban public space design such as park, water, green land, square, reflect the overall correlation.

(1) High-speed rail station housing North Square (5,6,7 plots) should use a large open-water surface to shape the public square and present the HSR view, on the one hand, interact with the "streamlined fish" housing shape to show a harmonious scene, on the other hand, fully show the characteristics of Changzhou, create a unique JiangNan style urban public space. Spring and water curtain movie will be integrated in design.

Fig. 5.12 Part of the Invitation of Attorney Excerpt for the conceptual urban design of the central area of the North HSR Station of Changzhou. *Source* Xinbei District Branch Office of Changzhou Urban Planning Bureau, *Bidding Invitation Letter*

5.4 Analysis of Development Stakeholders and Decision-Making of HSR New Town of Changzhou

5.4.1 Analysis of the Decision-Maker

In the planning and implementation process in the area surrounding the North HSR Station of Changzhou, the Changzhou municipal government, Xinbei District government and other relevant agencies play important roles, and the institutional organization and subordinate relations are shown in Fig. 5.13.

Like the Wuxi municipal government and the Xidong District government, the Changzhou municipal government and Xinbei District government are "growth-oriented governments." By simulating an enterprise system, the government conducts urban marketing to promote urban prosperity, attract investment, expand the local tax base and create employment. The government's administrative level and jurisdiction scope directly influences its demand. The Changzhou municipal government makes decisions on the basis of the most efficient development of the whole city; thus, how to coordinate the development of each urban area in the city and avoid invalid and disorganized competition are major challenges that the municipal government confronts. For the Xinbei District government, to achieve the most efficient development in its own jurisdiction, it must unavoidably compete with Tianning, Wujin, etc. in Changzhou City. Therefore, the municipal government and the district government have the joint objectives of promoting local development, but their specific decisions are not always consistent due to the differences in jurisdictional scope.

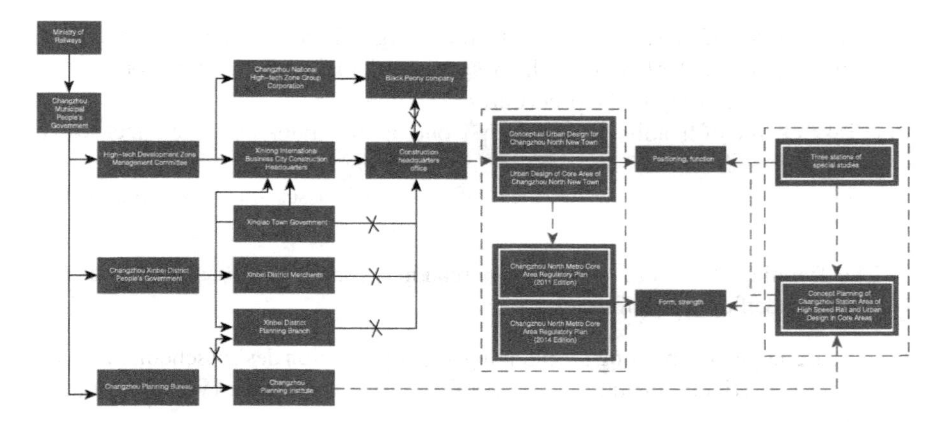

Fig. 5.13 The decision-makers and problems of development in the area surrounding the North HSR Station of Changzhou

The Changzhou people's government (municipal government) is the level I competent administrative department of Changzhou City, with a departmental administrative level, and manages the overall development of five municipal districts and two county-level cities. The Xinbei District people's government is the level I competent administrative department of Changzhou City Xinbei District and is established on the basis of the Changzhou National High-tech Development Zone. Its primary responsible persons concurrently serve as the responsible persons for the high-tech district management committee; for instance, the director (deputy departmental level) of the management committee serves concurrently as the Xinbei District mayor, and the deputy director of the management committee concurrently serves as the deputy mayor of the Xinbei District people's government. Since the state-level high-tech development zone adopts the "high-level officer distribution, independent management and separate expenditure," the administrative level of the main responsible person in the Xinbei District people's government is higher than that of the main responsible person in the general municipal district, and the district government often has more management power than the general municipal government. For example, regarding the functions and powers of planning management, the official Web site of the Changzhou Planning Bureau states that the Xinbei District Planning Sub-bureau has the following administrative functions and powers:

(1) Take charge of drawing up the planning conditions of Class I projects and approve according to procedures after reporting to the Municipal Bureau Technical Review Conference for approval;
(2) Take charge of drawing up the planning conditions of Class II and III projects in the area;
(3) Take charge of the campaign, review and approval of design schemes for the construction engineering of Class I, II and III projects in the area (Class I projects are reported and approved according to the procedures);

(4) Take charge of reviewing the roads (including rail transport) in Class I projects and the detailed planning schemes of municipal projects and conduct the approval according to the procedures;
(5) Take charge of handling all projects' "one opinion book and three licenses" in the area; and
(6) Take charge of organizing the traffic impact assessment of construction projects.[2]

The Tianning District Planning Sub-bureau in the same municipal district has the following specific responsibilities:

(1) Take charge of approving the planning and construction design schemes of Class III projects (including the outdoor landscape projects matched with the projects) in the area;
(2) Take charge of approving and issuing the "one opinion book and two certificates" of the Class III construction projects in the area;
(3) Take charge of the completion of planning acceptance of Class III projects in the area;
(4) Take charge of organizing the administrative licensing hearings of the projects being managed; and
(5) Participate in the hearings for administrative review, administrative litigation and administrative punishment cases of the projects being managed.[3]

Class I construction projects include municipal major infrastructure projects confirmed in the overall urban plan and major social public utilities projects concerning the urban functions and image of the whole city; real estate development projects with construction land areas of no less than 200 mu; public construction projects with building heights of more than 80 m or single building areas of more than 50,000 m^2; and real estate projects of public buildings with a building land area of less than 200 mu but a building height of more than 80 m or a single building area of more than 50,000 m^2. Class III construction projects include single public buildings with a building area of less than 5000 m^2 or a building height of less than 24 m; reconstruction projects of private houses; all industrial projects; and the accessory municipal engineering of the above projects. Class II building projects are all construction projects except Class I and III projects.

Through comparison, we can see that Xinbei District has far more power than Tianyu District in planning project management and administrative approval. In addition, the relevant personnel of the Changzhou City Planning Institute were interviewed and indicated that "Since Xinbei District has a good economic development

[2]Changzhou Xinbei District Planning Bureau executive authority. Source: Changzhou City Planning Network (http://www.czghj.gov.cn/).

[3]Changzhou Tianning District Planning Bureau executive authority. Source: Changzhou City Planning Network (http://www.czghj.gov.cn/).

foundation and a high administrative level, so the Xinbei District Planning Sub-bureau must follow the decisions of the Xinbei District government in specific decisions to a great extent, although it is a detached office of the Changzhou Planning Board."

The Changzhou High-tech Group Co., Ltd. (hereafter referred to as the Changzhou High-tech Group) belongs to the Changzhou High-tech District (Xinbei District). It was established and constructed in the Changzhou National High-tech Development Zone in 1992; it acquired the Black Peony (Group) Co., Ltd. (listed on the Shanghai Stock Exchange) in November 2008 and implemented enterprise restructuring in 2011. The Black Peony Group has become an investment stakeholder of the Xinbei District government, a carrier for development and construction, and an entity for capital operation. The first-level land development in the HSR core zone and the Longhu Park project are constructed by the Black Peony Group. The restructuring directly strengthens the enterprise property of the Changzhou High-tech Group, which will therefore gain more attention for its profit demands in the operations and decision-making processes. However, it is worth noting that the Changzhou High-tech Group is still a state-owned enterprise, and its board chairman also serves as the deputy director of the high-tech district management committee. The results are as follows: On the one hand, the Changzhou High-tech Group could still gain support from the Xinbei District government in relevant policies, such as admission in important fields and obtaining information relevant to certain decisions; however, since it is a state-owned enterprise, it cannot give complete priority to economic benefits in certain projects and must consider the social benefits and the vision of the decision-making group.

In the initial term of construction in 2011, the Xinbei District mayor, Xinqiao town chief and other officials served as the responsible persons for the construction headquarters of Xinlong International Business City. With a management scope of 1.6 km^2, it does not have specific responsibilities and permanent staff because it is not a permanent unit and does not undertake specific work. After 2013, the Xinlong International Business City is expanded to 24 km^2. The construction headquarters office of Xinlong International Business City is transferred to the permanent construction management department, which takes charge of specific development matters, such as demolition in the core zone. There are 4 officers until October 2014 who are civil servants transferred from Xinlong District and Xinqiao Town. They mainly take charge of coordinating, promoting and connecting the construction work in the area and take responsibility for investment promotion and capital introduction at the same time. In the actual operation, the headquarters office confronts many difficulties when carrying out the work. Unlike the management committee, the headquarters has no independent corporate capacity. Meanwhile, it has no subordinate relationship with the district department of Xinqiao Town and has no actual administrative functions and powers. It also does not have the subordinate relationship with the development company, Black Peony Group, so it cannot decide certain specific details of the investment promotion and negotiation. Finally, although this headquarters is responsible for investment promotion and funding, it has no decision-making power due to its organizational nature and lack of professional power. Although the district leaders instruct the Xinbei District Merchants Group to proactively coordinate the relevant work but lack a clear division of rights and liabilities, the progress of investment promotion, capital introduction and development work is slow. In fact, the investment promotion work mainly depends on the district leader for the consultation with the relevant enterprises in the area; then, the office follows up regarding the relevant work. In the interview, we learned that the decision cycle is very long, and progress is not smooth, as substantial problems are reported to the leaders for decisions.

5.4.2 Important Decision Analysis

(1) **Station selection**

The length of the Changzhou section of the Beijing-Shanghai HSR is 44.74 km. The internal line route of Changzhou City is similar to that of Wuxi City, and both are close to the edge of the built-up area in the central urban area. The original usage function in the area through which the route passes is farmland, and the total demolition amount of the route is approximately 260,000 m^2. The length of the Wuxi section of the Beijing-Shanghai HSR is 49.75 km^2, involving a total demolition of 480,700 m^2. By contrast, for the Ministry of Railways, the location of the Changzhou section of the Beijing-Shanghai HSR conforms to its goals of pursuing a high usage value for HSR and decreasing demolition costs.

In terms of siting, the North HSR Station of Changzhou is located at the north edge of the central urban area and is close to the Changzhou National High-tech Development Zone, which is undergoing rapid development at the time in line with the north–south expansion direction of the Changzhou urban area. The land surrounding the station is included in the total range of land use. By contrast, the East HSR Station of Wuxi is not located in the main direction of urban expansion, as planned in the original general plan. The location of the North HSR Station of Changzhou is a relatively ideal scheme for the Changzhou municipal government, and it could effectively drive the change of the surrounding land usage function, thus obtaining the exchange value of the space.

Since the Ministry of Railways and the Changzhou municipal government could both achieve their desired benefits in terms of the route and station siting, the Changzhou municipal government did not object to the scheme for the route selection of the Changzhou section of the Beijing-Shanghai HSR and the establishment of the North HSR Station of Changzhou after the Ministry of Railways proposed the scheme, and the scheme was confirmed rapidly.

(2) **Development orientation**

The development orientation of the HSR new town of the North HSR Station of Changzhou involves greater changes. The development status of the HSR new town is gradually transitioning from the *dominant area of the hub function* expected by the municipal government to *the business, commerce and public service center of the new town area in the north* in the 2009 *Conceptual Urban Design of the New town in the North* initially led and planned by the district government and then to the *business district integrating international functions and the new center highlighting an ecological image* raised in the 2011 urban design. The development status of the HSR new town is gradually improving.

The demand of the Changzhou municipal government is based on long-term development in the whole city. The demand for commercial office space in Changzhou City, especially relatively high-end commercial office space, is limited. If we emphasize the HSR station's function of stimulating business and commercial development, and building high-end business offices and the current business center in the area surrounding the HSR station is competitive because the construction receives much public investment, the area surrounding the HSR station may attract enterprises by virtue of cheaper land prices, more foreign traffic and a superior environment. Seemingly, the area surrounding the HSR station achieves the *exchange value* by changing the *usage value*, but the results are based on much public investment with limited appreciation. Meanwhile, the results may cause a decrease in the *usage value* of the original urban business and commercial center. Therefore, on the premise that the commercial business and other space requirements in the surrounding area are limited, blindly pursuing the high exchange value of space production in the area surrounding the station may cause an overall revenue decrease. "Therefore, the commercial office space of the Xinlong group in the *Study on Three Stations* is planned for the core development area, and the planned construction sequence is also arranged in the middle and later periods." As specified in the Study on Three Stations, *the area surrounding the HSR station may develop into the urban subcenter under the conditions of development demand in the future*. The commercial office space is regarded as a business and commercial gathering area. In terms of the planning concept for the HSR area, the municipal government emphasizes perfecting the supporting facilities of HSR to guarantee the usage efficiency of HSR. In addition, the efficient usage of HSR could drive a rapid flow of capital, information and talents in Changzhou, further promoting the overall development of Changzhou City. Meanwhile, the municipal government considers configuring certain public services, such as a theme park, commercial leisure functions, and research and development functions, thus promoting the residential development of the surrounding area.

As the direct stakeholder of the planning and implementation in the area surrounding the HSR station, the Xinbei District government is the independent stakeholder and obtains the exchange value of land by developing the area surrounding the HSR station, thus promoting regional prosperity and vitality. Unlike the Changzhou municipal government, the Xinbei District government, with its interests based on the Xinbei District administrative jurisdiction, aimed to attracts development resources from within and outside the city to promote regional development, which is a challenge confronting the *growth-oriented government* and is the necessary approach for a *growth-oriented government* to conduct development in competition with other municipal districts, especially when the current administrative performance evaluation content includes economic development, urban construction, etc. in the administrative jurisdiction. Therefore, giving full play to the promotional role of the area surrounding the HSR station and developing the area surrounding the station into the commercial center of Changzhou and even the Suzhou-Wuxi-Changzhou region is the approach that best meets the interest demand of the Xinbei District government. In the interview process, the chief planner of the Changzhou Urban Planning and Design Institute said, "As the municipal district formed at the deputy departmental level of Changzhou, Xinbei District and Wujin District have had very great development autonomy all the time. In approximately 2010, the two districts put forward the strategy of the new town construction, but the rigid demand for development was limited, and the coordination of the municipal government produced little effect." Based on the appeal for the optimal development of this district, Xinbei District government entrusts Debenham Thouard Zadelhoff with carrying out the development capacity forecast of the area surrounding the HSR station between 2009 and 2011 and prepares the 2011 *Urban Design and Regulatory Plan* on this basis. It intends to build a 380-m-high building, which will be the tallest in Changzhou; plan the super-high-rise *Changzhou financial business district* with 12 buildings in the area surrounding the HSR station; and is inclined to drive the development of business office activities through the relocation of the Xinbei District government, thus realizing the transformation and maximization of the *use value* of the land surrounding the station. Meanwhile, the district plans Xinlong Lake Garden, covering an area of 17.35 ha, at a cost of RMB 58 million with the goal of building a *new center with an outstanding ecological image*; its planning aims to draw attention to and improve the environmental quality and then promote the realization of land value. The planning objectives of the Xinlonghu Park, with a price of 58 million, are to gain popularity, improve environmental quality and therefore promote the value realization of the land. It is clear that the expectations of the district government for the HSR new town are much greater than the arrangements of the municipal government for the HSR new town.

(3) **Spatial arrangement**

The spatial arrangement of the area surrounding the North HSR Station of Changzhou is mainly embodied in the range of land use and development intensity. The adjustment of the range of land use is closely related to the development orientation of the area surrounding the station. In terms of development intensity, the change in the development intensity of the station core zone *is generally lower* in the 2008 *Study on Three Stations*, is *improved and partially prominent* in the 2011 version regulatory plan, *is higher and remains balanced* in the 2014 version regulatory plan and is *downregulated* in the 2014 development transfer.

In the *Study on Three Stations* dominated by the Changzhou municipal government in 2008, for the HSR area surrounding the *functional zone with a preferred transportation junction function*, the plan emphasizes the supporting *use value* of the transportation junction, giving full play to the advantage of the adjacent stations and effectively serving the various demands of station traffic organization and tourists. Therefore, the overall development intensity of the area surrounding the station, especially the 0.7 km^2 core zone of the adjacent stations, is relatively low, thus reducing the traffic pressure of other functional development.

In the 2011 *Urban Design and Regulatory Plan*, the Xinbei District government hopes to give play to the use value of the surrounding area as a portal area, to attract business activities and promote urban prosperity through the catalyst effect of the station. Meanwhile, the funding for municipal infrastructure construction is provided through the acquisition of the land exchange value. Therefore, the government expands the scope of the station core zone, improves the overall floor area ratio, reaching in some local areas a value of 9.0, and plans a 380-m-high new *urban landmark* building, thus further driving the development of the surrounding area, realizing the further transformation from the agricultural function to the urban function within the wider area and acquiring the *exchange value* of land development.

In the 2014 regulatory plan adjustment and subsequent actual land transfer process, the land floor area ratio decreases gradually. On the one hand, it is because the plan of the Xinbei District government moving to the core zone of Xinlong International Business City is abandoned, which leads directly to a decrease in the potential use value of the land for business offices within the scope of the core zone. On the other hand, it is because the Changzhou real estate market is trending downward overall, which leads to the withdrawal of enterprises that originally had investment intentions and a slowdown of the overall construction process. Meanwhile, the Changzhou municipal government proposes building the municipal financial business district at *east longitude 120* in Tianning District. This region is superior to the area surrounding the HSR station in terms of location, supporting facilities and overall business atmosphere, which weakens the attraction of the HSR new town for the original high-end business offices function. Private investors, especially developers of business buildings, all think highly of the land use value regardless of lease or sale after construction. There are two reasons that the area surrounding the HSR station can attract the business function. First, the location near the HSR station and district government can provide a unique value of convenience. Second, the area surrounding the HSR station changes from farmland to urban area, so the exchange value paid to acquire the new use value is lower. However, the use value of the area surrounding the station decreases sharply after the cancellation of the district government relocation. Meanwhile, due to better infrastructure, although the exchange value paid for the financial business district under development and construction within the original built-up area is higher than that paid for the area surrounding the station, the higher use value can cover these costs effectively.

Thus, the differences between the two stakeholders in the orientation of the Changzhou HSR new town leads to the change in the development intensity of the spatial sector (Figs. 5.14 and 5.15).

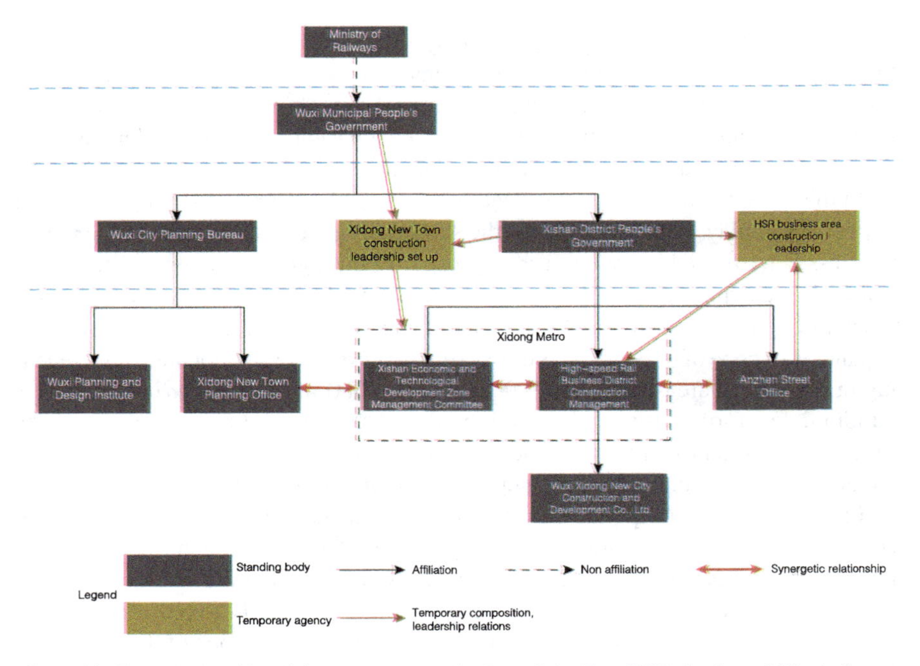

Fig. 5.14 The relationship of the management bodies of the East HSR Station of Wuxi. *Source* author, drawing based on research date and government website information

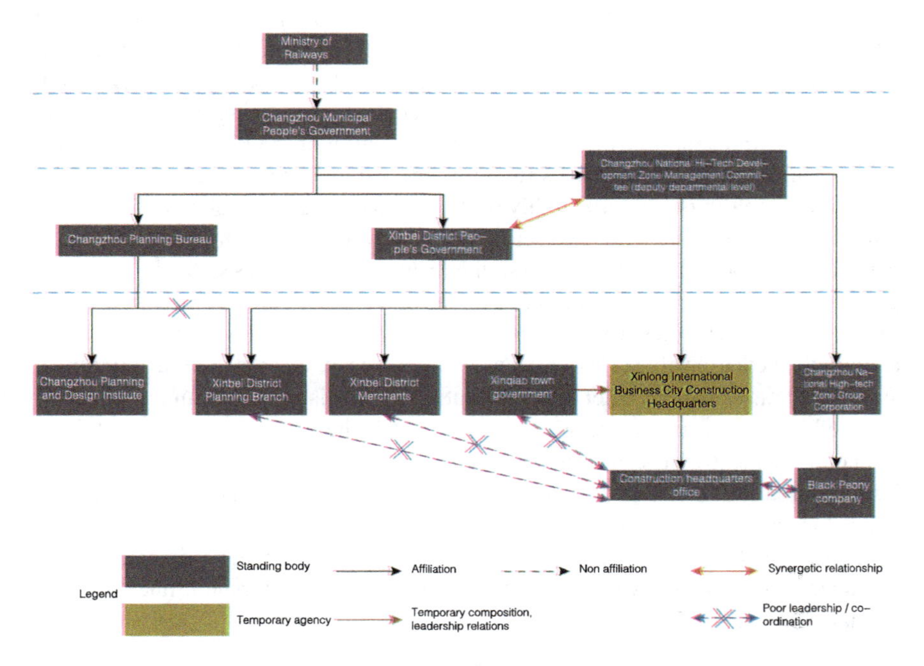

Fig. 5.15 The relationship of the management bodies of the North HSR Station of Changzhou. *Source* author, drawing based on research date and government website information

5.5 Comparative Analysis of Planning and Development of HSR New Towns of Wuxi and Changzhou

Based on the above detailed analysis of Wuxi and Changzhou, this section compares the organizational characteristics of the management and decision-making organs for the areas surrounding the two stations. It analyzes the specific process of planning and implementation, focuses on the decision-making similarities and differences regarding four key problems—station location, development orientation, spatial arrangement and implementation time sequence. Furthermore, it seeks the core factor that leads to different results against similar development backgrounds. In combination with the comparison conclusion, this section provides suggestions on the planning and implementation of the area surrounding the HSR station.

5.5.1 Management and Decision-Making Stakeholder and Its Relationship Comparison

In terms of the management setting and the functions of all the management departments in the area surrounding the HSR station, Wuxi and Changzhou have major differences, which are mainly embodied in three aspects: ① differences in the direct management organization of the area surrounding the station, namely, differences between Wuxi HSR Business District Construction Management Committee and the Changzhou Xinlong International Business City Construction Headquarters; ② differences in the relationship between the planning department and the municipal government; and ③ differences in the decision-making and coordinating agencies (Table 5.9 and Figs. 5.16, 5.17).

(1) **Differences in the direct management organization**

The setting of the direct management organization and its rights and responsibilities produce an important influence upon the development of the area surrounding the station. The Wuxi HSR Business District Construction Management Committee and the Changzhou Xinlong International Business City Construction Headquarters are the direct management organizations for the development and construction of the areas surrounding the East HSR Station of Wuxi and the North HSR Station of Changzhou. The two organizations are under the district governments, and the main responsible person is the responsible person of the district government and the local town government of the station. The differences are embodied in four aspects—organizational attributes, management object, constitution of the organization and responsibilities (Table 5.10).

The Wuxi HSR Business District Construction Management Committee is the department-level organization under the Wuxi Xishan District government. It consists of an office, a fiscal audit office, a construction office and other permanent

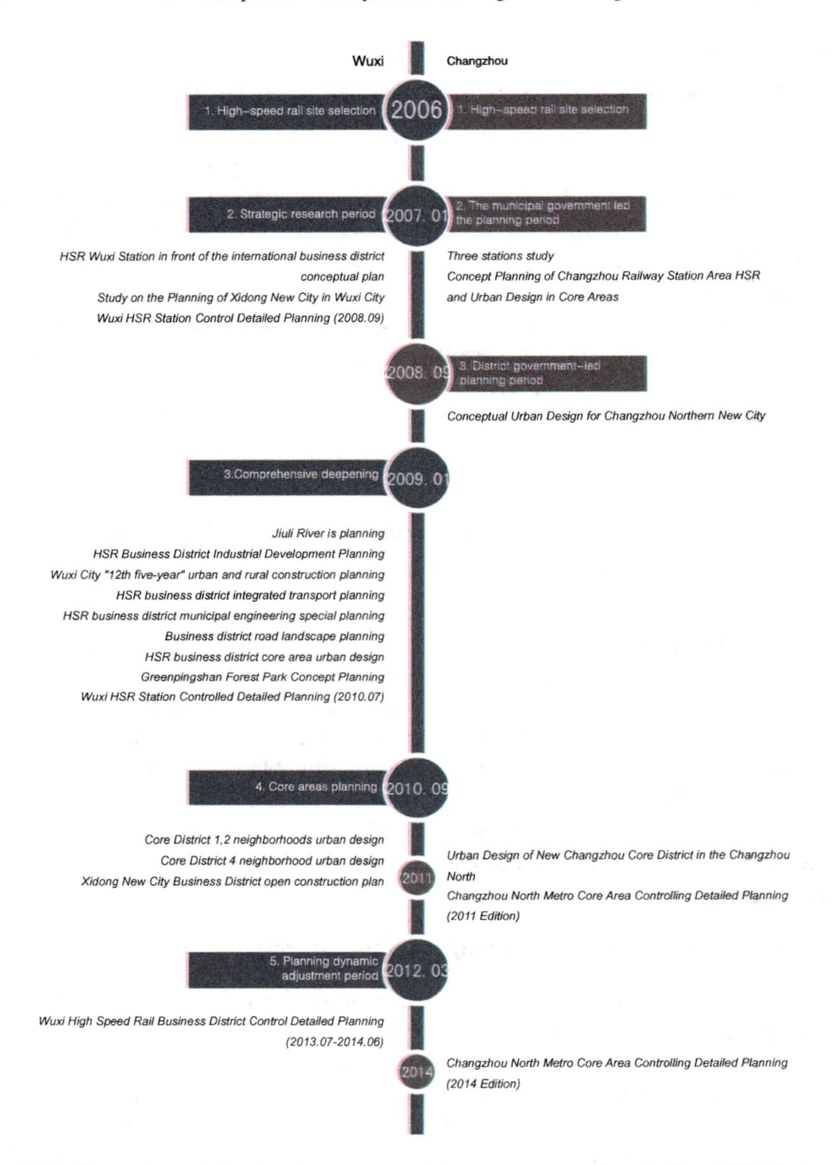

Fig. 5.16 Comparison of the planning process of the areas surrounding the East HSR Station of Wuxi and the North HSR Station of Changzhou

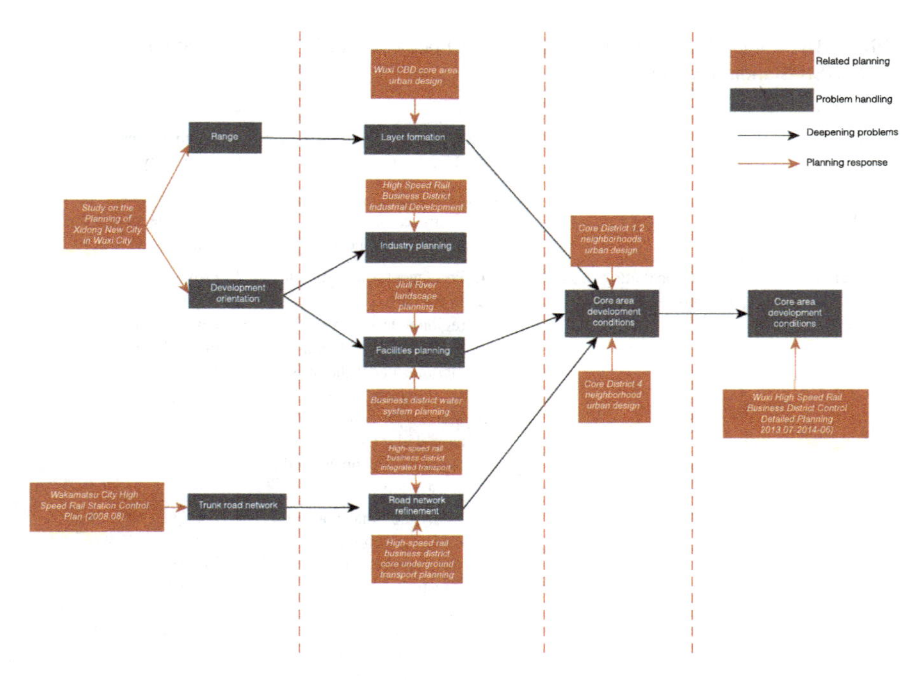

Fig. 5.17 Analysis of problems in each stage of planning of the area surrounding the East HSR Station of Wuxi

Table 5.9 Comparison of the management bodies of the East HSR Station of Wuxi and the North HSR Station of Changzhou

Department/object	East Wuxi HSR station	North Changzhou HSR station
Direct management agencies	High Speed Rail Business District Construction Management Committee	New Dragon International Business City Construction Headquarters
Planning authority	Xidong New Town Planning Office	New North District Planning Bureau
Policy coordination agency	Xidong new town Construction Leading Group and HSR business district construction leading group	None
Development company	Wuxi Xidong new town Construction and Development Co., Ltd.	Black Peony Group Co., Ltd.

Table 5.10 Comparison of the direct management bodies of the East HSR Station of Wuxi and the North HSR Station of Changzhou

Compare content/name		Wuxi	Changzhou
		HSR Business District Management Committee	New Dragon International Business City Construction Headquarters
Similarities		The same composition of leadership, with the district and the main leaders of the town holding the main responsibility	
Differences	Institutional attributes and management objects	1. Permanent establishment 2. In the early stage of regional development in 2009, the institutional facilities under the project are relatively complete 3. Management of business area, 45 km^2 of the overall development and construction	1. Provisional Institutions 2. 2009 years of early development setup, 2013 under the permanent office 3. 2009 1.6 km^2, 2013 expanded to 24 km^2
	Institutional composition	1. Under the Financial Audit Bureau, Economic Development Bureau, Construction Bureau, China Merchants 2. Under the New town Construction and Development Co., Ltd.	1. Set up an office (only 4 people in 2014) 2. In addition, often a high-tech group, Black Peony Investment Co., Ltd., has no affiliation
	Responsibility content	1. Financial revenue and expenditure 2. Social asset investment management 3. Planning organization and demonstration 4. Urban environment and environmental management content such as urban development and economic development responsibilities	1. There is no clear responsibility; the main responsibility of the office for the coordination of the district, town and bureau information 2. Financial revenues and expenditures, planning and management and other authority are still implemented by the district offices, and the various branches and construction headquarters have no affiliation

organizations and has a clear administrative stakeholder identity and administrative authority. The Changzhou Xinlong International Business City Construction Headquarters is a temporary agency, and its initial main purpose is to provide a management and coordination platform for the district, town and bureaus within the scope of the Xinlong International Business City; the office is set up in 2013 along with the expansion of the Xinlong International Business City. However, it still has no clear administrative stakeholder or administrative authority. For this reason, after the two cities propose the goal of development in the area surrounding the HSR station almost at the same time in 2009, the Wuxi HSR Business District Construction Management Committee could engage in work immediately, organize planning formulations and carry out investment promotion. Since the administrative power of Changzhou is still decentralized in the planning sub-bureau, the bureau of commerce and other departments of Xinbei District, the work is mainly promoted by the district government assigning work to the relevant bureaus. Among the levels and sectors of various departments, the plan formulation is dominated by the district government and entrusted to the Xinbei District Planning Sub-bureau, but coordination among the departments is difficult, and the overall advancement is slow.

In terms of the management object, the Wuxi HSR Business District Construction Management Committee defines the specific scope of the management object at the initial stage of development and has clear rights and responsibilities and an interest relationship with the related subdistrict, laying a good foundation for promoting the subsequent work. The Changzhou Xinlong International Business City Construction Headquarters is adjusting its management scope and has an unclear relationship with other administrative departments. This situation means the HSR station is an area without *jurisdiction*.

This comparison shows that the planning and development of the area surrounding the HSR station depend on a clearly defined and effective management organization. A unified management department with definite administrative power and clear management boundaries is an important foundation for guaranteeing the smooth implementation of the planning and development of the HSR new town.

(2) **Differences in the relationship between the planning department and municipal government**

The Xidong new town Planning Office and the Xinbei District Planning Sub-bureau are the main departments in charge of managing the planning for the areas surrounding the two stations. The Xidong new town Planning Office is established especially to support the Xidong new town construction strategy; its management scope focuses on Xidong new town, but the planning management of the external area of Xishan District is still in the charge of the Xishan District Planning Sub-bureau. The management scope of the Xinbei District Planning Bureau is the entire Xinbei District. In terms of management pertinence and efficiency, as the especially established department, the Xidong new town Planning Office has higher management efficiency. However, due to the wide management scope, the management efficiency of the Xinbei District Planning Bureau in the area surrounding the HSR station necessarily decreases (Table 5.11).

Nominally, the two departments are led by both the municipal planning bureau and the district government. However, we learned from the interview that the Xidong new town Planning Office is mainly led by the municipal planning bureau in the specific work, and its personnel are appointed by the municipal planning bureau. However, the Xinbei District Planning Bureau is basically led not by the municipal planning bureau but by the Xinbei District government, and its management rights and responsibilities of a higher level than those of other district sub-bureaus. With this management relationship, the Wuxi municipal government can produce a clear influence upon the planning and development of the area surrounding the HSR station through the municipal planning bureau. The planning of the area surrounding the

Table 5.11 Comparison of the planning and management bureaus of the East HSR Station of Wuxi and the North HSR Station of Changzhou

Compare content/city	Wuxi	Changzhou
Planning authority	Xidong new town planning office	Xinbei District Planning Branch
Time of establishment	2010	Earlier (no information support), no later than 2003
Management scope	Xidong new town, 125 km^2 within the scope of planning and management	Xinbei District (439 km^2) within the scope of planning and management work
Leadership department	Wuxi City Planning Bureau and Xishan District government jointly lead, actual City Planning Board has more influence	Nominally Changzhou Planning Bureau, actually main Xinbei District government leadership

Changzhou station is mainly influenced by the district government. This situation explains why the Wuxi HSR station business district planning and municipal-level planning are well connected. However, the planning related to Changzhou Xinlong International Business City is not connected to the development orientation of the municipal planning bureau in this region.

(3) **Differences in the settings of the decision-making and coordinating agencies**

The *Xidong new town Construction Leading Group* and the *HSR Business District Construction Leading Group* established by Wuxi are cross-level temporary agencies that play a role in coordinating the decisions of departments at various levels. The *Xidong new town Construction Leading Group* is mainly responsible for coordinating planning and development by the municipal government, district government and all committees, offices and bureaus at the municipal level within the scope of Xidong new town. The *HSR Business District Construction Leading Group* is mainly responsible for coordinating planning and development by the district government, the HSR Business District Management Committee, the Xishan Economic and Technological Development Zone Management Committee and related subdistrict offices. The leading group is led by the main leaders at a higher level and consists of the main leaders of related departments at a lower level. Its main role is embodied in two aspects: first, it can guarantee the effective implementation of decisions made at a higher level. Second, it can promote the coordination of the work of all the related departments. According to the interview, as a temporary agency, the leading group holds regular meetings to promote the work implementation of various departments and plays an important role in the specific work. For example, the HSR Business District Construction Leading Group coordinates problems related to conflict within the business district over projects within the east–west industrial sector and creates specific project categories to avoid homogeneous competition, and the Xidong new town Construction Leading Group coordinates various projects within Xidong new town at the city level to promote coordinated development at the city level. In contrast, there are the obvious disruptions between the municipal level and the district level in the planning and implementation process, due to the lack of such a coordinating agency in Changzhou, that influence the development speed of the area surrounding the station.

5.5.2 Comparison of the Planning and Implementation Process

The organization of the planning and implementation of the areas surrounding the two stations—the East HSR Station of Wuxi and the North HSR Station of Changzhou—it clearly has significant differences in two aspects—the planning process and the interaction between planning and implementation.

(1) **Comparison of the planning process**

Upon comparison of the planning process of the areas surrounding the Wuxi and Changzhou stations, the planning process of the area surrounding the East HSR Station of Wuxi shows the following two advantages:

(1) The planning scope is more stable. The scope of the HSR new town differs according to the station level and development vision, but the stable planning boundary realizes implementation consistency. Compared with the plan for the area surrounding the North HSR Station of Changzhou, the plan for the area surrounding the East HSR Station of Wuxi is based on a more stable spatial range. After the strategic research in the first stage, the Wuxi HSR new town planning defines a clear planning boundary and layer structure and follows that range in the subsequent planning stages. Regarding the related planning of the area surrounding the North HSR Station of Changzhou, the municipal government defines a spatial range of 0.7–4.5–9.0 km^2 in the planning stage, but that range is not adopted in the plan dominated by the district government. Then, the municipal government changes it to two ranges—4 and 1.6 km^2—being the core area zone stable until 2011. At the same time, the East HSR Station of Wuxi enters the key regional planning stage, while the planning schedule of the North HSR Station of Changzhou lags seriously.
(2) The logic of the planning organization is clearer. The plan for the area surrounding the East HSR Station of Wuxi follows the formulation sequence *strategic research—comprehensive planning—in-depth planning of key area—dynamic planning adjustment*, promoting a gradual and orderly planning formulation and implementation process. The planning formulation at various stages corresponds to certain actual development problems and can gradually promote problem-solving processes (Fig. 5.17). For the planning process of the Changzhou station, although the planning research is carried out at an early stage, the related

research conclusions are not reflected in the subsequent plan because the planning achievements are entangled in the functions, the development intensity is adjusted repeatedly, and the planning depth and width are restricted.

The clear and stable spatial range is the basis for the effective coordination of the planning formulation, while the definite planning organizational logic is a guarantee that the plan will be deepened continuously. With these two advantages in the planning process of the area surrounding the East HSR Station of Wuxi, the planning achievements have a variety of types and more systematic organization.

(2) **Planning and implementation process interaction**

Since the implementation schedule of the area surrounding the North HSR Station of Changzhou is relatively lagging, the planning and implementation interaction process require long-term observation. However, a comparison of the current situations indicates that the planning of the area surrounding the North HSR Station of Changzhou is forced to be adjusted after the implementation process encounters problems, but the interaction of the planning and implementation of the area surrounding the East HSR Station of Wuxi is more effective (Fig. 5.18). Its characteristics are mainly embodied in two aspects.

(1) Planning ranks first, namely, the planning priority of the previous stage is the implementation priority of the next stage. In terms of the planning process, the spatial level defined in the strategic research stage directly constitutes the basis for the management main stakeholder and mechanism construction in the construction start period. In the comprehensive planning deepening period, the deepening and certain adjustments are executed in planning the road network, but the road network construction that is implemented simultaneously is the definite and unchanged content from the previous stage, thus effectively avoiding the waste of time or investment brought by the conflict between the early and later planning content. For the development of the area surrounding the North HSR Station of Changzhou, the construction of Liaohe Road is forced to add an underpass due to the large-scale body of water in Xinlong Lake Park, which is a manifestation of a barrier to the interaction of planning and implementation.
(2) The planning formulation is used to organize the construction implementation. At the early stage of planning formulation in the key area, the Wuxi HSR Business District Construction Management Committee entrusts the Wuxi Planning Institute with organizing the planning and development; the same organization is also carried out at the early stage of dynamic planning adjustment. The staged organization of implementation content is beneficial in defining the basis of and barriers to subsequent work and can guarantee the smooth implementation of the subsequent planning content. For the North HSR Station of Changzhou, the residents refuse to move after the announcement of the new plan due to the lack of an organized demolition schedule in the process of the 2011 regulatory plan formulation; thus, the plan is forced to be modified, and it is difficult to realize the original planning intention.

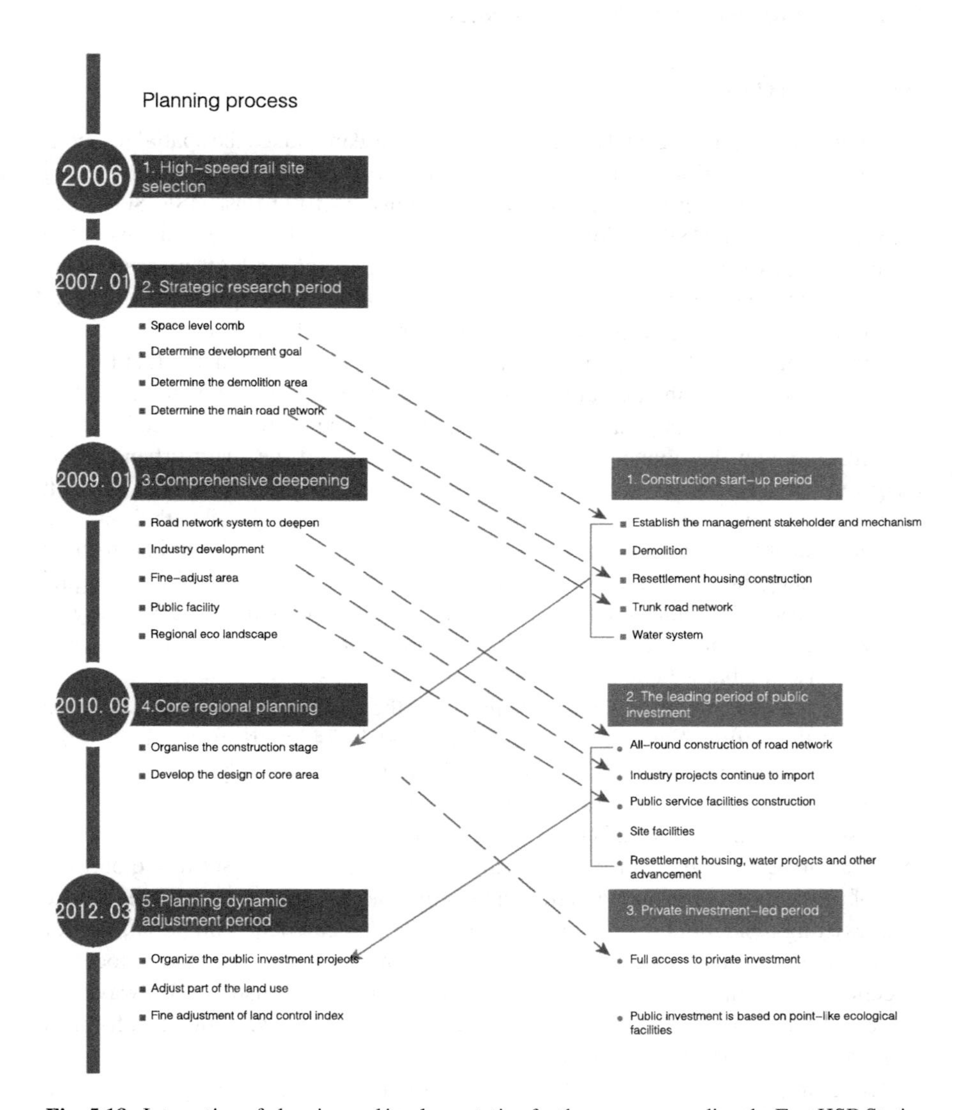

Fig. 5.18 Interaction of planning and implementation for the area surrounding the East HSR Station of Wuxi

5.5.3 *Comparison of Key Decisions*

(1) **Station selection**

Based on the comparative analysis framework, the research takes the spatial relationship between the station and the downtown area as one of the control factors during the case selection; therefore, the position relationship of the East HSR Station of Wuxi and the North HSR Station of Changzhou with the built-up areas of the central urban areas is very similar. Both are located at the edge of the built-up area of the central urban area and near the highway at the periphery of the original built-up area.

Though the location selection results are similar, we learned in the interview with the planning leaders who participated in the site selection decision that the site selection processes of the two cities had great differences. For Wuxi, since the line is too far from the urban area in the northern line scheme preferred by the Ministry of Railways, and the connection between the station and the main urban area is inconvenient, the scheme is rejected by the municipal government. After a difficult discussion, the two parties eventually obtain results that basically satisfy their interest demands but are not optimal. The chief engineer of the Changzhou Planning Institute said that Changzhou pays attention to reserving the channel for the Beijing-Shanghai HSR; therefore, the route scheme and station scheme proposed by the Ministry of Railways in the location selection process conform more closely to the development expectations of Changzhou.

In terms of the amount of route demolition, the Wuxi section and Changzhou section of the Beijing-Shanghai HSR are 49.75 and 44.74 km, respectively, and the route demolition amount is 480,000 and 260,000 m^2, respectively. In terms of the development foundation of the areas surrounding the stations, the area surrounding the East HSR Station of Wuxi is mainly the sites of original villages and towns, while the area surrounding the North HSR Station of Changzhou is closer to the built-up area of the central urban area, and the surrounding facilities are more suitable. There is an existing residential district east of the station that can help gain popularity for the development of the area surrounding the station. In terms of the station location selection, Changzhou reaches a smooth consensus with the Ministry of Railways due to the early reservation of development space, and the location selection results meet both parties' optimal development demand.

Therefore, the Changzhou HSR station location selection basically conforms to the urban development expectations of the city. Though it is difficult for the Wuxi municipal government to negotiate with the Ministry of Railways about the station location selection, the final location selection results are acceptable. In contrast, since the municipal governments of some cities on the route of the Beijing-Shanghai HSR, such as Suzhou, Cangzhou and Qufu, have limited power to negotiate with the Ministry of Railways about the selection of the station site. The HSR station can only be a transportation junction, and the traffic station will be built to realize the basic connection of the station with the urban area. Therefore, in terms of station location selection, the right of speech of local governments, especially those of cities with a small scale and low economic strength, should be improved. Meanwhile, a comprehensive assessment should be made of the costs and benefits of station location selection, and the short-term costs of demolition and construction should be considered and be integrated into the long-term benefits formed in the interaction between station and city, thus sharing the costs and benefits related to the location selection of HSR construction between the relevant departments and local governments and creating a foundation for the benign development of interaction between the area surrounding the station and the city.

(2) **Development orientation**

In terms of the planning condition of development orientation, the East HSR Station of Wuxi and the North HSR Station of Changzhou are very similar in two important aspects—the position relationship between the station and the main urban area and the urban economic background. The main difference lies in the setting of the decision-maker and the coordination of the relationship of the decision-makers at the municipal level and the district level.

The analysis of the planning process of the development orientation of the area surrounding the station indicates that the development orientation of the area surrounding the East HSR Station of Wuxi is based on the spatial level, and the development orientation of the various layers is definite. Over two years of research, the development orientation is not basically changed after being determined in 2009 but is further defined according to the spatial range. In contrast, the research support for the development orientation of the area surrounding the North HSR Station of Changzhou in the relevant planning is relatively poor. Therefore, adjustments of the development orientation are made repeatedly. The adjustment reasons lie in the differences in the development ideas at the municipal and district levels and the influence of national policy (it is difficult to realize the orientation of the Xinbei District Administrative Center in this region due to the suspension of the construction of the building, so it is forced to be canceled).

In terms of the specific content of the terminal decision, the development orientation of the area surrounding the East HSR Station of Wuxi is based on a spatial level. The first layer, namely, the surrounding area near the station, covers an area of approximately 0.9 km^2, is oriented around the "HSR property sector," and has high-end business offices as its core function. The second layer, namely, the area near the first layer with two Metro Line 2 stations as its core, covers an area of 1.54 km^2 and is oriented around the "waterfront creative sector" and "portal facilities sector." In terms of function, it not only has administrative management and hotel services matching the high-end business offices of the first layer but also is equipped with business, leisure and cultural facilities providing services for the surrounding residents in combination with the landscape node. The third layer, namely, the range of land used for urban construction within the HSR business district, covers an area of approximately 30 km^2, is oriented around the urban functional area of Xidong new town, and drives the development of the surrounding area with schools, hospitals, district-level business centers and other facilities. The adjustment of the development orientation of the area surrounding the North HSR Station of Changzhou is greater in the different planning stages: In 2008, in the planning research stage dominated by the municipal government, a range with an area of 0.7 km^2 in the area surrounding the station is oriented around the transportation junction area, with the main functions of traffic collection and distribution and supporting the construction of businesses and parks. In 2009, the area surrounding the station is expanded to approximately 4 km^2 and is oriented around an urban comprehensive area with HSR driving the development, and the functions of the core zone with an area of 1.6 km^2 are administrative and high-end business offices. In 2011, the HSR core zone with an area of 1.6 km^2 is oriented around the business finance gathering area, and its core functions are administrative offices, financial offices and enterprise headquarters offices. In the actual plan implementation process, the core zone with an area of 1.6 km^2 is mainly built according to the "high-end business office district" determined in 2011. The surrounding area is mainly the range of "Xinlong International Business City" with an area of 24 km^2 set in 2013, and its main functions are residences and education. For various sectors that have not entered the substantial implementation stage, the development orientation is temporarily not defined.

Upon comparison, there are mainly the following two aspects of difference.

(1) The Wuxi municipal government and district government require a high level of consistency of the development orientation of the area surrounding the station, and the development orientation and functional partition are stable. In terms of the development orientation of the area surrounding the North HSR Station of Changzhou, the differences between the decisions of the municipal government and the district government are significant, and the development orientation and partition range are adjusted frequently (Table 5.12).
(2) The content of the development orientation of the area surrounding the East HSR Station of Wuxi is detailed. It not only includes the main functions but also involves the development and operation strategy for realizing the functions within various spatial ranges. Due to the frequent changes in development orientation, the plan for the area surrounding the North HSR Station of Changzhou has defects in depth and comprehensiveness.

A comparison of the decision implementation results of the two areas indicates that the consistency and stability of the development orientation are of great significance for the development of the area surrounding the HSR station. *Consistency* mainly refers to the unification and coordination of the related supporting policies of the municipal government and district government in terms of the development orientation of the area surrounding the HSR station. *Stability* mainly means that the development orientation should not change frequently in the different stages of planning and implementation. The Wuxi Municipal Party Committee and municipal government issued Several Opinions about Accelerating the Development and Construction of Xidong new town in January 2009. The opinions not only define the development orientation of Xidong new town but also determine the central area in principle, namely, the scope, orientation and development goals of the HSR business district, and identify the core zone producing the HSR influence in the area surrounding the station. In March 2009, the Xishan District Party Committee and district government issued the Implementation Opinions about Accelerating the Development

Table 5.12 Comparison of the development positioning of the areas surrounding the East HSR Station of Wuxi and the North HSR Station of Changzhou

Development orientation comparison				
Similarities/differences			Wuxi	Changzhou
Comparison of planning conditions	The overall similarities		1. The municipal and district governments are the main decision-makers for the development and positioning of the areas surrounding the site	
			2. The macro development background, such as city size and economic strength, is similar	
			3. The site is located at the edge of the built-up area	
			4. The original function around the site is mainly agricultural or rural land	
	Significant differences	1. Direct management agencies	HSR Business District Management Committee	New Dragon International Business City Construction Headquarters
		2. Planning authority	Xidong new town Planning Office mainly affected by the intent of the Municipal Planning Bureau	Xinbei District Planning Branch is mainly affected by the intention of the government of New North District
		3. City, district government relations	Urban government coordination smooth, cooperation between the municipal government and district government	District government decision-making more dominant, municipal government may be unable to implement decisions

(continued)

Table 5.12 (continued)

Development orientation comparison			
Planning process	Specific planning points in each period	1. In 2009, the scope of the peripheral area delineated, with overall development orientation as "regional transportation hub transfer center, high-end service industry gathering area, Wuxi city center"	1. In 2008, the municipal government designated 0.7 km^2, located as a transport hub, mainly to traffic distribution functions, supporting the construction of commercial parks
		2. In 2010, fine-tuning of the core area as "HSR property plate," "waterfront creative plate" and "gateway facilities plate." The corresponding functional content is refined, and the development positioning in the subsequent plan is stable	2. In 2009, approximately 4 km^2 of the site were redrawn and positioned as "the city comprehensive district driven by HSR." The core area of 1.6 km^2 is "administrative and high-end business office functions"
			3. In 2011, the core area of 1.6 km^2 was adjusted to "Xinbei Administrative Center, a new business center with a gathering of international functions"

(continued)

Table 5.12 (continued)

Development orientation comparison			
			4. In 2014, the outer area was adjusted to 24 km^2, and the "Xinbei Administrative Center" in the core area was canceled
Planning content comparison	Similarities	Site outside the "first layer" development orientation is "high-end business office area"	
	Differences	"The first layer" positioning based on the scope of the service area	"The first layer" positioning based on competition within the city
		"The second and third layers" have a clear development orientation. The second layer is urban integrated services driven by rail transit and landscape nodes. The third layer is a comprehensive residential area driven by regional public facilities	"The second and the third layers" lack development orientation
Planning results	Differences	The "first layer" is supported at the municipal level and has been growing steadily	The "first layer" positioning conflicts with the municipal "east longitude 120," and development is hindered
		The rapid development of "the second and the third layers" provides a foundation for the development of "the first layer"	"The second and third layers" have been slow to develop; their interaction with the "first layer" is weak

and Construction of the HSR Business District based on Several Opinions about Accelerating the Development and Construction of Xidong new town and provided detailed explanations of the orientation, development goals, management systems, supporting policies and other content of the HSR business district development. The development orientation of the HSR business district in these two documents is highly consistent, and the core content is *regional transportation junction transfer center, gathering area of high-end service industry and subcenter in eastern Wuxi*. Then, the Urban Design of the HSR Core Zone completed by RTKL in 2010 deepens the core zone function of the HSR station on the basis of following the previously established development orientation of this area. The subsequent planning formulation and implementation process also follow the previously established development orientation of this area. In contrast, the development orientation of the North HSR Station of Changzhou is not unified between the municipal government and district government. The municipal government's dominant planning places more emphasis on the *pure transportation junction* orientation, while the district government's dominant planning emphasizes the *high-end financial business district* orientation, which leads directly to the repeated repositioning of *Xinlong International Business City* and *Tokyo 120 Financial Business Central District*, difficulty in investment promotion and slow construction. Then, the district-level orientation of the HSR core zone changes frequently due to the changes in development conditions and leadership intention. On the one hand, it leads to the cancellation of the district-level administrative center and the withdrawal of enterprises with original investment intention. On the other hand, the road that has been built is forced to build an underpass due to the construction of the *ecological center* characterized by the large-scale body of water, and the BRT station is canceled after completion, leading to serious construction waste. The change of orientation under internal and external influences causes the repetition and dislocation of implementation.

For HSR stations built at the edge of a built-up area, such as the East HSR Station of Wuxi and the North HSR Station of Changzhou, their development should interact and be coordinated with the built-up area and has many possible development spaces. Therefore, the development orientation of the surrounding area requires all management levels to be coordinated and remain relatively stable while managing the overall development of resource input, thus effectively organizing the subsequent planning formulation and implementation process.

(3) **Spatial arrangement**

Against the background of spatial arrangement planning, the location and function of the areas surrounding the two stations are similar, and Xinbei District, where the North HSR Station of Changzhou is located, has more prominent economic strength. The main difference lies in the direct management stakeholder. Compared with the Changzhou Xinlong International Business Construction Headquarters, the Wuxi HSR Business District Construction Management Committee has a more definite scope of responsibility, more professional planning technology support departments and a more independent administrative position.

The spatial arrangement mainly has the following differences.

(1) Spatial level. The three spatial levels of *Xidong new town HSR Business District (Xidong new town middle area)—HSR Business District Core Zone* are determined for the area surrounding the East HSR Station of Wuxi in the initial period of construction. The range of the core zone is slightly adjusted in 2009, but the overall spatial level does not change. In contrast, the range division of the area surrounding the North HSR Station of Changzhou changes from *Xinlong cluster—HSR core zone (0.7 km^2)* in the Study on Three Stations prepared by the municipal planning institute in 2008 to *Northern New town—HSR Complex (4.0 km^2)* in the Conceptual Urban Design of the Northern New town prepared by the Xinbei District government in 2009, to 1.6 km^2 proposed in the *Xinlong International Business City* orientation in 2011, and finally to *Xinlong International Business City (24 km^2)—Business City Core Zone (1.6 km^2 in the area surrounding the station)* in 2013. The core significance of defining the spatial level lies in defining the spatial framework and mutual relationship for the subsequent planning formulation and implementation (Table 5.13).
(2) Land use mode. After the comprehensive planning deepening period in 2009–2010, the land use mode on a large scale for the periphery and a small scale for the core is formed in the area surrounding the East HSR Station of Wuxi. The peripheral land area plots are 2.5.4 ha, while the land of the core zone is divided into small plots with an area of 0.6 ha. The planning of the peripheral area of the North HSR Station of Changzhou (not the built-up area) is not covered, and the land area of the core zone is approximately 1.3 ha.
(3) Design control content. The East HSR Station of Wuxi refines the design control content of the core area constantly through the urban design of three levels, *overall urban design—urban design of core zone—important neighboring urban design*, and has detailed guidance for the street interface, building form and other content, thus providing good spatial quality. The content of the planning adjustment of the North HSR Station of Changzhou is mainly *floor area ratio* and *height control*, and its binding effect upon the spatial quality is obviously inferior.
(4) Peripheral area planning. The peripheral area planning of the East HSR Station of Wuxi precedes the core area planning, and the various planning achievements comprehensively cover public facilities arrangement, landscape system organization and other content. However, the planning of the North HSR Station of Changzhou concentrates mainly on the core area, and the peripheral area planning is temporarily limited to a rough functional orientation.

From the specific planning content of the spatial arrangement, the areas surrounding the East HSR Station of Wuxi and the North HSR Station of Changzhou basically follow the “three layers structure” model (Tables 5.14 and 5.15) in terms of function and development intensity. For the East HSR Station of Wuxi, the first layer mainly focuses on high-end business offices, with a land area that occupies more than 35% of the total land area, and the land for mixed business and offices functions occupies 17% of the total land area. The plan considers various functions of passenger flow

Table 5.13 Comparison of the spatial distribution of the areas around the Wuxi East Railway Station and the Changzhou North Station

Space layout			
Similarities/differences		Wuxi	Changzhou
Comparison of planning conditions	The overall similarities	1. The macrolevel development background, such as urban scale and economic strength, is similar 2. The site is located at the edge of the built-up area 3. The original function of the site is mainly agricultural or village land 4. The HSR site construction drives the urbanization of similar demands	
	Significant differences	There is a permanent management stakeholder, "HSR business district management committee," and its management scope is clear	Management agencies "Xinlong International Business City Construction Headquarters" as a temporary agency, no clear scope of management
Planning process	The main planning nodes and content	1. In 2008, the spatial structure of the space was determined, and the nature of the land and the strength of development were determined	1. In 2008, a spatial structure of 0.7–4.5 km^2 was formed, and the content of land use and development intensity were clear
		2. From 2008 to 2010, the major public facilities in the peripheral areas were determined; the overall urban design and development in the core area and the spatial layout were refined	2. In 2009, the spatial level was redefined as "1.6–4 km^2," and all the land properties were adjusted

(continued)

Table 5.13 (continued)

Space layout			
Similarities/differences		Wuxi	Changzhou
		3. In 2011, the design of the key neighborhood cities in the core area was deepened, and the spatial layout of the core areas was further optimized	3. In 2011, the development intensity of the core area of 1.6 km^2 was determined, and there was less spatial layout in the peripheral areas; temporarily put forward the building of the site with a large body of water
			4. In 2014, the land area of 1.6 km^2 was adjusted in nature, and the intensity of land development was changed
Planning content comparison	Similarities	The core area of the land features is mainly business offices with a floor area ratio of 5.0	
	Differences	The sphere structure is obvious, with a layer of range stability	The process of repeatedly adjusting the scope of the layer, poor planning stability
		Large land plots (between 2.5 and 4 ha) and core land plots (average 0.7 ha)	No external planning, the core block plots are still more extensive (an average of 1.3 ha)
		The research on design control content is more in-depth, with detailed content of urban design control in the core area	Design control content is relatively infrequent, many planning adjustments focused on the content and development strength

(continued)

Table 5.13 (continued)

Space layout			
Similarities/differences		Wuxi	Changzhou
Planning results	Differences	The spatial layer is clear and stable, providing a synergistic foundation for planning and implementation	Frequent adjustments of the scope and level of the space and difficulties in compiling and implementing the plan
		Peripheral areas, with determined public service facilities as the core point, have effectively attracted private investment and promoted the peripheral space model to expand the urban design results in the core area as the "design control" conditions for land transfer, effectively supplementing the content of the regulatory code and unifying the overall environmental conditions in the core area	A driving force is lacking in the peripheral areas, the basic urban design depth is undeveloped and poor, the land cannot be formed in the process of "design control" conditions, and the intensity of development is also transferred in the process of adjustment

Table 5.14 Function and floor area ratio of area surrounding the East HSR Station of Wuxi

Surrounding areas of the East Wuxi Station				
Scope		Area (km^2)	Functional content	Volume rate
Core area	The first layer	0.85	Business office (0.3 km^2), commercial mixed function (0.14 km^2)	3.5–7.0
	The second layer	1.54	Commercial (0.21 km^2), Commercial Mixed (0.27 km^2), Residential (0.25 km^2), Civic Activities (0.12 km^2)	2.5–6.0
External area	The third layer	45 (construction area of 31 km^2)	Residences, public services, industrial activities	2.0–3.5

Table 5.15 Function and floor area ratio of area surrounding the North HSR Station of Changzhou

Surrounding areas of North Changzhou Station					
Layer/content	Planning documents		Area (km^2)	Function	Volume rate
Core area (first and second layers)	2008	Three stations study	0.7	Business, convention and exhibition, parks	2.0–4.0
	2009	Northern Metro	1.6	Business, administrative offices, business offices	–
	2011	Control regulations1	1.6	Business offices (0.44 km^2), administrative offices (0.15 km^2), parks (0.18 km^2)	5.0–9.0
	2014	Control regulations2	1.6	Business offices (0.52 km^2), commercial (0.16 km^2), parks (0.18 km^2)	4.0–6.5
Peripheral area (third layer)	2008	Three stations study	4.5	Living, technology industry, entertainment	2.0–3.5

collection and distribution of the HSR station, the overall development intensity is the highest, and the floor area ratio can be up to 5.0 and even higher. The proportion of land use for business, business and offices, residences and public services in the second layer is 14%, 17%, 16% and 8% respectively. The floor area ratio of this layer is higher and is only slightly lower than that of the first layer. The function of the third layer is mainly urban area. It is less closely associated with the HSR station and mainly focuses on residential, public services and industrial activities. The floor area ratio of this layer is determined according to the urban standard and is lower than that of the first and second layers. For Changzhou, the first and second layers constitute the core area. Business offices are the main function, with land that occupies 32% of the total land area; business land occupies 10% of the total land area, and parkland occupies 11% of the total land area. The floor area ratio of these two layers is the highest at 4.0–6.5. The function of the peripheral area (the third layer) is mainly decided by the urban functional zone and is mainly residential, entertainment and scientific and technological industries, with a floor area ratio of 2.0–3.5.

The TOD pattern of the HSR station is similar to that of the common public transport station. The HSR has a stronger external radiation ability, can reach the larger urban area within a short time, and brings more passenger flow volume, thus causing the larger development scale in the area surrounding the HSR station. Meanwhile, there may be more diversified and separate facilities around the HSR station than in the TOD development pattern of the common public transport station HSR. With the East HSR Station of Wuxi as an example, the spatial structure of the metro is adjusted from the typical layer structure to the *point-axis + layer* structure for the HSR station and metro station due to the influence of the station.

According to the implementation results of the spatial arrangement, the development in the peripheral area of the East HSR Station of Wuxi is good and effectively supports the construction of the core area. In contrast, the development in the peripheral area of the North HSR Station of Changzhou is slow, and the development control content of the core area is adjusted many times.

The above comparison shows that a clear spatial level is of positive significance for the planning formulation and implementation. The development of the peripheral layer around the HSR station has a significant influence upon the development of the whole HSR area. The clear spatial level facilitates the planning research and formulation for the different spaces. In the East HSR Station of Wuxi area, the industrial development planning, comprehensive traffic planning, green space system planning and municipal infrastructure planning are formulated successively with the *HSR business district* as the object. The urban design, underground ring road planning and urban deepening design of the block streets of the core area are formulated with the "core area" as the object. This series of planning determines much of the space development content. For the North HSR Station of Changzhou, the urban design content of the area surrounding the station goes through three stages—the 2008 *Study on Three Stations*, the 2009 *Conceptual Urban Design of Northern New town* and the 2011 international bidding of the urban design of the area surrounding the HSR station. The work achievements of the three stages have great differences in coverage area, and the continuity of the early and later achievements is weak, so the depth and width of the achievements are limited.

The separate planning research of the different spatial levels differs in control guidance content and priority, and the clear spatial level division can help effectively unify the work object of the subsequent planning, providing a foundation for the overall coordination of the subsequent planning. The division of the spatial level is beneficial to promote the plan implementation and establish a clear spatial development organization. For the East HSR Station of Wuxi, after the spatial level division of the area surrounding the HSR station, the *Xidong new town Construction Leading Group*, *HSR Business District Construction Leading Group* and *HSR Business District Management Committee* are established, and the rights and responsibilities of these groups in relation to the original *Xishan Economic and Technological Development Zone Management Committee* and all subdistrict offices are defined. The construction leading group as the temporary agency will implement the decision intentions and coordinate the work of all the relevant departments. The Xidong new town Construction Leading Group and HSR Business District Construction Leading Group will enable communication and coordination between the municipal government and district government and among functional departments within the district government. As a permanent organization, the management committee is responsible for organizing the planning formulation, construction planning review, project investment and other specific work. The construction leading group provides guidance, and the management mechanism of the management committee promotes the implementation of the core planning content at various levels. For example, in terms of development orientation and function organization, the Xidong new town Construction Leading Group effectively coordinates the industrial development orientation of the *HSR business district* and the *east–west industrial sector* and determines that the *HSR business district* will form a dislocation development with the *east–west industrial sector* with a core urban function. For the North HSR Station of Changzhou, the development of the area surrounding the HSR station lacks the obvious spatial level. The area of *Xinlong International Business City* is only 1.6 km^2 in 2009, and the temporary agency *Xinlong International Business City Construction Headquarters* consists of all department leaders but has no specific management theme, so it is difficult to execute the actual functions. In 2013, the area of *Xinlong International Business City* is adjusted to 24 km^2, and the permanent *management office* is set up under the *headquarters*. The relevant personnel of the management office said, "Since there is no document defining the spatial range, the headquarters is still a temporary agency and has no legal qualification, and the boundary of responsibilities of the management office is unclear; it is difficult to carry out the work." After determination of the spatial arrangement, the establishment of a spatial development management agency will be beneficial to the development within the specific boundary range.

The *third layer* (or peripheral layer) of the area surrounding the HSR station is an important factor in promoting the connection of the area surrounding the station with the urban function and supporting the development of the core layer. The "three layers structure" theory of development of the area surrounding a HSR station is a mature theory. The development orientation of the area surrounding the HSR station should not only pay attention to the influence of the HSR station but also investigate the relationship between the station and overall urban development from a macroscopic perspective. For Wuxi and Changzhou, the Beijing-Shanghai HSR stations are new stations located at the edges of built-up areas in the central urban areas. The development of the areas surrounding such stations differs substantially from that of the internal rebuilt stations in the built-up areas. The original function of the area surrounding the station is mainly agricultural production and rural settlement, and public service facilities are lacking. To realize the development orientation of the first and second layers in the "three layers structure," support of the urban function should be provided within a macroscopic range. Therefore, though the development of the "third layer" receives a low level of radiation from the HSR station, it is a necessary condition for the benign development of the "first and second layers" and an important medium for the fusion of the area surrounding the HSR station with the original urban function. In the development process of the area surrounding the East HSR Station of Wuxi, the relocation and rehousing developments, schools, hospitals and other public facilities construction as well as the Auto-Park industrial park, Country Garden, housing development of Ever Grande, Long for Properties and other companies are located within the third layer. As of the end of 2013, the investment attracted within the range of the third layer is far more than the total private investment attracted in the first and second layers. For the development of the area surrounding the East HSR Station of Wuxi, the HSR station actually serves as the catalyst that first stimulates the development of the third layer, realizing the transformation of this area from a nonurban function to an urban function and producing the spatial requirement for the businesses, offices, leisure and other urban activities within the "first and second layers" and finally realizing the high use value of the businesses and offices in the internal layers. For the North HSR Station of Changzhou, on the one hand, its core area plan abandons the orientation of a pure transportation junction in the *Study on Three Stations* and emphasizes the planning of a high-end business office function according to the "three layers structure." On the other hand, the construction of the HSR station does not stimulate the new development of the peripheral layer and does not play a supporting role in the planning function of the core area. In addition, the development of area surrounding the North HSR Station of Changzhou is expected to take the relocation of the Xinbei District government and the construction of Xinlong Lake Park as catalysts that will directly drive the development of the "first and second layers (core area)" of the area surrounding the station in combination with the HSR station, thus attracting the original usage requirements in the urban area. However, since this orientation fails to obtain the support of the municipal government, the "municipal-district" and "district-district" homogeneous competition in the subsequent development is serious, which directly leads to the lagging development of the area surrounding the station.

(4) **Implementation sequence**

The planning formulation is the basis of the plan implementation. To define the construction sequence of the areas surrounding the two stations, the research categorizes the planning achievements formed in various stages and the corresponding planning ranges (Tables 5.16 and 5.17). The comparison shows that the related planning formulation of the East HSR Station of the Wuxi HSR business district goes through 4 stages and obtains 16 planning achievements from 2007 to 2014. The spatial range covers the *Xidong new town*, *HSR Business District* and *core area*, and the related planning with the business district as the planning range completely covers the *core area*, so there are 14 planning achievements involving the *core area* range. In this period, the planning of the area surrounding the North HSR Station of Changzhou goes through 2 stages and has 5 planning achievements, and the main spatial range is mainly the *core area* surrounding the HSR station. For Wuxi and Changzhou, the greatest difference in the plan of the area surrounding the station is that the special plan of the core area (first and second layers) of the East HSR Station of Wuxi follows the related plan for the *HSR business district*, and the core area plan of the North HSR Station of Changzhou precedes the peripheral area plan. Therefore, the HSR new town in Wuxi focuses on the core area after the overall arrangement, whereas Changzhou prioritizes the planning design of the core area.

On this basis, the research compares the implementation sequence for the area surrounding the two stations. In terms of the decision background, the two stations are similar in terms of the urban area development background and development demand, and the planners of both hope to drive the urbanization of the surrounding area through the station construction. The difference in the decision background includes the direct management and spatial level division in the early stage of planning. The difference in the management is that the Wuxi HSR Business District Construction Management Committee as the direct manager of the area surrounding the East HSR Station of Wuxi has clear and centralized administrative power, while the Changzhou Xinlong International Business City Construction Headquarters has no direct administrative power, and the related administrative power is decentralized to various district departments. In terms of the spatial level division in the early stage of planning, Wuxi forms a definite and stable spatial range and level division, while the spatial range and level of the North HSR Station of Changzhou change repeatedly and do not become stable until 2013.

In terms of the specific content of the plan implementation sequence, the basic logic of the development of the area surrounding the two stations prioritizes public facilities and then promotes private investment. However, the two have certain differences in the specific construction project and spatial range. The public investment project content of the area surrounding the East HSR Station of Wuxi is richer, including the road network, municipal facilities, water system engineering and other infrastructure facilities as well as schools, hospitals, civic centers and other public projects; in contrast, there is only infrastructure in the area surrounding the North HSR Station of Changzhou. In terms of the spatial sequence of the plan implementation, the sequence for the area surrounding the East HSR Station of Wuxi is *peripheral area first, followed by the core area supported by the peripheral area*; almost all private investment projects are distributed in the peripheral area in the second implementation stage (2010–2012), and the project investment and land transfer of the core area are carried out in the second half of 2012. For the North HSR Station

Table 5.16 Main completion time and scope of planning for the area surrounding the East HSR Station of Wuxi

Number	Stage	Completion time	Plan name	Plan range
1	Stage 1 (2007–2008)	2007	Planning and Research of Xidong new town District	Business area
2		2008	Wuxi HSR station control detailed planning (2008.09)	Business area
3	Stage 2 (2009–2010)	2009.10	HSR Business District Industrial Development Plan	Business area
4		2010	Wuxi City, "12th Five-Year Plan" urban and rural construction plan	Xidong Metro
5		2010.7	HSR business district integrated transport plan	Business area
6		2010	HSR business district municipal engineering special plan	Business area
7		2010	Business district road landscape plan	Business area
8		2010.7	HSR business district core area urban design	Core area
9		2009	Jiuli River landscape plan	Business area
10		2010	Greenpingshan Forest Park Concept Plan	Business area
11		2010	HSR business district core area underground transport plan	Core area
12		2010.7	Wuxi HSR Station Control Detailed Plan (2010.07)	Business area
13	Stage 3 (2011–2012)	2011.2	Xidong new town Business District Development and Construction Plan	Xidong Metro
14		2011.4	Neighborhoods urban design of Core District 1, 2	Core area
15		Beginning of 2012	Neighborhood urban design of Core District 4	Core area
16	Stage 4 (2013–)	2013–2014	Wuxi HSR Business District Control Detailed Plan (2013.07–2014.06)	Business area

Table 5.17 Main completion time and scope of planning for the area surrounding the North HSR Station of Changzhou

Number	Stage	Completion time	Plan name	Plan range
1	Stage 1	2008.5	*"Three Stations" special study, including HSR Changzhou Station Area Concept Plan and Key Areas Urban Design*	Core area
2	Stage 2	2009	*Conceptual Urban Design for Changzhou Northern New town*	Northern metro
3		2011	*Urban Design of New Changzhou Core District North of Changzhou*	Core area
4		2011	*Changzhou North Metro Core Area Control Detailed Plan (2011 Edition)*	Core area
5		2014	*Changzhou North Metro Core Area Control Detailed Plan (2014 Edition)*	Core area

of Changzhou, the road construction and water park construction concentrate on the 1.6 km^2 of core area.

According to the implementation results, the areas surrounding the East HSR Station of Wuxi are better developed, and public investment drives the input of private projects effectively. The peripheral area attracts the residential projects of Country Garden, Long for Properties, Ever Grande and other large groups beginning in 2011. The core area attracts many business building projects such as the Huaxia Business Building, Hongdou Business Building and Tianyu Building. As of 2013, the commercial residential property and business building project areas under construction in the HSR business district exceed 1.5 million and 850,000 m^2, respectively. Of the three goals of *becoming famous within one year, taking shape within three years, and building into a city within ten years* set for the Wuxi HSR business district, the first two goals are basically achieved, and the third goal is proceeding in an orderly manner. For the area surrounding the North HSR Station of Changzhou, after the disinvestment of several intentional investment projects in 2013, the first transfer of 5 land plots and 8 ha of business land is completed in 2014, and the construction goals of *starting business project construction in 2012, starting science and technology research and development function construction in 2013 and completing the core area construction in 2015* have not been achieved (Table 5.18).

The comparison shows that the planning sequence is the basis for the implementation sequence, and the high-performance implementation sequence should be

Table 5.18 Comparison of the plan implementation of the area surrounding the East HSR Station of Wuxi and the North HSR Station of Changzhou

Timing comparison of planning and implementation			
Similarities/differences		Wuxi	Changzhou
Comparison of decision-making background	The overall similarities	1. They are all nonurbanized areas that are significantly affected by urban areas 2. The use of HSR site construction, driving the urbanization of similar demands	
	Significant differences	Planning has formed a clear spatial level	The planning process in the spatial hierarchy undergoes many adjustments, and the scope of the periphery is not clear
		The management stakeholder is clear and stable, and the relationships among the various decision-makers are clear	The main stakeholder of management is not clear and is mainly district-level leadership
Comparison of decision-making content	Similarities	Both are government-funded public projects	
	Differences	First build peripheral areas, and then develop the core area	Focus on the core area
		First nurture industry and living, then develop commercial and business functions	Development focuses on business
Decision result	Differences	Public investment has effectively boosted private sector investment	There is a lack of private investment in the region, resulting in a situation of concentrated divestment midway
		The peripheral areas continued to develop, the core areas were steadily advancing and the construction targets for all phases were basically completed	The development of the peripheral area is relatively small, and the promotion of the core area is slow. All stages of the targets are lagging

supported by a reasonable planning sequence. For the current development of the areas surrounding HSR stations in China, advance public investment is the main way to drive private investment, and public investment should be coordinated in terms of spatial range and project type. The core of public investment is to form an attraction at various spatial levels in the area surrounding the station. In addition, the priority in improving the use value of the *peripheral layer* is the current common point of more successful development in the areas surrounding several stations.

5.6 Case Analysis Conclusion

In sum, based on the comparison of the planning and implementation of the areas surrounding the 22 stations of the Beijing-Shanghai HSR, the selected cases and on-site investigations and in-depth interviews, this study compared and analyzed the planning and implementation processes of the areas surrounding the East HSR Station of Wuxi and the North HSR Station of Changzhou. The following suggestions are made regarding the four key issues concerning the planning and development of the areas surrounding the stations.

(1) Site selection. First, coordination between local interests and the interests of the railway construction sector should be fully considered. The location of HSR stations cannot emphasize the reduction of construction costs and ignore local development demands. Local governments should actively communicate with HSR construction departments and strive to meet the site selection requirements of urban development by sharing site development rights and construction costs. Second, the government at the city and district levels should participate in the site selection process. During the site selection period, all interested entities should participate in the demonstration of important content, such as the development and positioning of the site area, and reach a consensus on development ideas for the areas surrounding the station in order to build a foundation for future development.

(2) Development orientation. The development orientation of the area surrounding the station must ensure *consistency* and *stability*. *Consistency* means that the decision-makers at all levels should be unified in their assumptions and judgments regarding the development orientation of the areas surrounding HSR stations. For example, regional-level positioning shall not conflict with municipal-level positioning. *Stability* refers to the basis for thoroughly researching the issues and the judgment that a development position, once made, should not be arbitrarily changed. *Consistency* can ensure that the development resources invested by decision-makers at all levels within this scope can be effectively coordinated, and *stability* can ensure that the investment and construction before and after planning will not conflict with each other and cause repeated investment.

(3) Spatial arrangement. First, the spatial level should be clear, forming an important basis for follow-up planning and plan implementation. The corresponding planning and development stakeholders can be established accordingly, which is conducive to the continuous refinement of the planning content and the effective organization of planning and implementation. Second, in the arrangement of functions, the emphasis should be on mutual support between the development power and each layer. For example, the core layer of the East HSR Station of Wuxi is driven by HSR stations and subway stations, and the peripheral layers are driven by regional high schools, hospitals, and community centers. The business functions of the core layer and the peripheral layer serve each other.

(4) Implementation timing. According to the development foundation and development path, a reasonable implementation sequence should be established. For cities such as Wuxi and Changzhou, which are located on the edges of built-up areas, the sequence of *first overall core, first public and then private* is based on scientific timing. First, the overall emphasis on the *peripheral layer* concerns the *peripheral layer* as the key to the integration of the areas surrounding the site and the original urban areas and also as an important basis for the development of the *core zone*. First, the benign development of the areas surrounding the station should be completed with the *stitching* of the original city functions, and second, the service needs of the *core area* should be met. In the public and private sectors, public investment is the main driving force behind private investment. Mature public investment is conducive to attracting high-quality private investment. However, under the premise of immature public investment, the public investment is reduced. The lower entry threshold will cause harm to the overall development quality of the region.

At the same time, the study found that the realization of a superior interaction between the planning and implementation of the areas surrounding HSR stations requires the following key steps.

(1) A progression was based on certain spatial levels. A clear spatial level provides a framework and foundation for the orderly advancement of planning and implementation in the area surrounding the station. By relying on a fixed spatial level and organizing the planning and implementation of each period, it is possible to match the planning priorities with the implementation content.

(2) Regular integration of planning results and implementation. The regular integration of planning results can ensure the consistency of implementation and reduce the repetition caused by unclear planning conditions in the implementation process. The dynamic combination of the implementation situation can effectively feed back into the planning process, clarifying the current problems and problems existing in the previous stage of planning. Additionally, the focuses for the next phase should be noted.

(3) Set up special management agencies for the area around the site. The key point of the interaction between planning and implementation lies in the unity and coordination of the time sequence. The greatest challenge is the disconnection between the decisions of various departments. By arranging the management

authority of the area surrounding the station separately through the administrative departments of the Bureau of Commerce, Construction Bureau, and Finance Bureau, it is possible to form a management organization that specializes in managing this specific area (such as the *Wuxi HSR Business District Management Committee*). In this way, the scope of management can be strengthened, management objects can be clarified, decision-making can become faster, communication efficiency can be improved, and effective interaction between planning and implementation can be fully promoted.

References

Lauria M (1997) Reconstructing urban regime theory: regulating urban politics in a global economy. Sage Publications

Wang L, Liu G (2007) The evolution of public-private partnership: American downtown redevelopment in the second half of the twentieth century. Urban Plann Int 22(4):21–26

Zhao Y (2014) Land finance in China: history, logic and choice. Urban Dev Stud 21(01):1–13

Zhou F (2009) The study on the administrative subject qualification of administrative organization of development zone. Human Normal University

Chapter 6
Planning Analysis of the Surrounding Area of HSR Station

The analysis of the factors influencing the HSR functional layout is conducted on the basis of the planning compilation of the area surrounding the HSR or the new town around the HSR at the macro-, meso- and micro-spatial levels in combination with the urban development stage and the existing conditions of the adjacent and surrounding areas, which will be beneficial in defining the possible passenger sources for the HSR station and promoting industrial development and the required service functions. This chapter discusses the planning compilation research implemented at different spatial levels and the associated promotional approaches and analysis methods, which are explained through a case study. This chapter shows that specifically designed research can provide a basis for the spatial planning of the new town around a HSR station, with Wuhu City as the example.

6.1 Analysis Content and Methodology at Macrolevel

6.1.1 Analysis of Regional Industry

The research priority at the macrolevel is the possible opportunities brought to the city where the station is located after the space-time distance between cities are shortened by the HSR, including industry transfer, investment possibility and tourism. Therefore, the planning compilation research analyzes the urban areas that the HSR can reach within 1, 2 and 3 h. By analyzing the urban industries and resources included in such coverage, we can judge the complementarity and competitiveness of those facilities with the city where the station is located.

L. Wang and H. Gu, *Studies on China's High-Speed Rail New Town Planning and Development*, https://doi.org/10.1007/978-981-13-6916-2_6

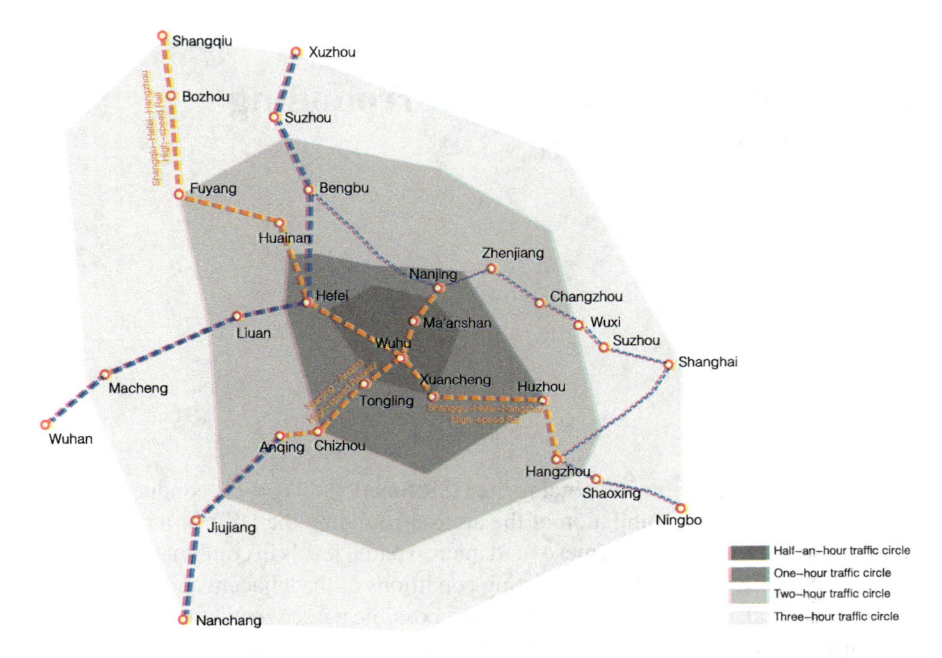

Fig. 6.1 Urban districts accessible in 1, 2 and 3 h from the HSR station of Wuhu

In the joint planning research conducted by the author's research team and the Research Center of Wuhu Urban and Rural Construction Strategy and Planning, the correlation analysis of industry (Fig. 6.1) is conducted for the cities that can be reached by HSR from Wuhu City within 1, 2 and 3 h against the HSR station plan in Wuhu City.

There are many industrial analysis methods, including the industrial structure, location quotient and deviation share models. The analysis mainly defines the comparative advantage of a specific urban industry in relation to other cities. This research uses the industrial structure and location quotient methods and makes a basic comparison of cities that the HSR can reach within 3 h after the completion of the station. In terms of the type of dominant service industry, there is a great similarity between Wuhu and all the main cities of the Yangtze River Delta. The industries with great similarity include technical research and development, finance and insurance, modern logistics, cultural creativity, commercial and trade services, and tourism (Table 6.1). In terms of industrial development power, the development of various industries of Wuhu has a certain gap compared with those in other cities of the eastern Yangtze River Delta through the contrasting analysis of two indicators—employees and fixed assets investment amount. The industries with a certain development potential include logistics, commercial and trade services, and tourism. In the comparison of the economic development of the demonstration area of selected industries in the Wanjiang City belt, the economic aggregate, GDP growth rate and per capita GDP of Wuhu have advantages and rank in the top three. However, Wuhu is far lower than

Table 6.1 Analysis of the leading service industry in major cities in the Yangtze River Delta

Reachability	City	Software	Information service	R&D of science and technology	Financial insurance	Modern logistics	Business services	Cultural innovation	Business services	Tourism	Real estate	Service outsource	Intermediary services	Exhibition
	Wuhu	●		●	●	●	●		●	●		●		
1 h	Nanjing	●	●	●	●	●	●	●	●	●	●			
	Huzhou					●		●	●	●		●		
2 h	Zhenjiang		●	●	●	●		●	●	●				
	Changzhou	●	●	●	●	●		●	●	●	●	●		
	Hangzhou			●	●	●		●	●	●	●		●	
3 h	Wuxi	●		●	●	●	●	●		●		●		
	Suzhou	●	●	●	●	●	●		●	●	●	●		
	Shaoxing			●	●	●		●	●	●	●	●	●	●
	Ningbo		●	●	●	●		●	●	●			●	●
	Shanghai		●		●	●		●	●	●	●			●

Hefei in terms of fiscal revenue and fixed assets investment. The industrial structure of various cities in the Wanjiang City belt is similar, with the second industry as the dominant industry. Except in Liu'An and Chizhou, where the proportion of the primary industry is higher and the proportion of the second industry is less than 50%, the proportion of the second industry in most cities is higher than 50%. The development of the tertiary industry is relatively slow, and the city with the highest proportion of the tertiary industry in the industrial structure is Hefei, at 39%. The proportion of the second industry and tertiary industry in Wuhu is 66 and 28%, respectively. The HSR construction will provide the developed second industry with better support in terms of traffic and funding, drive the development of the tertiary industry and promote the industrial upgrading and structural optimization of Wuhu (Table 6.2). Meanwhile, in the internal comparison of cities in Anhui Province, the commercial and trade services industry, logistics industry and tourism industry of Wuhu have a regional competitive advantage, with the location quotient exceeding 1 and the commercial and trade services industry reaching 3.2 (Table 6.3). This finding shows that the HSR station has a basis and demand in Wuhu, and the HSR may promote the further development of industry owing to the regional competitive advantage. The location quotient of the financial industry is less than 1, which shows that the current development of this industry is limited. In addition, the distribution analysis of Wuhu financial business office buildings indicates that the construction quantity of financial business office buildings near the HSR station should be moderate. The planning compilation research suggests targeting the advantageous commercial and trade services industry, logistics industry and tourism industry and constructing the corresponding buildings and facilities.

6.1.2 Analysis of Passengers' Demands

The investigation of the demand of HSR passengers can help planners define what appeals to the potential space user and then respond in the spatial planning. In the planning of the Wuhu HSR station conducted by the author's research team and the Research Center of Wuhu Urban and Rural Construction Strategy and Planning, the investigation and analysis based on the questionnaire are conducted to determine the demands of the related HSR passengers. The research indicates that the potential passenger group after the construction of the Wuhu HSR station is mainly divided into two types, namely business and tourism. The questionnaire is prepared according to the people's demand for travel and space. The questionnaire for businesspeople is distributed by an email that is sent to provincial and urban enterprises. The survey for tourism is based on a field questionnaire distributed at Wuhu Fangte and other scenic destinations. The number of questionnaires distributed is 320, with 240 business questionnaires and 80 tourism questionnaires. The number of recovered valid questionnaires is 300. Among the survey subjects, the respondents living in Wuhu or within a one-hour economic circle represent 60%, those living within a three-hour economic circle represent 29%, and those living beyond a three-hour economic circle

Table 6.2 Comparison of the number of employees in service industries in major cities in the Yangtze River Delta

Area	2012 Regional GDP (100 million yuan)	Growth rate (%)	The added value of the primary industry (100 million yuan)	Secondary industry added value (100 million yuan)	The tertiary industry added value (100 million yuan)	Per capita GDP (yuan)	Above-scale industrial added value (100 million yuan)	Growth rate (%)	Financial revenue (100 million yuan)	Investment in fixed assets (100 million yuan)	Growth rate (%)
Hefei	4 164.3	13.6	229.1(6%)	2 303.9 (55%)	1 631.4 (39%)	54 997	1653.5	17.4	694.4	4 001.1	23.7
Wuhu	1 873.6	14.9	117.6 (6%)	1 234.2(66%)	521.8(28%)	52 365	1080.2	17.3	337.1	1 700.79	27.4
Ma'anshan	1 232.0	12.0	71.5 (6%)	818.9 (66%)	341.6 (28%)	56 127	538.2	13.7	210.6	1 201.15	25.4
Anqing	1 359.7	11.5%	196.2 (14%)	759.4 (56%)	404.1(30%)	25 559	476.4	16.2	170.3	972	21.1
Tongling	621.3	11.0	11.8 (2%)	456.3 (73%)	153.2 (25%)	84 646	408.7	11.2	127.3	534.3	29.5
Chuzhou	970.7	13.1	192.8 (20%)	507.6 (52%)	270.4(28%)	24 607	401.3	17.2	153.2	882.6	27.9
Luan	918.2	11.0	198.8(22%)	423.4 (46%)	296.0 (32%)	16 248	354.8	17.7	112.4	687.2	23.1
Xuancheng	757.5	12.6	111.7 (15%)	395.1 (52%)	250.7(33%)	29 635	297.5	17.0	137.6	805.3	26.7
Chizhou	417.4	12.3	62.2 (15%)	204.2 (49%)	151.1 (36%)	29 418	118.5	18.5	71.7	374.7	29.3

Table 6.3 Comparison of industrial development of major cities in Anhui Province

Reachability	1 h							
City	Wuhu		Hefei		Xuancheng		Tongling	
Industry (2012)	Added value (ten thousand yuan)	Growth rate	Added value (ten thousand yuan)	Growth rate	Added value (ten thousand yuan)	Growth rate	Added value (ten thousand yuan)	Growth rate
Logistics	75.36	10.8%	165.1	8.9%	40.0	10.3%	21.1	8.2%
Business services	102.97	22.1%	320.6	11.5%	8.8	11.0%	28	11.8%
Tourism	40.14	13.8%	55.4	9.3%	19.9	10.2%	12	9.5%
Financial industry	46.18	15.3%	214.6	17.6%	33.3	18.9%	22.6	14.6%
Reachability	2 h				3 h		Added value of various indus-tries in Anhui Province (100 million yuan)	Location business
City	Fuyang		Bengbu		Bozhou			
Industry (2012)	Added value (ten thousand yuan)	Growth rate	Added value (ten thousand yuan)	Growth rate	Added value (ten thousand yuan)	Growth rate		
Logistics	46.5	8.0%	33.5	10%	37.6	10.4%	650.21	1.3
Business services	88	9.4%	51	13.8%	62.5	11.8%	351.87	3.2
Tourism	19.9	6.5%	23.8	9.9%	14.1	9.0%	268.95	1.6
Financial industry	33.3	22.9%	24.2	17.7%	11	20.9%	617.62	0.8

Source Anhui Statistical Yearbook (2013)

represent 11%. Most of the HSR passenger respondents are aged 25–34, representing 60%. Most of the HSR passenger respondents are enterprise employees, representing approximately 50%, including private business owners, teachers, students and other professionals (Fig. 6.2).

The survey finds that the average time per year spent taking HSR is 1–3 h for 34% of the respondents and 4–6 for 25% of the respondents; 9% of respondents say that they never take HSR. Of the respondents, 93% think they will take the HSR in Wuhu after its completion. The survey shows that the main purpose of taking HSR is tourism, representing 46%. Business travel and visiting relatives represent 28 and 15%, respectively (Fig. 6.3).

In planning the area surrounding the HSR station, it is very important to define the time spent by passengers in the station and the intention to use the facilities. The survey on Wuhu HSR indicates that the average time spent by 57% of the respondents and 34% of the respondents stopping or waiting in the HSR station is within one hour and within half an hour, respectively. Generally, the activity of 54% of the respondents

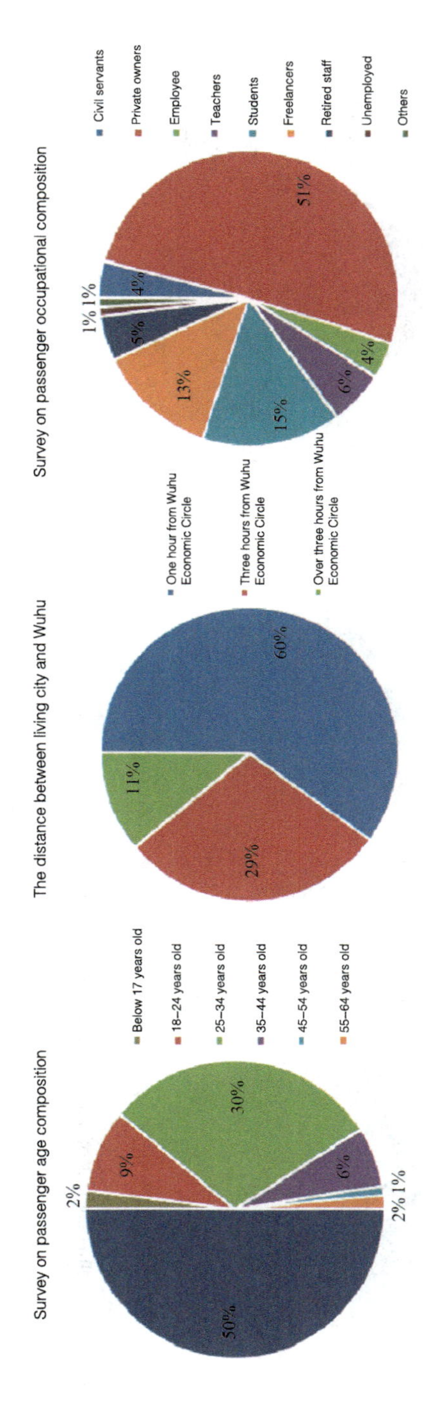

Fig. 6.2 Basic information from the questionnaire survey

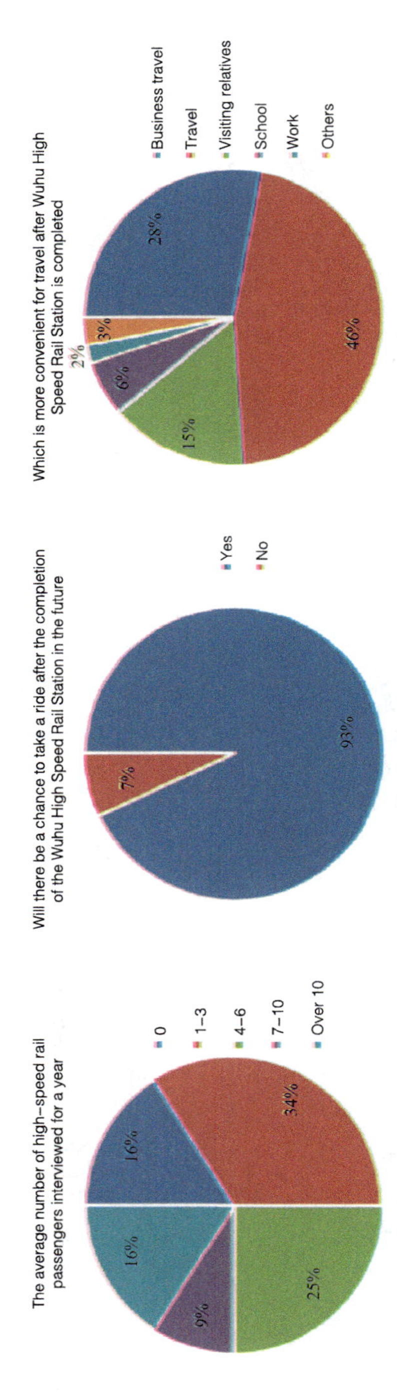

Fig. 6.3 Traveling choices of potential passengers of HSR

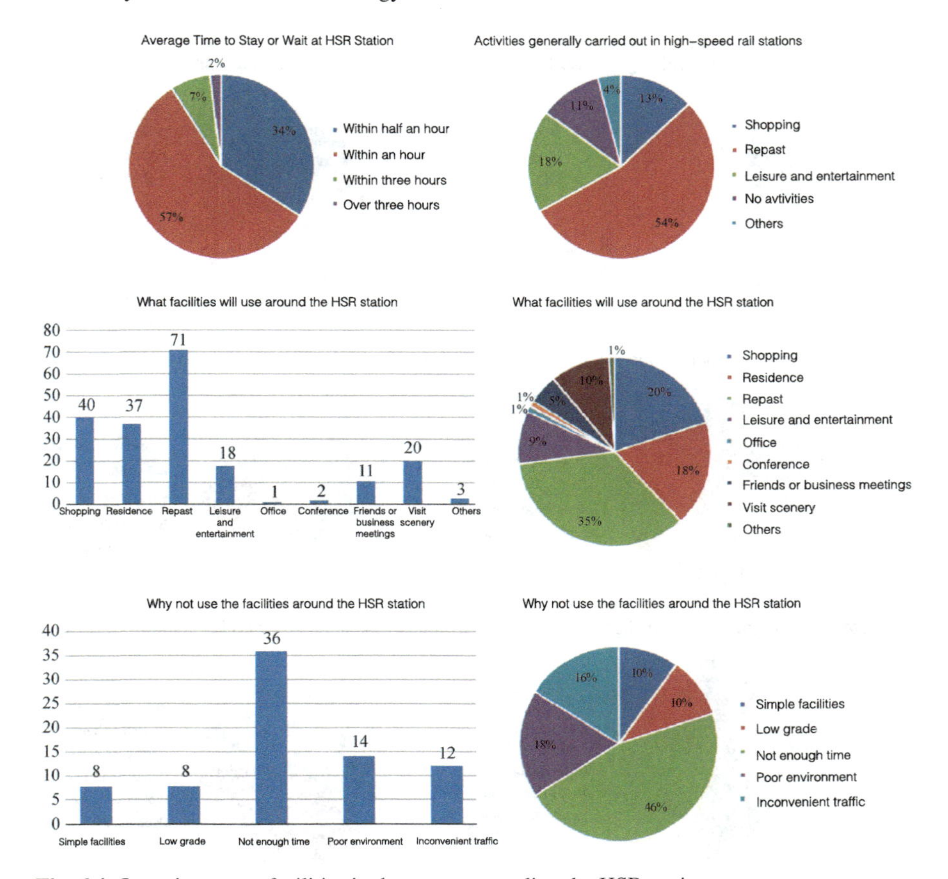

Fig. 6.4 Intention to use facilities in the area surrounding the HSR station

in the HSR station is dining, followed by leisure, entertainment and shopping. Of the respondents, 13% do not undertake any activity in the HSR station except waiting. In terms of their reason for being unable to use the facilities surrounding the HSR station, 46% of respondents choose lack of time, 18% choose poor environment, 16% choose inconvenient traffic, and 10% choose the simple facilities and low grade of the facility (Fig. 6.4).

The survey summarizes different types of facilities that the potential HSR passengers wish to find in the surrounding area: restaurant (20%), supermarket (19%), shopping mall (13%), hotel (10%), coffee shop (9%), park (9%) and bookstore (7%). The further survey in a refined format finds that 53% of the respondents hope that a restaurant with fast service, affordable prices and a good dining environment will be built, and 31% of the respondents suggest building a middle-grade restaurant. In terms of accommodations, 47% of the passengers believe business chain hotels should be built rather than luxury hotels. In terms of retail format, 70% of the passengers choosing to make purchases in the area surrounding the HSR station prefer

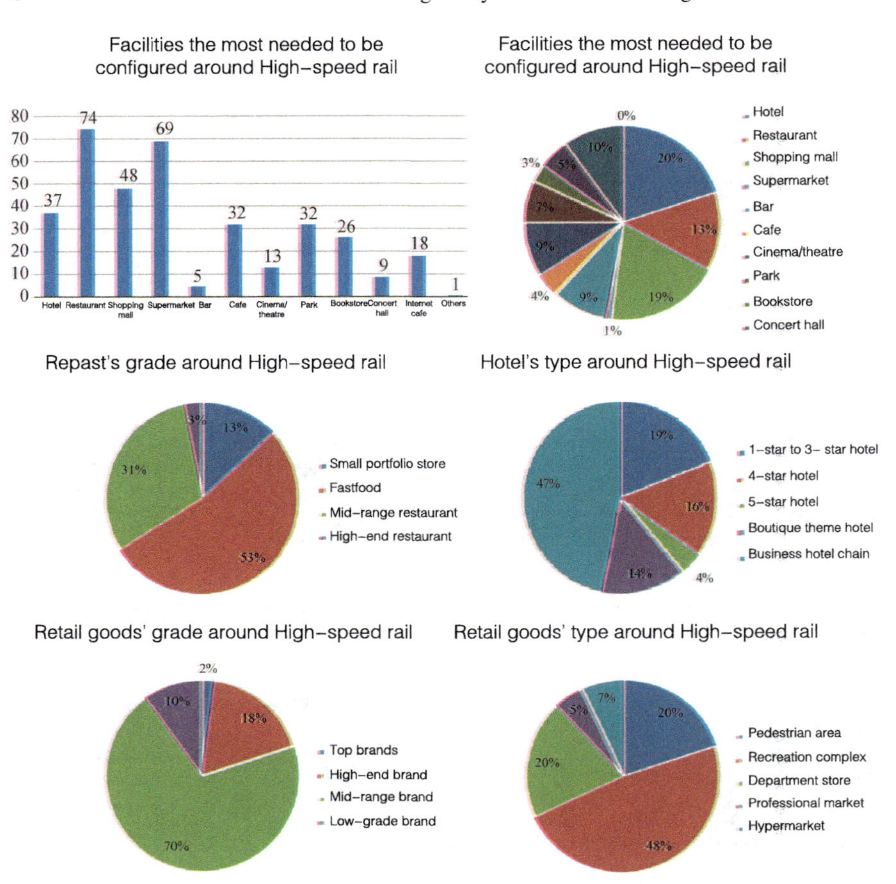

Fig. 6.5 Demands for facilities, infrastructure and format in the area surrounding the HSR station

middle-grade brands, and 18% and 10% of the respondents will choose high-grade and low-grade brands, respectively. Of the respondents, 48% suggest that the retail area in the area surrounding the HSR station should be mainly a leisure and entertainment complex, with 20% of the respondents choose a pedestrian street and department store (Fig. 6.5).

The demand survey of the potential passengers of the Wuhu HSR station reflects the space demand of the area surrounding a HSR station of this type and grade. The main purposes of taking HSR are tourism and business travel. The time that most passengers spend waiting in the HSR station is short, within about 1 h. There is a demand for dining, business and lodging facilities in the area surrounding the HSR station, but there are also problems such as poor environmental quality and inconvenient traffic. In terms of specific facilities, the demands for restaurants, shopping malls, supermarkets, hotels, coffee shops and parks are the greatest. Among these facilities, the middle-grade facilities and services are highly required. The preferred business model is a leisure and entertainment complex. The questionnaire survey conclusion provides clues for the planning of the area surrounding the HSR station, from service facilities to the format type, to a certain extent.

The problem to be defined through the macrolevel analysis is the resources brought by the HSR to the city where the station is located at the regional level and the desire of passengers to visit this city. There are differences among cities on different scales and stations of different grades, and the research approaches and methods of the Wuhu HSR station can provide some references regarding this issue.

6.2 Analysis Content and Methodology at Mesolevel

Focusing on the specific space surrounding the HSR station, a mesolevel analysis is conducted of the function to be supplemented for the overall urban development as well as the specific space demand of the surrounding residents living around the station. For the functional demand of the overall urban development, the service scope analysis of the functional layout can be conducted through a geographical information system (GIS). The questionnaire survey is a direct and effective way to comprehend the specific space demand of residents.

In the planning of the area surrounding the Wuhu HSR station, the research analyzes the current distribution of the related industrial and functional zones of the Wuhu central urban area and then proposes the functional allocation requirements of the area surrounding the station. Meanwhile, the research delimits the spatial range of the area surrounding the station and performs the questionnaire survey to determine the residents' demands for facilities and functions.

6.2.1 Analysis of Functional Sector and Distribution of Facilities

The demand survey at the urban level establishes a database of public service facilities using GIS software. It analyzes the service coverage of the facilities based on the road network and then defines the missing functions of urban service within the planning scope. The spatial analysis of the service scope of facilities should follow three steps. First, the system carries out a general survey of the current distribution of various functional facilities, integrates multiple plans and defines the functional facilities to be built in order to establish data integration, including different types of facilities and their locations, in the GIS. The following functions are included in the analysis: commercial and business facilities (commercial complex, business and office buildings, various types of hotel, etc.), cultural and leisure facilities (cultural display and experience facilities, such as museums, cultural centers), sports facilities and parkland. Second, it defines a reasonable service radius for various functional

Table 6.4 Development of the agglomeration district of finance, business and commerce in the central urban area

Financial gathering area	Development orientation	Planning target industries
Beijing West Road and Cultural Road financial gathering area	Business services gathering area Regional Business Center District	Headquarters economy, building economy, finance, insurance
Zheshan-Jinghu regional business center area	Business services gathering area Regional Business Center District	Legal advice, accounting, engineering consulting, advertising exhibition, product development, intermediary services, marketing, education and training, web services, creative enterprises
Eastern financial business services area	Wanjiang and southern Anhui development services Financial industry concentration area	Corporate headquarters of financial institutions, regional headquarters, backstage service centers

facilities and delimits the multilevel service radius for the spatial analysis and calculation. It determines the coverage of the service radius of the existing functional facilities at different levels based on the urban road data in the GIS and its spatial analysis tool. Finally, it obtains the current services of functional facilities in the plan base and the surrounding area through analysis and suggests functions within the planning scope that need to be added. Meanwhile, it judges the position of the specific function in the plan base through the superposition of coverage and planning scope. The distribution analysis of the industry and functions of the Wuhu central urban area includes financial business, tourism, retail commerce, hotel, residential, cultural and sports facilities. The distance relationship between these areas and the area surrounding the station can be analyzed through the GIS, and the plan can define the functional setting of the area surrounding the HSR station in combination with the development direction and target industry.

(1) Financial Business

The GIS analysis shows that the distance of the three financial business clusters of Wuhu from the area surrounding the HSR station is relatively short (Fig. 6.6). The convenience of regional traffic accessibility brought by HSR may improve the service level for financial business. The specific industrial planning of the financial business cluster indicates that the main functional orientation is beneficial, and the conceived industrial chain is relatively complete (Table 6.4). The service radius of these three clusters is 400 m, 800 m and 1600 m, respectively, and the calculation of the service scope is carried out through the road network calculation in the GIS.

Banking is the main financial business development in Wuhu, and banks are densely distributed. In terms of spatial allocation, banks are mainly distributed in Yinhu Road, Beijing Road, Changjiang Road, Zheshan Road and other main roads.

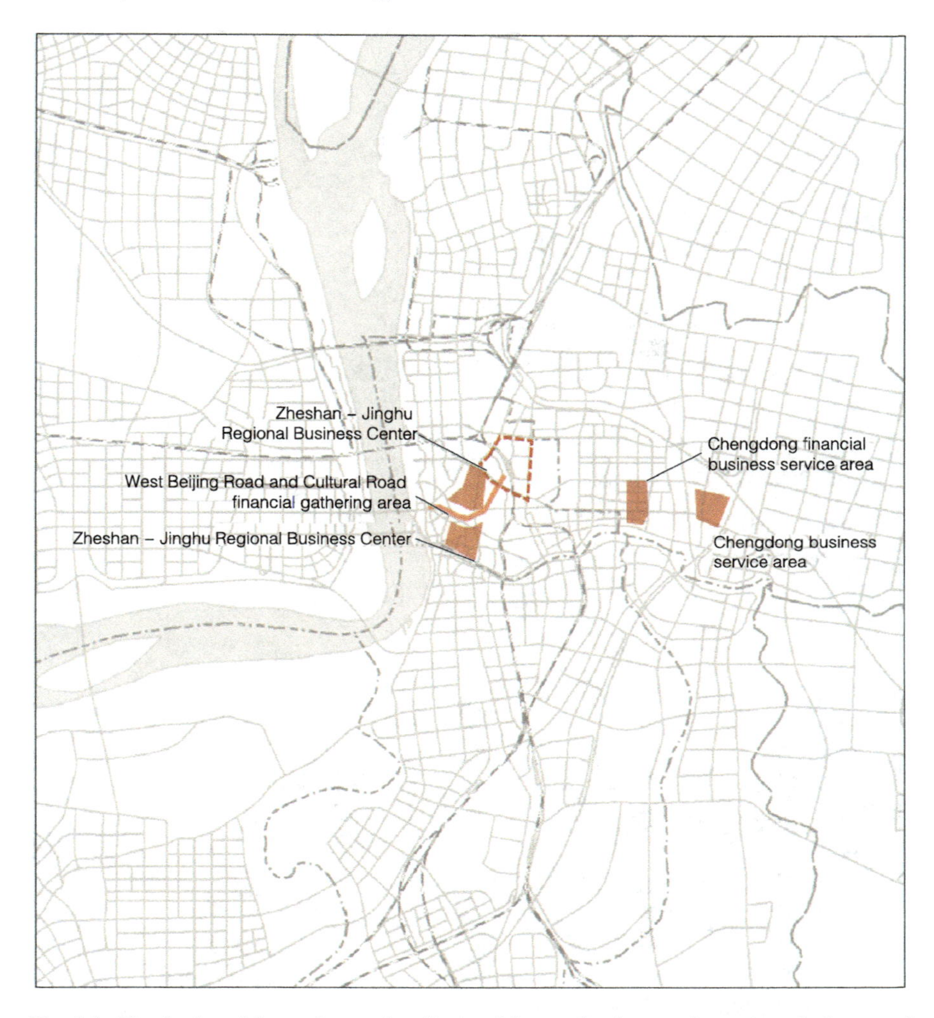

Fig. 6.6 Distribution of the agglomeration district of finance, business and commerce in the central urban area

They are also distributed in Jiuhua Middle Road and Zheshan Middle Road in the area surrounding the HSR station. Insurance service facilities are intensively distributed near the Beijing Road and Wenhua Road. The relationship with the HSR station of those locations and their service scopes proves the following views. The area surrounding the HSR station is in the peripheral area of the service scope of the current financial business facilities, and it is necessary to provide a proper amount of financial business in this area. The space required for the financial industry can be provided for the business circulating group brought by the HSR and further attract banks and securities, insurance and guarantee facilities (Figs. 6.7 and 6.8).

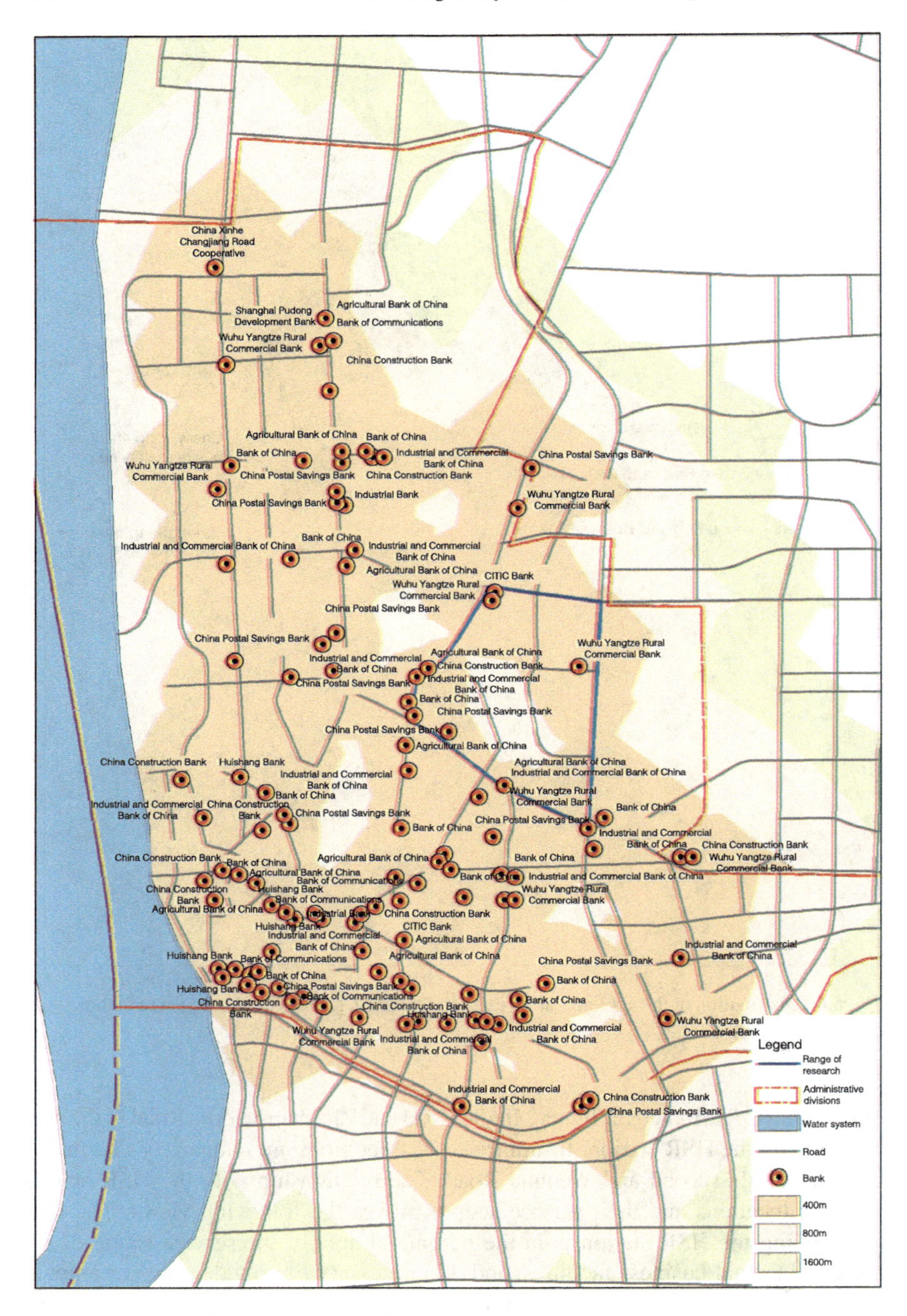

Fig. 6.7 Analysis of the service scope of banks

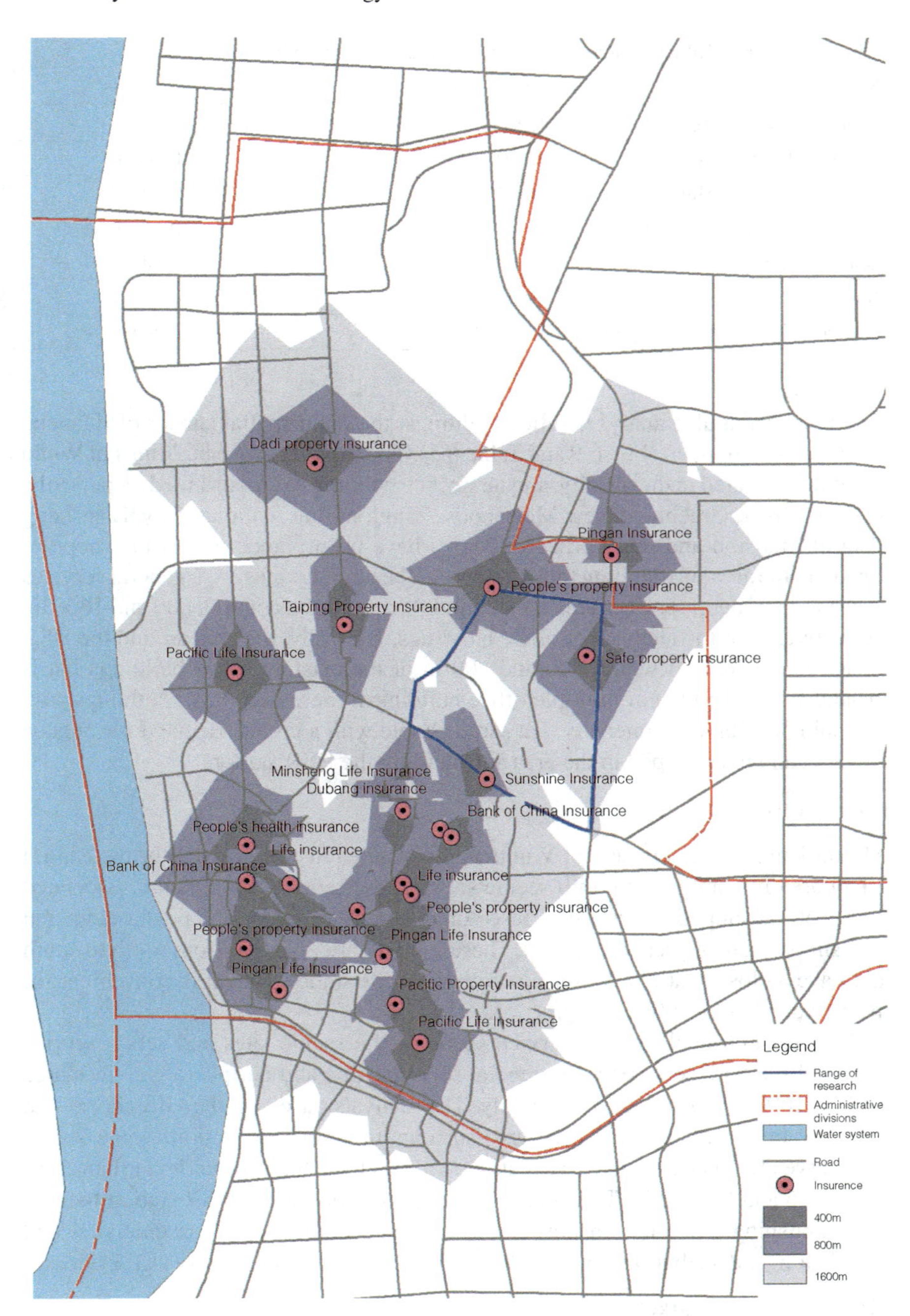

Fig. 6.8 Analysis of the service scope of insurance companies

Table 6.5 Status of office buildings in the area surrounding the HSR station

Name	Scale (m^2)	Grade	Remarks
Wanda Center (Phase II)	138,000	5A	In stock
Greenfield Xinduhui	130,000	5A	In stock
Liansheng Central Building	50,000	–	Sold out
Star International City	50,000	5A	In stock
Qiao Hong International Mall	29,000	–	Sold out
Jinding Street	30,000	–	Sold out
Qiaohong Riverside Century Plaza	120,000	5A	In stock

The research also analyzes office buildings, the main spatial carrier of financial business development (Fig. 6.9 and Table 6.5). The existing office buildings of Wuhu are mainly located in the old city, and new office buildings are distributed dispersedly. Wanda Center, Greenland New Metropolis, Xinglong International City, Liansheng Central Mansion and other office buildings have been planed within the scope of 1.6 km in the area surrounding the rail station. The core services area, general services area and peripheral services area are delimited with a radius of 400, 800 and 1600 m, respectively, for the analysis of office buildings. Basically, the area around the HSR station is not located within 400 and 800 m of the existing office buildings but is located within 1600 m of multiple office buildings. Therefore, office buildings with a certain development intensity that serve people with a greater demand for outside contact can be developed in the area surrounding the HSR station.

(2) Tourism Services

All tourism services clusters of Wuhu are distributed in various parts of the central urban area in combination with scenic sites. Those tourism clusters are associated with well-developed theme parks based on different resources and positioning, and tourism projects of great prosperity. These tourism areas spread from north to south and have a close spatial relationship with the HSR station, which is convenient for the arrival of tourists (Fig. 6.10).

The research analyzes the service scope of the scenic sites and sets a service radius of the core services area, general services area and peripheral services area of 800, 1600 and 3200 m, respectively. The analysis based on the GIS shows that there are more scenic sites in the area surrounding the HSR station. Some scenic sites serve the inner city, and others attract tourists from the surrounding urban areas (Table 6.6 and Fig. 6.11). The plan of the area surrounding the HSR station focuses on establishing a public transport connection with multiple urban scenic sites and establishing a distribution center of tourism, dining and hotels for tourist services.

(3) Specialized Market

Currently, the specialized markets of Wuhu are mainly distributed in the central urban area, leaded by automobile products, building materials and agricultural products markets. The service radius of the specialized markets is divided into the core

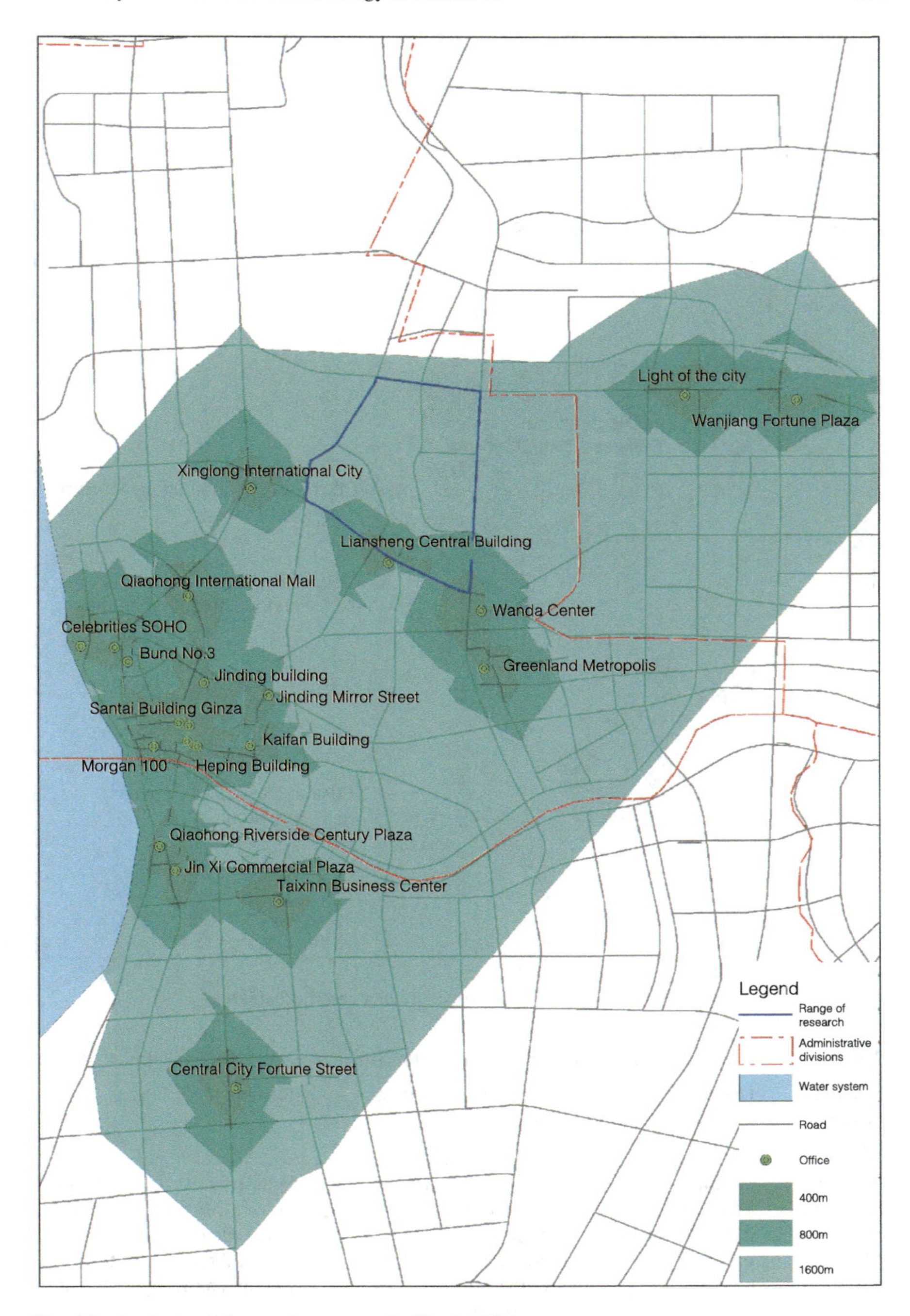

Fig. 6.9 Analysis of the service scope of office buildings

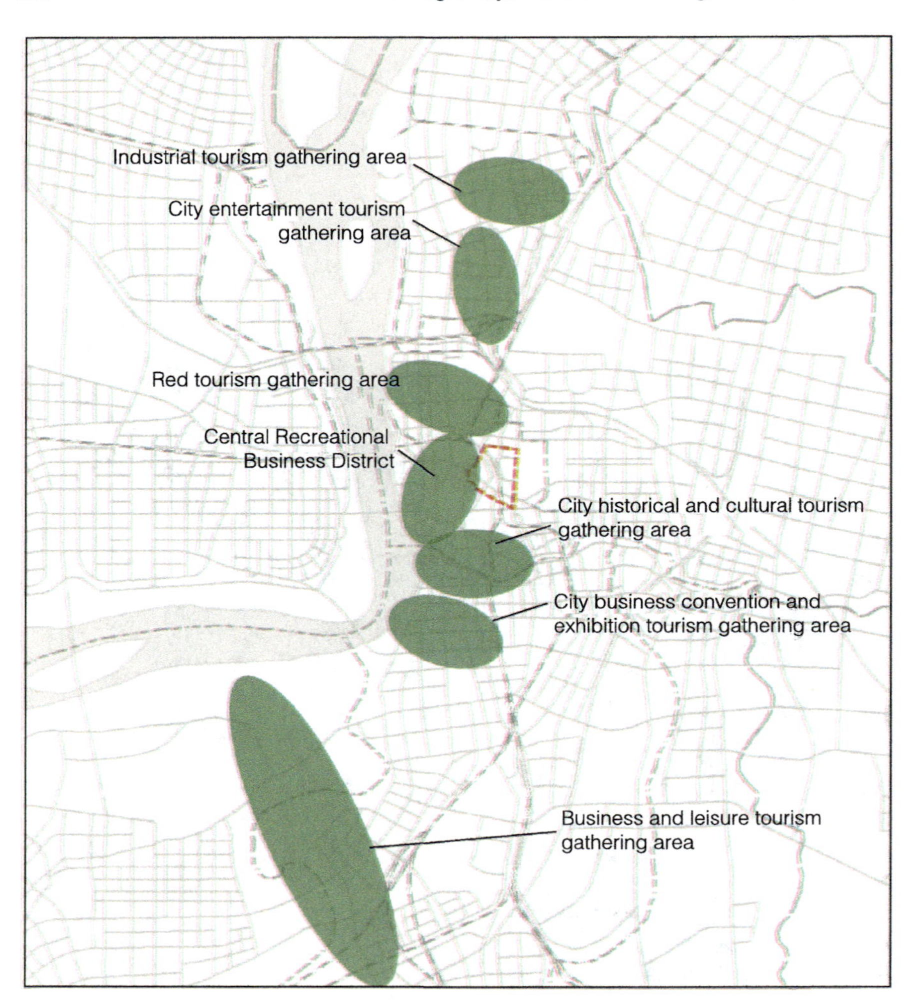

Fig. 6.10 Distribution of tourism and service zones in the central urban area

Table 6.6 Development of agglomeration district of tourism and services in the central urban area

Tourist services area	Development orientation	Industry type or tourism project	Relying on resources
Central Tourism Gathering Area	Wanjiang City Tourism RBD Construction of first-class tourism environment	Recreation and business services (leisure and entertainment, special dining, cultural activities, film and television performances, fitness and beauty, boutique shopping)	Pedestrian Street, Jiuzhaigou Square, Mirror Lake Park, Riverside Park, Zheshan Scenic Area, Food Street, Zhongjiang Ronghui Square

(continued)

Table 6.6 (continued)

Tourist services area	Development orientation	Industry type or tourism project	Relying on resources
Business Exhibition Tourism Gathering Area	Business travel Shopping travel MICE travel	Tourism public facilities (business hotels, dining, entertainment, shopping, business services, financial services, etc.), travel agencies, MICE tourism products, convention and exhibition tourism and leisure systems	Wuhu Culture and Drama Park, Wuhu International Conference and Exhibition Center, Olympic Plaza, etc.
City Entertainment Tourism Gathering Area	Theme park travel Cartoon travel	Wuhu theme park clusters, animation design, animation shopping, animation science and education, animation and amusement, animation experience	Huaqiang Tourism City, based on the theme parks of Fantawild World and Fantawild Kingdom, plans to build 6.8 high-tech cultural theme parks with world-class standards
City History and Culture Tourism Gathering Area	Cultural experience Leisure travel boutique	Travel services, Folklore presentations and participatory activities	Wuhu Ancient City, Jiuzhaowan Scenic Area, Fanluoshan folk collection Exposition Garden, modern Western architecture, etc.
Red Tourism Gathering Area	Red travel	Develop red tourism products, forming a series of red tourist attractions	Wang Jiaxiang Memorial Park, the site of the New Fourth Army seventh division, Sun Yat-sen, Dai Anlan, Tan Zhenlin, crossing the first ship, etc.
Industrial Tourism Gathering Area	Industrial travel	Production visits, cultural awareness, environmental appreciation, learning to watch, experience entertainment	Industrial Park, Wuhu Arts and Crafts Factory, Chery Company
Business Casual Tourism Gathering Area	Longwo Lake ecological travel Leisure travel Business casual travel	Lake business and leisure projects, business eco-clubs, business eco-restaurants, business health and leisure, water recreation and entertainment projects, Binhu sports and leisure fitness park, popular clubs, Lakefront	Sanshan Longwo Lake Ecological Sports Park, ring water as the representative of the business and leisure tourism gathering area

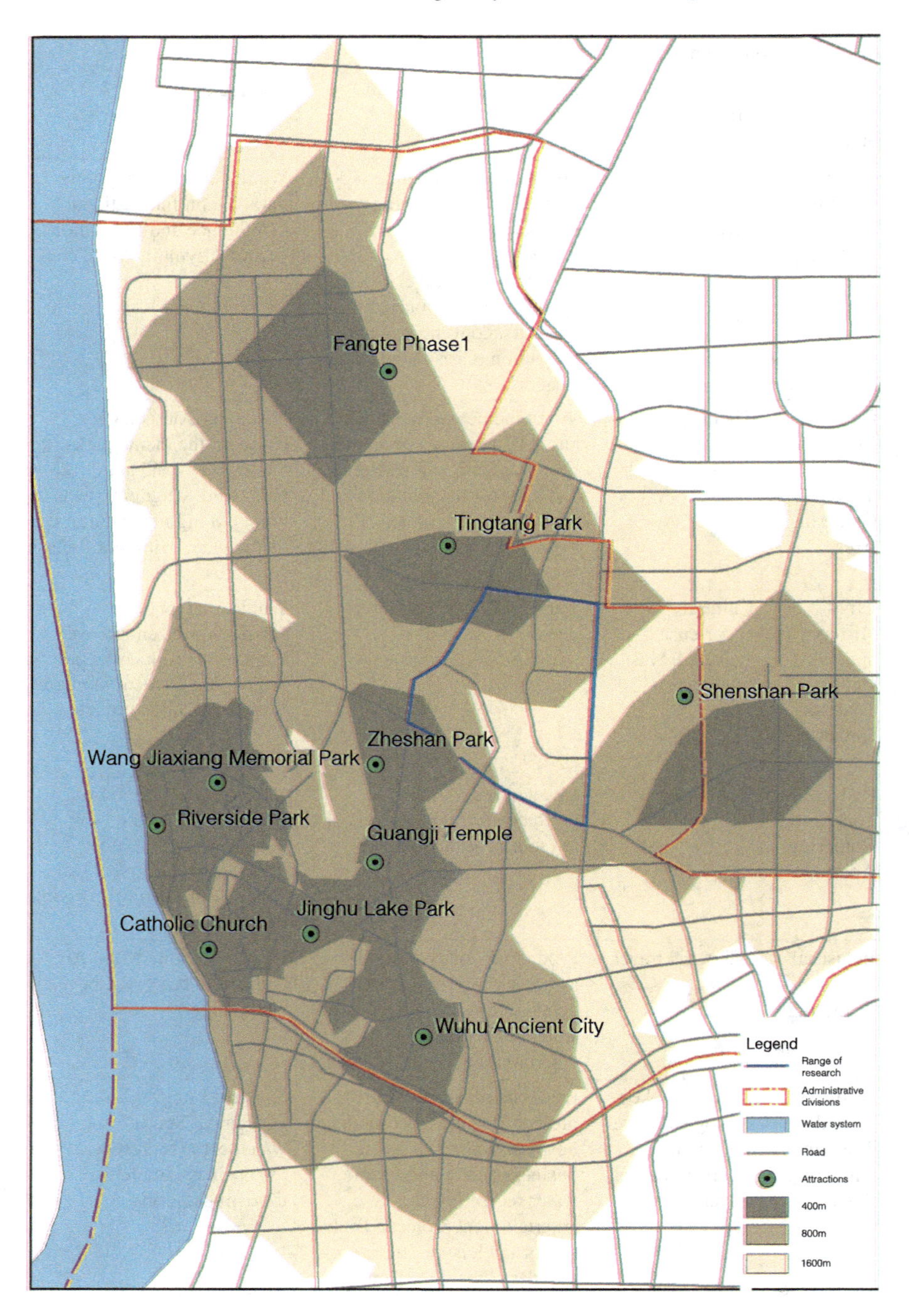

Fig. 6.11 Analysis of the service scope of tourism sites

Table 6.7 Status of specialized markets in the area surrounding the HSR station

Project name	Nature	Land area (ha)	Construction area (m^2)	Remarks
Wenzhou trade city	Commodity market	1.48	Business area of more than 20,000	Mainly apparel, shoes and hats, knitwear, wholesale department stores
Changjiang market park	Building materials wholesale market	32.4	Business area of more than 268,000	Integrated business logistics market
Yaxia auto city	Automotive and auto parts market	6.66	Business area of more than 206,000	Mainly car sales and auto accessories
Wuhu auto parts city	Automotive and auto parts market	–	Business area of more than 18,000	Under demolition
Wuhu Dadi agricultural and sideline products wholesale market	Agricultural products trading market	13.0	Business area of more than 69,600	Mainly agricultural products
Ouyada furniture market	Furniture trading market	2.0	Business area of more than 120,000	Mainly furniture sales

services area, general services area and peripheral services area, with a radius of 800, 1600 and 3000 m, respectively. The HSR station and its surrounding area are located within the services area with a radius of 3000 m, including many specialized markets such as Yangtze River Market Garden, Ouyada Furniture Market and Yaxia Motor City. Additionally, the southern and northern specialized markets are located within the service scope with a radius of 1600 m, so it is not necessary to add a specialized market in the area surrounding the HSR station (Table 6.7 and Fig. 6.12). Improvement and expansion can be carried out in combination with the existing markets. A logistics coordination platform can be established in the area surrounding the station to integrate the resources of multiple specialized markets, forming a beneficial interaction with the regional traffic accessibility provided by the HSR station.

(4) Hotel

The hotels of Wuhu are mainly distributed in the central urban area. The current hotels are mainly starred hotels, and the majority are high star-level hotels. At present, Conch International Hotel, Hanjue Plaza Hotel and Tiantai International Hotel are within a radius of 1600 m in the area surrounding the HSR station.

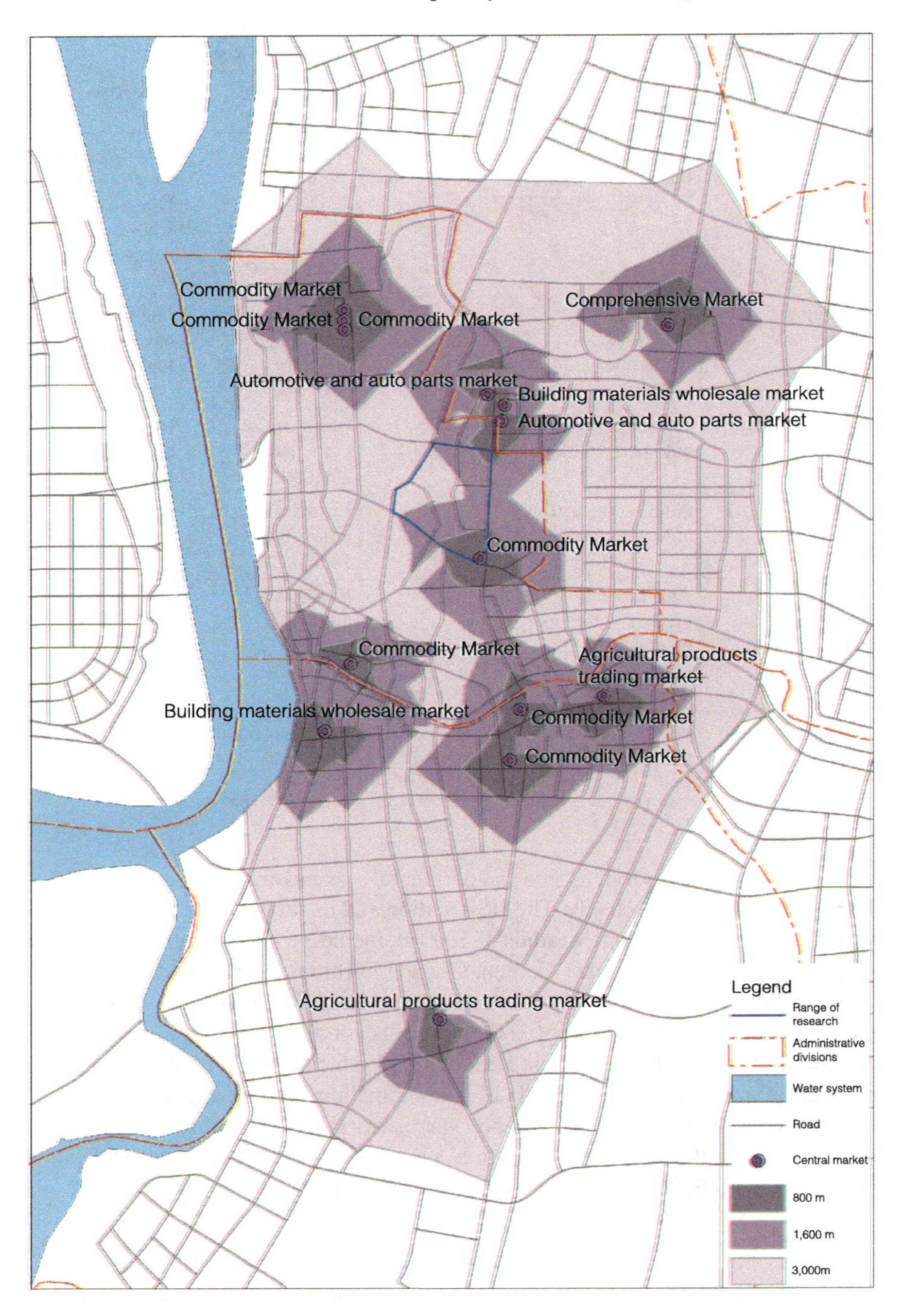

Fig. 6.12 Analysis of the service scope of specialized markets

The radius of the core services area, general services area and peripheral services area for hotels is 400, 800 and 1600 m, respectively. A quasi-five-star hotel and a three-star hotel are within the area of the station, and they are located within the scope of the service radius of 1600 m (Table 6.8 and Fig. 6.13). With the construction of the HSR, more passengers may visit Wuhu, thus bringing a greater demand for hotels. According to the questionnaire survey of potential passengers, the middle-grade business hotel is the most popular type, and correspondingly, the plan may consider the establishment of such hotels. Meanwhile, the characteristic hostel has become popular among young people and is especially suitable for backpackers. Its combination with the HSR will promote the urban tourism industry and hotel services industry.

(5) Retail Commerce

At present, the municipal-level, district-level and community-level commercial and trade service systems have basically been formed in the Wuhu central urban area. The city plans to optimize and improve the high-end commercial and trade services themed by Zhongshan Road Commercial Pedestrian Street as well as Fenghuang Food Street, Qingshan Leisure Commercial Street, New Era Commercial Street and other characteristic commercial streets and to accelerate the development and construction of Food Street Phase II, Xiaojiuhua Tourist Commodity Street, Yinhu Commercial Street, Binjiang Commercial Street, Fangte Theme Park Commercial Street, Ancient City Travel and Leisure Commercial Street and Yanhe Road Leisure Commercial Street. Those approaches are meant to enrich the commercial resources and improve the urban central business district. The current retail commerce of Wuhu is mainly distributed in the central urban area, and the business model is mainly commercial streets and commercial complexes. At present, Wanda Plaza, Liansheng Plaza, Greenland New Metropolis, Xiaojiuhua Pedestrian Street and many other large commercial streets and commercial complexes are within the radius of 1,600 m in the area surrounding the rail station (Table 6.9).

The radius of the core services area, general services area and peripheral services area for retail commerce is 400, 800 and 1600 m, respectively. The rail station and its surrounding area are located within a radius of 1600 m of multiple commercial streets but beyond the scopes with a radius of 400 and 800 m, which is more than a comfortable walking distance (Fig. 6.14). Therefore, the retail commerce in the area surrounding the HSR station can serve the HSR circulating group, and the community commerce can serve industrial workers and residents. Additionally, the characteristic commerce can be designed to attract people from other urban areas.

(6) Culture and Sports

There are various cultural facilities in Wuhu that are mainly distributed within the scope of the central urban area; Huizhou Merchants Museum, Wuhu Radio and Television Center and other facilities are in the area surrounding the railway station. The main sports facility is Wuhu Olympic Park (Tables 6.10 and 6.11).

The radius of the core services area, general services area and peripheral services area for cultural and sports facilities is 800, 1600 and 3200 m, respectively. According

Table 6.8 Status of hotels in the area surrounding the HSR station

Hotel name	Star rating	Distance from planned plot (km)	Number of rooms	Meeting capacity	Occupancy rate		Customer source	Function
					High season (%)	Low season (%)		
Hanjue Plaza Hotel	Three stars	Within the scope	220	1 room for 50 people	60	60	Business	Business
Tiantai International Hotel	Quasi-three stars	Within the scope	492	20 rooms for 200 people	40	40	Business, local	Local
Conch International Hotel	Four stars	0.5	220	2 rooms for over 200 people, a total of 14 rooms	60	50	Business, local	Business, local
Wuhu Xin Ocean Hotel	Three stars	1.2	218	1 room for 100 people, a total of 3 rooms	50	45	Local	Local
Tieshan Hotel	Four stars	3.5	166	3 rooms for over 300 people, a total of 18 rooms	70	70	Government reception, local	Local
Orton Hotel	Quasi-five stars	1.8	118	1 room for 70 people	50	50	Local	Local
Qiaohong Crowne Plaza Hongqiao	Five stars	3.2	391	3 rooms for over 300 people, a total of 8 rooms	60	60	Business, local	Business, local
New material Building Hotel	Three stars	2.2	96	2 rooms for over 100 people, a total of 3 rooms	50	50	Local	Local

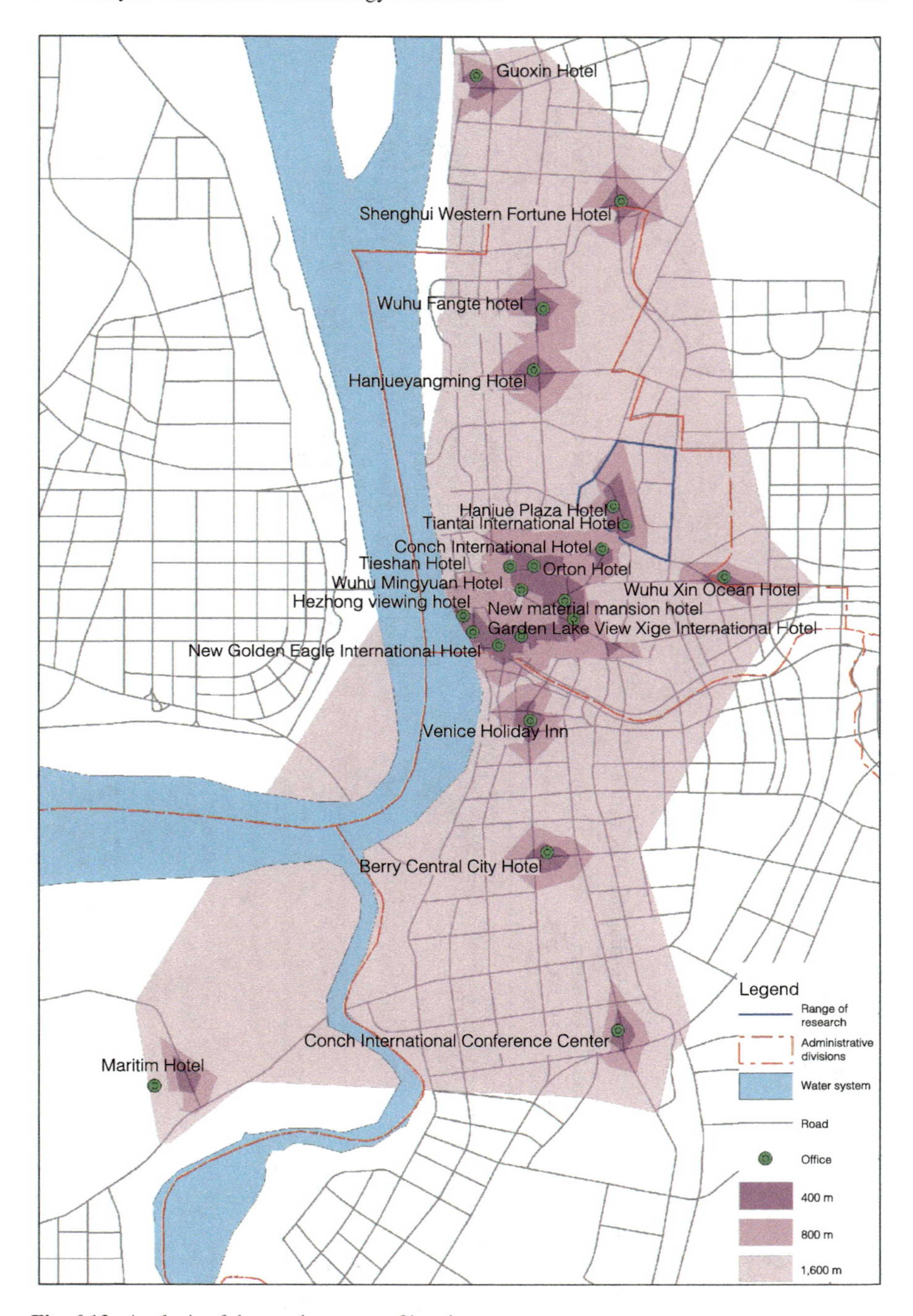

Fig. 6.13 Analysis of the service scope of hotels

Table 6.9 Status of retail stores in the area surrounding the HSR station

Project name	Total land area (10,000 m^2)	Construction area (10,000 m^2)	Volume rate	Year of development	Property type	Distance from the planned plot (km)	Parking capacity
Wanda (phase I)	8.4	24.9	3.7	2011.01	Complex	0.5	1 561
Liansheng	3.8	18.0	4.5	2011.08	Complex	0.5	800
Greenfield	19.0	45.0	2.4	2011.06	Complex	2.0	2 785
Little Jiuhua Pedestrian Street	1.68	2.33	1.39	2007	Commercial street	2.0	200
Zhongshan Road Pedestrian Street	–	60.0	–	1999	Commercial street	2.8	–

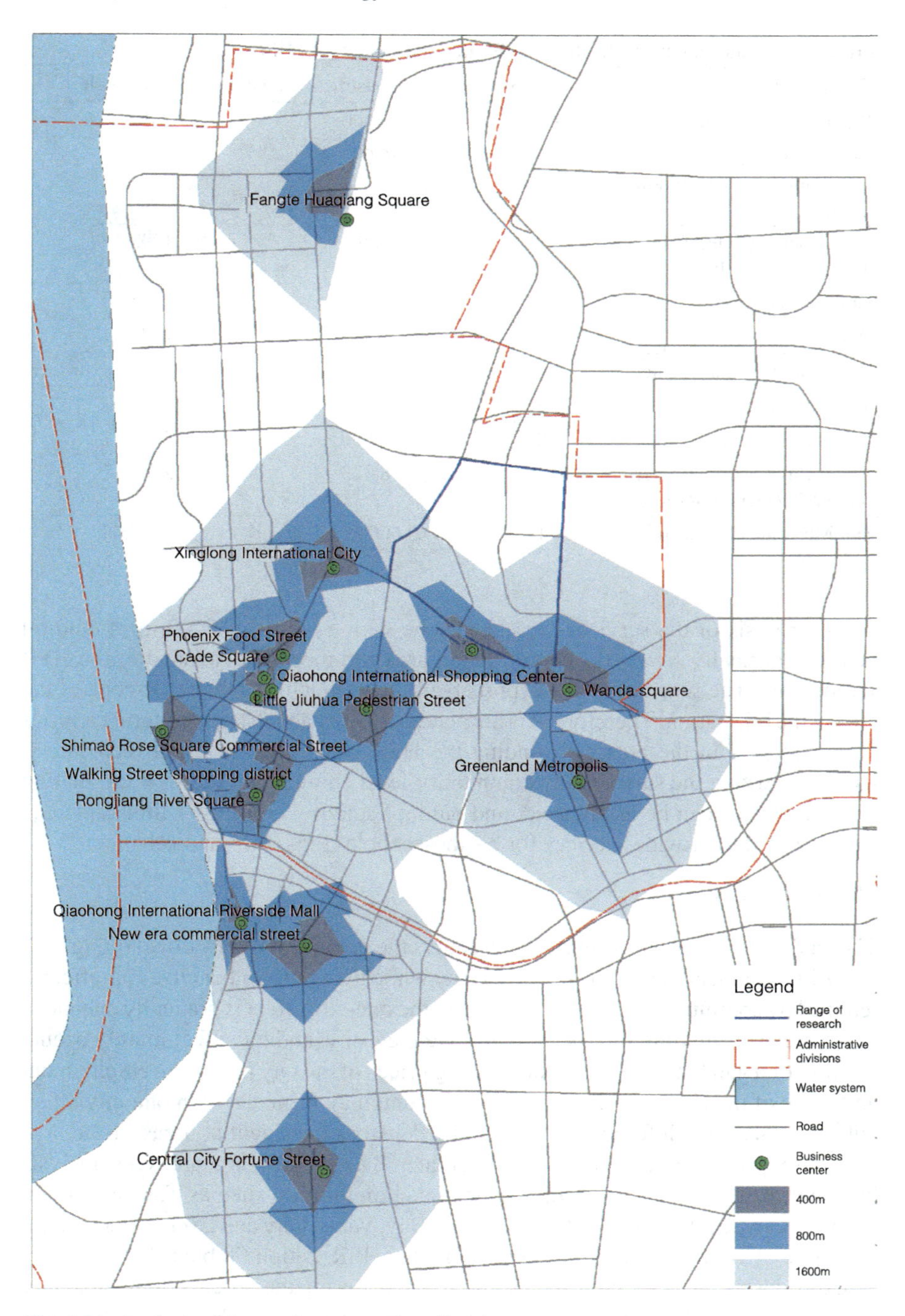

Fig. 6.14 Analysis of the service scope of retail stores

Table 6.10 Status of cultural facilities in the area surrounding the HSR station

Facility name	Land size (ha)	Construction area (m^2)	Time built
Wuhu Grand Theater	3.5	15,000	2012.05
Wuhu International Convention and Exhibition Center	14.5	56,000	2011.12
Art Museum (Wuhu Painting and Calligraphy Institute)	0.13	1600	1993
Children's Palace	0.88	5000	1995
Wuhu Daily	1.9	18,843	2008.04
Wuhu Broadcasting Center	2.0	31,000	1999
Wuhu Archives	0.8	1.5	2011.09
Science Museum	2.0	16,600	2008.11
Wuhu City Cultural Center (cultural centers, libraries)	0.75	18,123	2008.12
Hui Merchants Museum	1.1	6000	2010.04

to the analysis of the services area, the scope with a radius of 3200 m of cultural facilities basically covers the central urban area and meets the demand of the HSR station area. The service scope of sports facilities is much less than the scope of the area of the rail station. Therefore, it is not necessary to establish large cultural facilities separately within the area surrounding the HSR station. Leisure and recreational facilities serving the surrounding community-level cultural and business circulating group demands can be established, and outdoor activity facilities or fitness centers can be added to provide services for businesspeople (Figs. 6.15 and 6.16).

(7) Residences

The analysis of residences generally considers the possible one-city effect brought by the HSR system, namely working in one city while living in another city. This effect is realized by commuting via HSR. At present, the one-city effect is gradually enhanced in the area surrounding the special intensive megacity, and the effect mainly occurs because of a gap between the urban housing prices of the city providing employment and those of the city providing residences. Wuhu may form a certain one-city effect with Nanjing and Hefei in the future. The density of residential areas in the area surrounding the Wuhu HSR station is greater. The project data and GIS analysis indicate that there are many large residential communities, such as Tianhe Garden, Weixing City, City Lights, Left Bank, Baijin Bay, Vanke City and Evergrande, within the scope of 1600 m in the area surrounding the HSR station (Table 6.12).

Considering the walking distance, the research during the process of the planning compilation delimits a spatial range with a radius of 400, 800 and 1600 m for the area surrounding these residential communities (Fig. 6.17). There are more residential

Table 6.11 Status of sports facilities in the area surrounding the HSR station

Facility name	Grade	Land size (ha)	Construction area (m^2)	Type	Time built	Remarks
Wuhu Olympic Park	City level	30.8	–	Sports, leisure	2012	The second and third galleries and the corresponding integrated facilities, namely the 40,000-seat main stadium and track and field training ground, the 5500-seat integrated gymnasium, the 2000-seat swimming pool and the shooting hall with competition standards

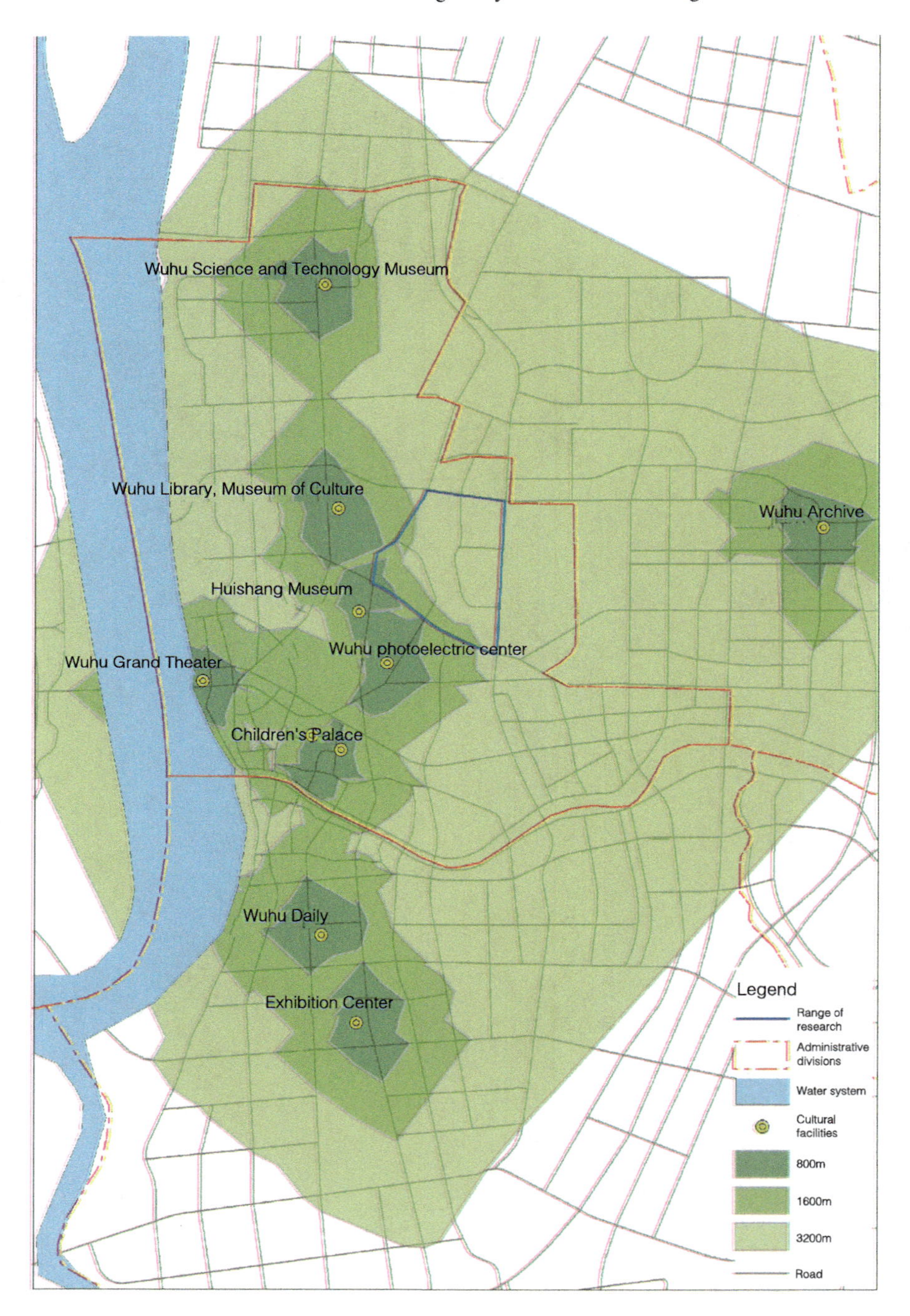

Fig. 6.15 Analysis of the service scope of cultural facilities

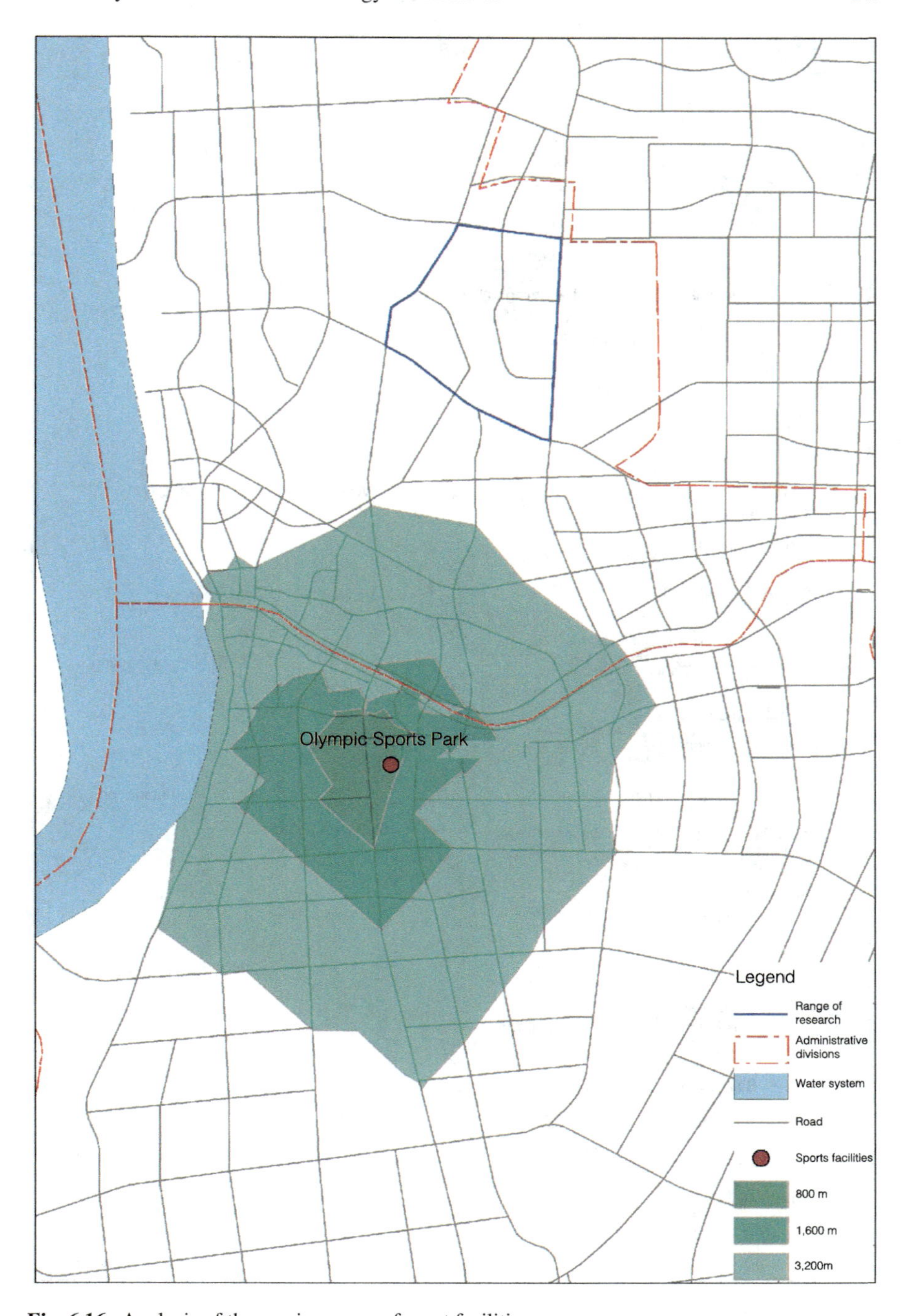

Fig. 6.16 Analysis of the service scope of sport facilities

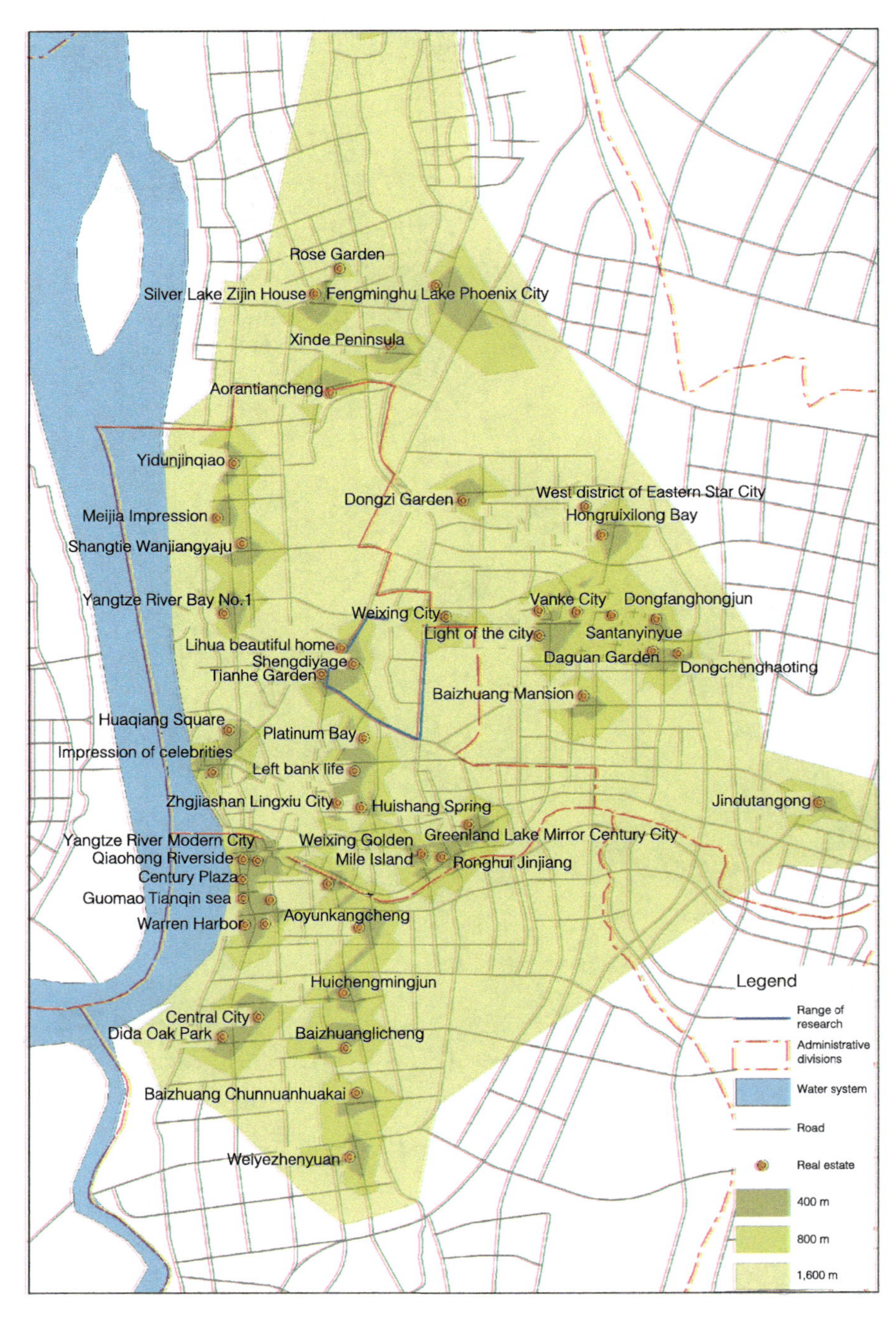

Fig. 6.17 Analysis of the service scope of residences

Table 6.12 Status of residential districts in the area surrounding the HSR station

Project name	Land area (ha)	Construction area (10,000 m^2)	Year of development	Property type	Price (yuan/m^2)	Parking capacity	Current sales ratio (%)
Lihua Beautiful Home	5.0	10	2011.04	High rise, multistory	6000	100	90
Great Star City	34.6	62.3	2010	High rise, multistory	5300	5745	80.9
San Diego	21.6	27.0	2010	High rise, multistory	8200	2500	70
Tianhe Garden	4.68	16.0	2008	High rise	6700	300	85
Platinum Bay City Lights	61.5	112.0	2013	High rise, bungalow	5500	5000	40
Vanke City	21.5	47.0	2011	High rise	6700	5400	65
Hengda Huafu	21.8	40.0	2011	High rise	7000	2841	65
Zhangjiashan Lingxiu City	16.6	48.3	2009	High rise, multistory	6800	3000	50

communities in the area surrounding the HSR station, so it is not suitable to set the land use type for this area, but the apartments that serve businesspeople can be taken into consideration.

6.2.2 Questionnaire Survey for Residents

The questionnaire survey is an effective way to collect data from specific groups and obtain answers to specific questions. The questionnaire design is very important, including aspects such as question content, answer design, overall length and question seriation. The questions should be direct and clear, and the question content should be focused on the spatial demand to be determined by the planning. The answer design generally should provide multiple choices, with a preference for multiple choices or sorting. Meanwhile, the predetermined answers should be based on adequate literature research or site analysis so that the proper choices can be provided for the respondents. For the proper overall length, respondents should be able to complete the questionnaire within 10 min, and the layout should be compact to avoid the sense of a questionnaire that is too long. In terms of question order, it should start with simple questions, gradually proceed to the key question and then end with secondary questions and open questions.

In the case of Wuhu, the research team conducted a questionnaire survey for the space demand of the residents in the area surrounding the HSR station. The number of questionnaires distributed is 300, and the number of recovered valid questionnaires is 215. The distribution site is mainly the current residential community within 1000 m in the area surrounding the station, including Santiago Community, Julong Garden, Tingyuan Community, Xindu Garden and Tiantai Garden (Fig. 6.18). The questionnaire content mainly includes basic information about the respondents, their way and frequency of reaching the area surrounding the HSR station and their preferred activities in the area surrounding the HSR station.

The gender, age and occupation of the respondents are balanced, which fully reflects the spatial demand and preferences of different types of residents in the area surrounding the HSR station (Fig. 6.19), thus providing a certain basis for spatial planning. The survey in Wuhu finds a preference for specific commercial forms (Fig. 6.20). Of the respondents, 46% which that restaurants with fast service, affordable prices and a good dining environment can be built in the area surrounding the station. Of the respondents, 48% choose a large supermarket as the main retail form because they believe its variety is complete and the prices are affordable. Of respondents, 33% (mainly the elderly) choose a small commodity market because they think that transportation is convenient and the prices are low. A commercial complex (54%) and commercial pedestrian street (20%) are favored by the respondents who are residents, followed by a specialized market (15%). Meanwhile, in terms of refined retail commerce, 77% of the respondents choose a middle-grade brand, and 54% of the respondents suggest that the main combination mode of retail commerce in the area surrounding the rail station should be a complex.

Fig. 6.18 Scope of the questionnaire survey

The research performed a survey of the selection of specific consumption options through the presentation of proper questions (Fig. 6.21). For the consumption of snacks, beverages and daily necessities, 66% of the respondents choose a large supermarket, and 16% choose a small and medium-sized supermarket. For the consumption of home appliances and furniture, 59% of respondents choose a large shopping mall, and 15 and 14% choose a brand-name store and specialty market, respectively. For the consumption of clothing, shoes and cosmetics, 53% of the residents choose a large shopping mall, and 21% choose a brand-name store. Therefore, large shopping malls and supermarkets are the preferred shopping options of Wuhu residents. Such types can be determined when considering adding commercial facilities in the area

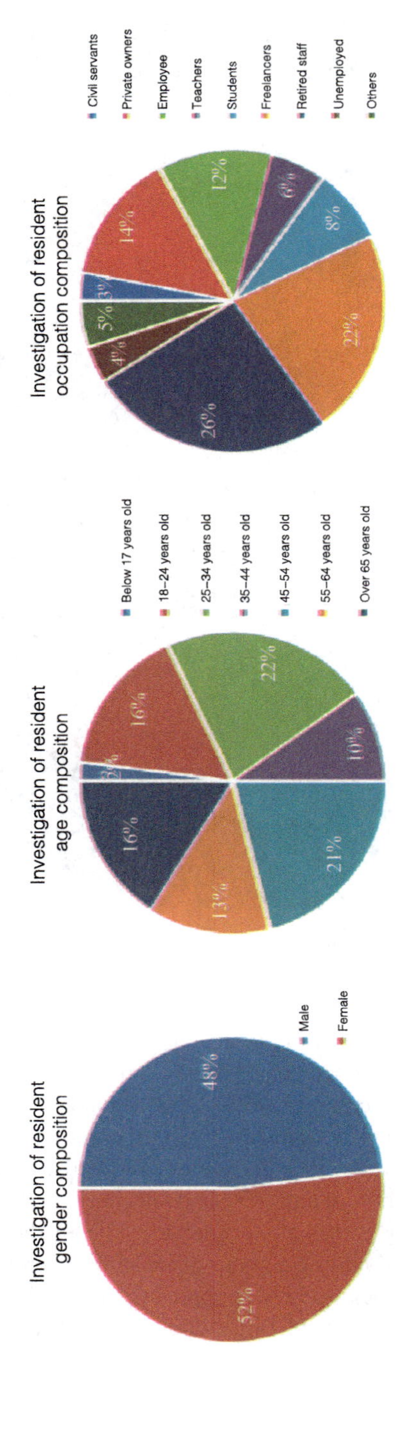

Fig. 6.19 Basic information of the objects of the questionnaire survey

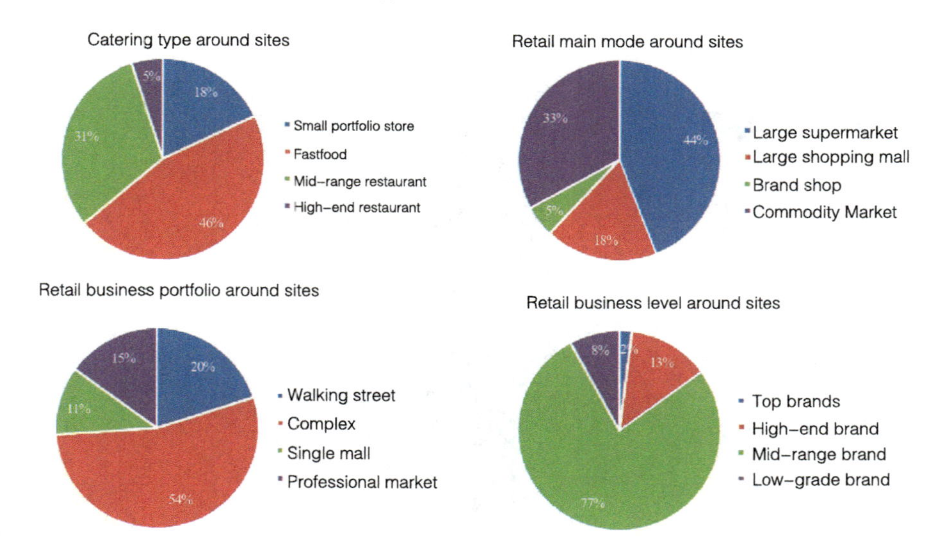

Fig. 6.20 Preference for commercial formats in the area surrounding the HSR station

surrounding the station. By conducting similar surveys in different urban districts, different preference conclusions are obtained.

The questionnaire surveys the surrounding residents' demand for cultural, leisure and recreational facilities (Fig. 6.22). The demand for various facilities is quite equal, including libraries (18%), movie theaters (16%), mass art centers (14%), bookstores (13%), youth activity centers (10%), etc. The recreational facilities with the most obvious demand are indoor and outdoor playgrounds for children (18%), small and medium-sized activity squares (17%), coffee shops (14%), book bars (13%) and elderly activity centers (13%). The diverse demands reflect the lack of regional cultural and recreational facilities. Despite the differences between the demands of residents and passengers brought to the HSR station, some of the passengers' demands can be met at the same time that residents' demands are met. In this way, the continuity and vitality of consumption and activity in the area surrounding the HSR station can be guaranteed.

The factors influencing regional development can be integrated into the research during the planning compilation process for the area surrounding the HSR station. For example, in the planning survey of the area around the Wuhu HSR station, 32% of the respondents consider that an important factor influencing the development surrounding the rail station is public security, 27% choose sanitary conditions, 25% focus on the convenience of public transport, and 16% focus on parking (Fig. 6.23).

The macrolevel survey of passenger demand can be supplemented by the questionnaire survey of residents in the area surrounding the HSR station. Additionally, a combined analysis of urban functional sectors and the service scope of facilities would provide support for the planning in locating the demand for the functions and facilities of the area surrounding the HSR station to a certain extent. For example, the

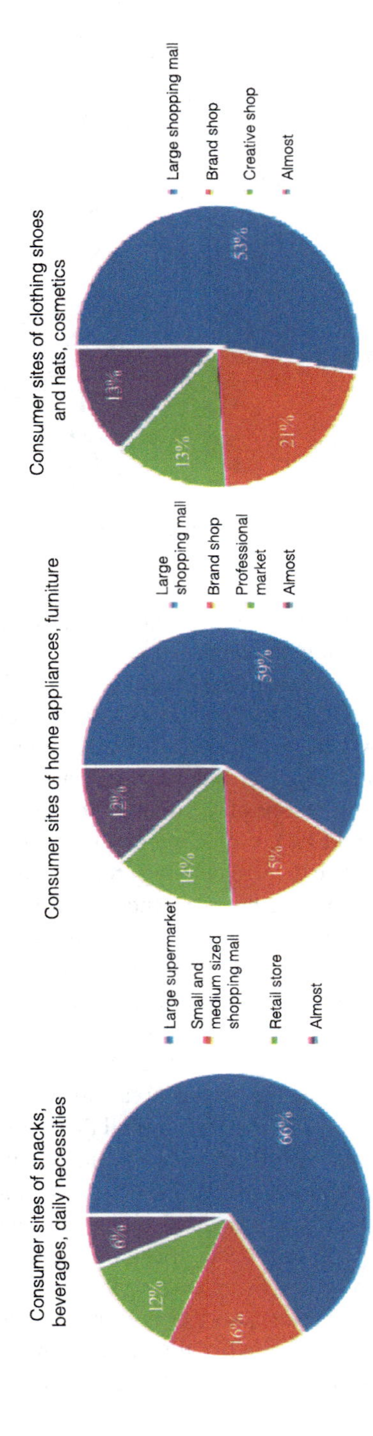

Fig. 6.21 Preference for specific consumption space

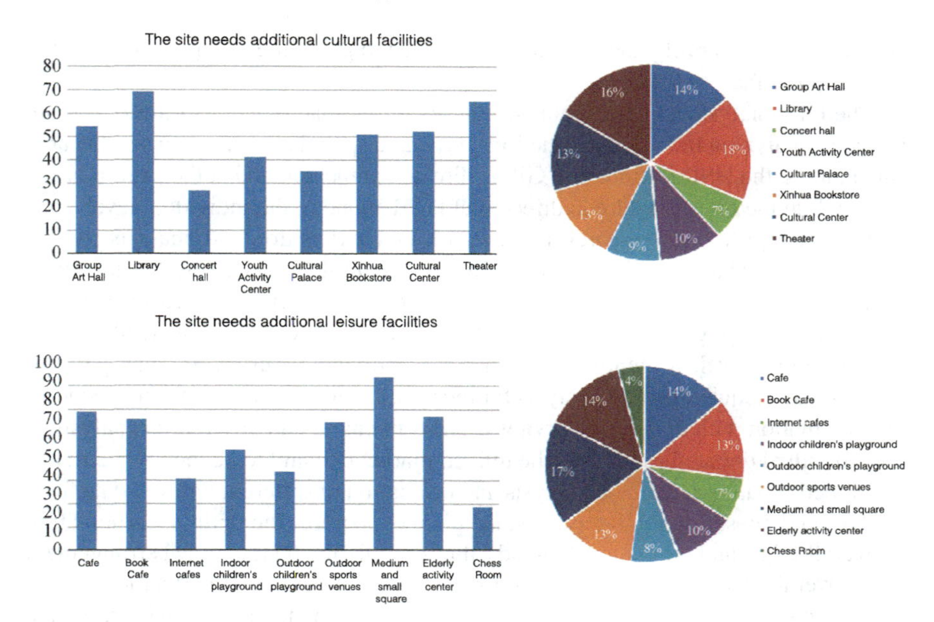

Fig. 6.22 Preferences for cultural and leisure facilities in the area surrounding the HSR station

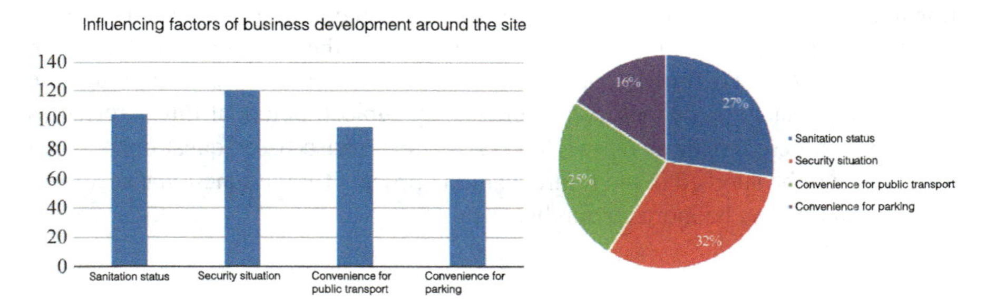

Fig. 6.23 Factors influencing the development of the area surrounding the HSR station

survey in Wuhu finds that the passengers and residents have a demand for shopping malls, supermarkets, dining facilities and open space. The passengers belong to the group with a shorter duration of stay and strong fluidity, while residents belong to the group with high space usage frequency. The groups complement each other. According to the follow-up of the questionnaire, the residents hope to remain separated from the passengers in terms of daily activities to a certain extent, avoiding the mutual interference of crossing a rapid flow of people or being stopped by a flow of people. The conclusion of the GIS analysis and questionnaire survey provides the planning of the area surrounding the HSR station with clues and a basis. It makes the designer deploy and design the corresponding space in combination with the local demand

and consumption ability and provides a reference to guide the designer in providing consumption facilities.

The GIS analysis and questionnaire survey at the macrolevel and mesolevel are aimed at analyzing the specific functional type and spatial characteristics of the area surrounding the HSR station. The GIS is directed at the demand at the overall urban level, while the questionnaire is directed at the demand at the individual level. This section exemplifies the research methods of data acquisition and analysis. At the urban level, it takes the GIS as the main analysis tool for acquiring the layout of functional facilities, calculating different service radius ranges based on the road system and judging the necessity of establishing specific functional facilities. At the individual level, it mainly relies on the questionnaire survey and can perform detailed data acquisition and analysis for the spatial demand of tourists and residents in combination with in-depth interviews. The refinement and differential analysis of the individual demand can derive the refined market demand to be met by the newly developed urban space. In the analysis and evaluation of different types of data from the GIS and questionnaire interviews, the planner acquires the optimal combination of overall urban and individual demands, thus optimizing the functional arrangement at the overall urban level and the functional satisfaction at the individual level.

The analysis and method at these two levels can refine the functional format to a certain extent and provide a basis for further detailed planning and urban design. Yet, there are certain deficiencies. At the overall urban level, the service scope of functional facilities can provide a basic clue to whether a function should be set. However, the influence of the scale of functional facilities upon the functional configuration for the target set needs further detailed analysis. At the individual level, although the surrounding residents are included as the survey subject, potential future residents are not integrated into the analysis. However, the enterprise-based questionnaire and interview will be adopted in the future planning process for the functional sectors to obtain data that can be more comprehensive.

6.3 Analysis Content and Methodology at Microlevel

At the microlevel, the planning of the area surrounding the HSR station focuses on comparing and selecting the land use types according to the land conditions based on the macrolevel and mesolevel analyses, thus defining the functional arrangement, land use types, development conditions and other planning compilation content. The planning of the area surrounding the HSR station can conduct the following planning analysis: determine the possibilities of land development according to the current land use condition and building condition; analyze the land use potentiality; and define the functional configuration of the surrounding area according to the comprehensive analysis at three spatial levels.

With the area surrounding the Wuhu HSR station as an example, the planning research first conducts the current land use analysis. The HSR station is built by upgrading and expanding the original railway station. The surrounding land use types

Table 6.13 Chart of composition of current land use

The nature of land use	Land area (ha)	Ratio (%)
Transport hub site	38.8	16.2
Residential land	111.7	46.5
Commercial land	16.5	6.9
Public service facilities land	6.3	2.6
Road and water system land	9.9	4.1
Other sites	56.8	23.7
Total land	240	100

are much more complicated: the majority are road traffic, commercial and residential, while the minority are industrial, public facility and vacant. Overall, the function is mixed, the environmental quality is poor, and public green space is missing in the urban area. The commercial land use mainly includes Ouyada Furniture Plaza, Huxin Plaza, Wenzhou Commerce City, Hongxiang Business Center and some commercial land along the streets. The internal residential land use on site is complicated as well. The majority is second-class residential land (R2), and the minority are the first-class (R1) and third-class residential land (R3). In terms of the specific land use structure, the area of residence is 111.7 ha, occupying 46.5% of the total land use area, and the area of commercial land is 16.5 ha, occupying 6.9% (Fig. 6.24 and Table 6.13).

The urban public supporting facilities at the site are mainly traffic facilities, educational facilities and administrative service facilities. The traffic facilities are mainly the railway station and bus station, providing external traffic services for Wuhu and the surrounding areas. The administrative service facilities and municipal facilities are distributed unevenly, and their sanitary conditions are poor. There are water supply and drainage, power and telecommunication infrastructure in some areas, but various municipal infrastructures are lacking in the shantytown, with bad sanitary conditions awaiting overall improvement. The educational facilities mainly include primary school and kindergarten, and some residential communities with complete infrastructure have their own internal kindergarten.

The research in the planning compilation process analyzes the building type, quality and height based on the GIS (Fig. 6.25). In terms of building type, the majority of the current buildings at the site are residential, commercial and transportation, and some industrial buildings. In terms of building quality, the buildings are mainly first class and third class. Among them, the first-class buildings are mainly newly built residential and commercial buildings, and the third-class buildings are mainly old residential community and commercial streets. Buildings at the site are mainly low-rise (1–3 floors) and multistory (4–6 floors), and the high rises are mainly residential buildings distributed singly.

Based on the above analysis, the research can conduct a potentiality analysis of land use in the area surrounding the HSR station and further classify the current land into three types: reserved land, land for reconstruction and land for redevelopment

Fig. 6.24 Current land use status

(Fig. 6.26). Reserved land refers to land that maintains its original state. Land for reconstruction refers to land that preserves the building but inserts different functions based on the new land use type. Land for redevelopment refers to land where the original buildings were demolished for future construction. There are only reserved land and land for redevelopment on the site in this case. The reserved land is mainly residential (Julong City Garden, Tingyuan Community, Shengdi Yage, Taihua Garden, Wenbo Garden, Tiantai Garden, Xindu Community and other new residential communities) and commercial (Ouyada, Hanjue Plaza Hotel, Huixin Plaza and other

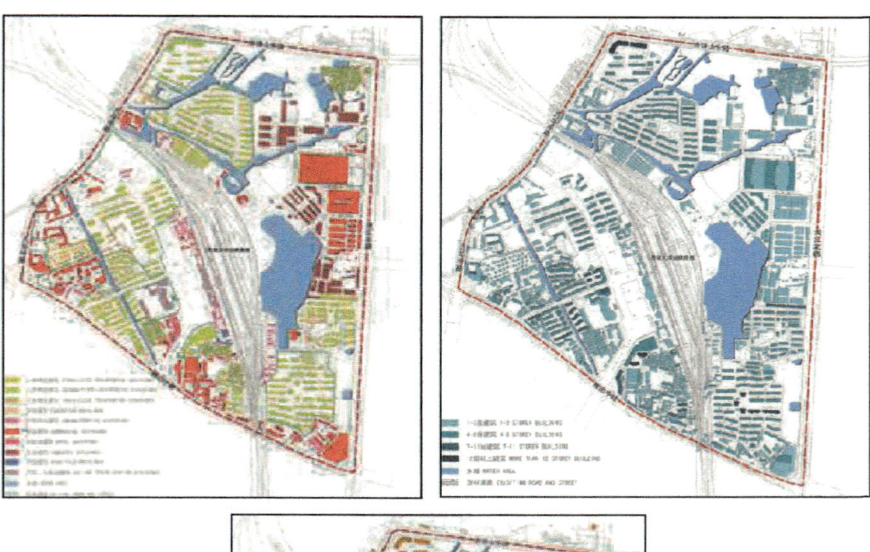

Fig. 6.25 Analysis of the type, quality and height of current buildings

commercial buildings). After determining the availability of the land development, the planning research will conduct the corresponding comparison analysis of the land type.

The land function suitability evaluation can be made in the area surrounding the HSR station. The evaluation method includes the following steps. First, numbers are allocated for each development land parcel, and the evaluation factor of the land use function is determined. Second, the refinement of functions is conducted according to the proper functions determined by the macro- and mesoanalysis, including nine functional directions, namely retail, dining, hotel, residence, business office, admin-

Fig. 6.26 Distribution of land for redevelopment

istrative office, exhibition, cultural facility and logistics storage. Then, the evaluation is made of each land parcel on the basis of the nine different use functions, and the evaluation standard of each evaluation factor is determined and a score is given. The maximum score is five points, the minimum score is one point, and three points are the point of reference that represents that the condition of the land is basically suitable for the function. If the condition is beneficial to the functional implementation, a point will be added or deducted. Finally, the score for the functional development direction of each land parcel is assigned. The full score for the suitability of a single function is 80 points, with 60 points (scoring rate equal to or more than 75%) representing very suitable, 48–59 points (scoring rate between 60 and 75%) representing suitable and 47 points (scoring rate less than 60%) or below representing unsuitable.

The evaluation factor can be divided into three main classes (location, urban facilities and environmental conditions) and six medium classes (prosperity degree, traffic, public service facilities, public facilities, environmental quality and natural conditions). In combination with the specific conditions of the land, these classes can be refined into 16 elements. Location is divided into two medium classes and six subclasses: prosperity degree (class of trade, commercial cluster and surrounding residential district) and traffic (road grade, bus stop and public parking). Urban facilities are divided into two medium classes and four subclasses: public facilities (railway station or bus station, park green space, school and cultural and recreational facilities) and public infrastructure (gas station). Environmental conditions are divided into two medium classes and six subclasses: environmental quality (water environment, sound environment, atmospheric environment and sanitary environment) and geographic conditions (land size and land shape). The evaluation factor system gives a score on the basis of the different types of land use and forms a matrix (Table 6.14). Under this framework, the planning compilation research provides a specific scoring basis for various scoring factors corresponding to the different functions, thus making a scoring classification for the specific land (Table 6.15).

The comparison and selection of the applicability of the land use type in the planning research for the area surrounding the Wuhu HSR station follow the above method. According to the analysis of the current land conditions and development potential, the study of the planning compilation process determines the reserved land and its function. On the basis of meeting the standards of facilities configuration of the relevant plan, the ecological environmental protection requirements and the cadastral boundaries, the planning research defines the location and boundaries of land for redevelopment and allocates numbers to each of them. The area surrounding the Wuhu HSR station is divided into 14 development land parcels (Fig. 6.27). Take the #5 land parcel as an evaluation example (Fig. 6.28). The research gives the land a score according to the rules, and land with a score of more than 60 points is suitable for development. With this scoring system as a basis, it can be concluded that the #5 land parcel is suitable for the development of business offices, administrative offices, retail, dining, hotel and exhibition facilities (Table 6.16). Furthermore, the planning research evaluates the functions of development one by one and summarizes the suitable function for each land parcel for development (Fig. 6.29 and Table 6.17).

The land with functions to be adjusted is organized based on the scoring standards of land use suitability. The suitable land use type of the land to be developed is obtained according to the score of each land. The #1, #2 and #3 land parcels are suitable for the development of business service facilities; the #6 land parcel is suitable for the development of hotels; the #5, #7, #8, #9, #10 and #11 land parcels are suitable for the development of business service facilities; and the #12, #13 and #14 land parcels are suitable for the development of public service facilities or commercial and business facilities (Fig. 6.29). On this basis, the functional arrangement is carried out in combination with the macro- and mesoanalysis according to the different format types and key space requirements (Fig. 6.30). This quantitative method provides functional planning on a rational basis. It is applicable to the updated planning of the upgrading and transformation of the existing station as well as the planning compilation of the new town around the HSR station located at the edge of the built-up area.

Table 6.14 Scoring matrix of land use applicability on each block

Rating content/function			Retail	Dining	Hotel	Residence	Business office	Administration	Exhibition	Cultural facilities	Logistics and warehousing
Location	Business level	Business level									
		Business gathering									
		Surrounding residential area									
	Transport	Road level									
		Bus station									
		Public parking									
City facilities	Public facilities	Train station, bus station									
		Parks, green space									
		Schools, cultural and recreational facilities									
	Public facilities	GAS									
Environmental conditions	Environmental quality	Water environment									
		Acoustic environment									
		Atmospheric environment									
		Health environment									
	Natural conditions	Plot size									
		Lot shape									

Table 6.15 Comparison of land use applicability on each block

Rating content/points			Retail	Dining	Hotel	Residence	Commercial office	Business office	Exhibition	Cultural facilities	Logistics and ware-housing
Location	Business level	Business level	The higher the rating, the better the suitability	The higher the rating, the better the suitability	The higher the rating, the better the suitability	Community level, suitability is strong	Not sensitive	Not sensitive	Not sensitive	Not sensitive	The higher the rating, the better the suitability
		Business gathering	Business office, suitability is strong Nonbusiness district, suitable in general	Suitability is strong when near business offices Suitability is general when not near business offices	Business offices, suitability is strong Nonbusiness district, suitable in general	Suitability is strong when near business offices Suitability is general when not near business offices	Business offices, suitability is strong Nonbusiness district, suitable in general	Business offices, suitability is strong Nonbusiness district, suitable in general	Business offices, suitability is strong	Not sensitive	Not sensitive
		Surrounding residential area	Suitability is strong when adjacent Suitability is general when not adjacent	Suitability is strong when adjacent Suitability is general when not adjacent	Not sensitive	Suitability is strong when adjacent Suitability is general when not adjacent	Suitability is strong when adjacent Suitability is general when not adjacent	Suitability is strong when adjacent Suitability is general when not adjacent	Not sensitive	Suitability is strong when adjacent Suitability is general when not adjacent	Not sensitive

(continued)

Table 6.15 (continued)

Rating content/points			Retail	Dining	Hotel	Residence	Commercial office	Business office	Exhibition	Cultural facilities	Logistics and ware-housing
	Transport	Road level	Suitability is strong when close to main road Suitability is general when near expressway, slip road	Suitability is strong when close to main road Suitability is general when near expressway, slip road	Suitability is strong when close to main road Suitability is general when near expressway, slip road	Suitability is strong when near secondary roads Suitability is general when near main road Not suitable when near expressway	Suitability is strong when close to main road Suitability is general when near expressway, slip road	Suitability is strong when close to main road Suitability is general when near expressway, slip road	Suitability is strong when close to main road Suitability is general when near expressway, slip road	Suitability is strong when close to main road Suitability is general when near expressway, slip road	The higher the rating, the better the suitability
		Bus station	The greater the number, the closer the distance, the more suitable	The greater the number, the closer the distance, the more suitable	The greater the number, the closer the distance, the more suitable	The greater the number, the closer the distance, the more suitable	The greater the number, the closer the distance, the more suitable	The greater the number, the closer the distance, the more suitable	The greater the number, the closer the distance, the more suitable	The greater the number, the closer the distance, the more suitable	Not sensitive

(continued)

Table 6.15 (continued)

Rating content/points			Retail	Dining	Hotel	Residence	Commercial office	Business office	Exhibition	Cultural facilities	Logistics and ware-housing
		Public parking	The closer the distance, the better the suitability	The closer the distance, the better the suitability	The closer the distance, the better the suitability	The closer the distance, the better the suitability	The closer the distance, the better the suitability	The closer the distance, the better the suitability	The closer the distance, the better the suitability	The closer the distance, the better the suitability	Not sensitive
City facilities	Public facilities	Train station, bus station	Suitability increases	Suitability increases	Suitability increases	General suitability	Suitability increases	Suitability increases	Suitability increases	Not sensitive	Not sensitive
		Parks, green space	Not sensitive	Not sensitive	Suitability increases	Suitability increases	Not sensitive	Not sensitive	Not sensitive	Not sensitive	Not sensitive
		Schools, cultural and recreational facilities	Suitability increases	Suitability increases	Suitability increases	General suitability	Suitability increases	Suitability increases	Not sensitive	Not sensitive	Not sensitive
	Public facilities	GAS	General suitability	Not suitable	Not suitable	Not suitable	General suitability	General suitability	Not sensitive	Not sensitive	Not sensitive
Environmental conditions	Environmental quality	Water environment	Suitability increases when near water	Suitability increases when near water	Suitability increases when near water	Suitability increases when near water	Suitability increases when near water	Suitability increases when near water	Not sensitive	Not sensitive	Not sensitive

(continued)

Table 6.15 (continued)

Rating content/points			Retail	Dining	Hotel	Residence	Commercial office	Business office	Exhibition	Cultural facilities	Logistics and warehousing
		Acoustic environment	The lower the noise, the better the suitability	Not sensitive	The lower the noise, the better the suitability	The lower the noise, the better the suitability	Not sensitive	Not sensitive	Not sensitive	Not sensitive	Not sensitive
		Atmospheric environment	The better the air quality, the better the suitability	The better the air quality, the better the suitability	The better the air quality, the better the suitability	The better the air quality, the better the suitability	The better the air quality, the better the suitability	The better the air quality, the better the suitability	The better the air quality, the better the suitability	The better the air quality, the better the suitability	Not sensitive
		Health environment	The better the sanitation, the better the fitness	The better the sanitation, the better the fitness	The better the sanitation, the better the fitness	The better the sanitation, the better the fitness Train stations, expressways strongly influence suitability	The better the sanitation, the better the fitness	The better the sanitation, the better the fitness	The better the sanitation, the better the fitness	The better the sanitation, the better the fitness	Not sensitive

(continued)

Table 6.15 (continued)

Rating content/points			Retail	Dining	Hotel	Residence	Commercial office	Business office	Exhibition	Cultural facilities	Logistics and ware-housing
	Natural conditions	Plot size	Not sensitive	Not sensitive	Suitability decreases when too little	Suitability decreases when too little or too much	Suitability decreases when too little or too much	Suitability decreases when too little or too much	Suitability decreases when too little	Suitability decreases when too much	Suitability decreases when too little
		Lot shape	Not sensitive	Not sensitive	The more regular, the more suitable	The more regular, the more suitable	The more regular, the more suitable	The more regular, the more suitable	Not sensitive	Not sensitive	The more regular, the more suitable

Table 6.16 Example of comparison for land use applicability on a specific block

Lot number	Rating content/points			Feature description	Retail	Dining	Hotel	Residence	Business office	Administration	Exhibition	Cultural facilities	Logistics and ware-housing
5	Location	Business level	Business level	High end	5	5	5	5	5	5	5	4	2
			Business gathering	Liansheng	4	4	5	5	5	5	5	3	2
			Surrounding residential area	East living status	5	5	4	5	5	5	3	4	2
		Transport	Road level	Near main road + secondary road	5	5	5	3	5	5	3	4	3
			Bus station	Status of two bus stations	5	5	4	4	5	5	5	4	2
			Public parking	Near the bus terminal, parking lot	4	4	4	3	5	5	5	4	3
	City facilities	Public facilities	Train station, bus station	Railway station square	5	5	5	3	5	4	5	2	3
			Parks, green space	None	3	3	4	3	3	3	3	3	3
			Schools, cultural and recreational facilities	None	3	3	3	3	3	3	3	3	3
		Public facilities	GAS	None	3	3	3	3	3	3	3	3	3
	Environmental conditions	Environmental quality	Water environment	Not near water	3	3	3	3	3	3	3	3	3
			Acoustic environment	Main road	3	3	3	3	3	3	3	3	3
			Atmospheric environment	Expressway	3	3	3	2	3	3	3	3	3
			Health environment	Influence of expressway	3	2	3	2	3	3	3	3	3
		Natural conditions	Plot size		5	3	3	5	5	5	5	5	3
			Lot shape	More regular	5	5	5	5	5	5	5	5	3
Total score					64	61	62	57	66	65	62	56	44

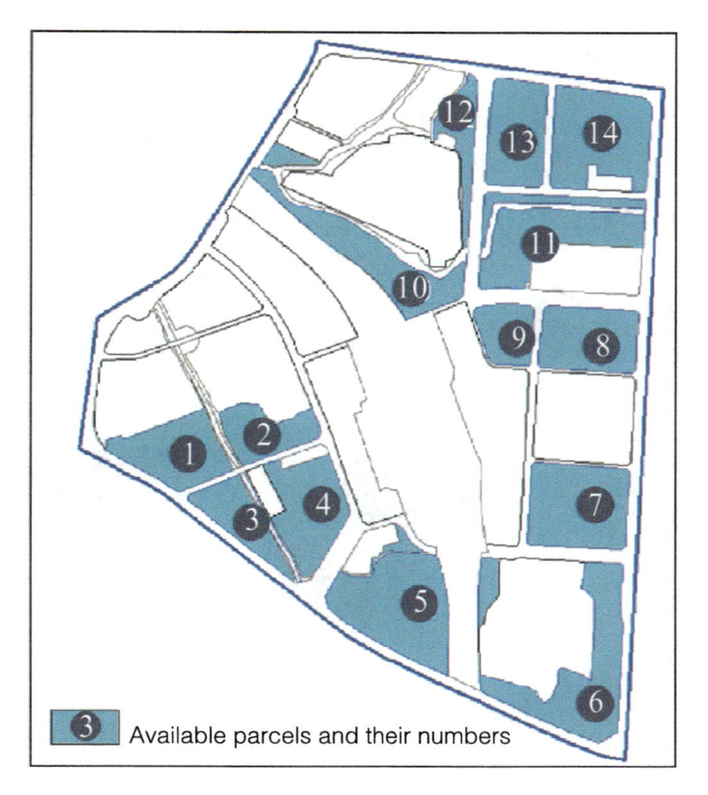

Fig. 6.27 Numbering of land for redevelopment

6.4 Conclusion

The HSR has increasingly become people's travel choice because of its high speed, large transport volume and guaranteed punctuality. Meanwhile, it is gradually changing the development patterns of the regions, cities and areas surrounding the stations. According to the latest railway network news in June 2016, the *Medium- and Long-term Railway Network Plan (2016–2030)* was adopted at the executive meeting of the State Council. Originally, in 2004, the HSR line in China had a length of 12,000 km with *four longitudinal lines and four transverse lines* structure. However, it has currently lengthened to 72,000 km with a new structure of *eight longitudinal lines and eight transverse lines*. The rail network covers cities with a population of more than 200,000; the express railway network connects all cities with a population of more than 500,000; and the HSR network connects all cities with a population of more than 1 million; moreover, the intercity HSR connects cities within urban agglomerations. The total length of China's railway network will eventually reach 204,000 km. This increase means that more cities will be connected to the HSR network system, and more stations will be built or updated. Both the development of the new town around

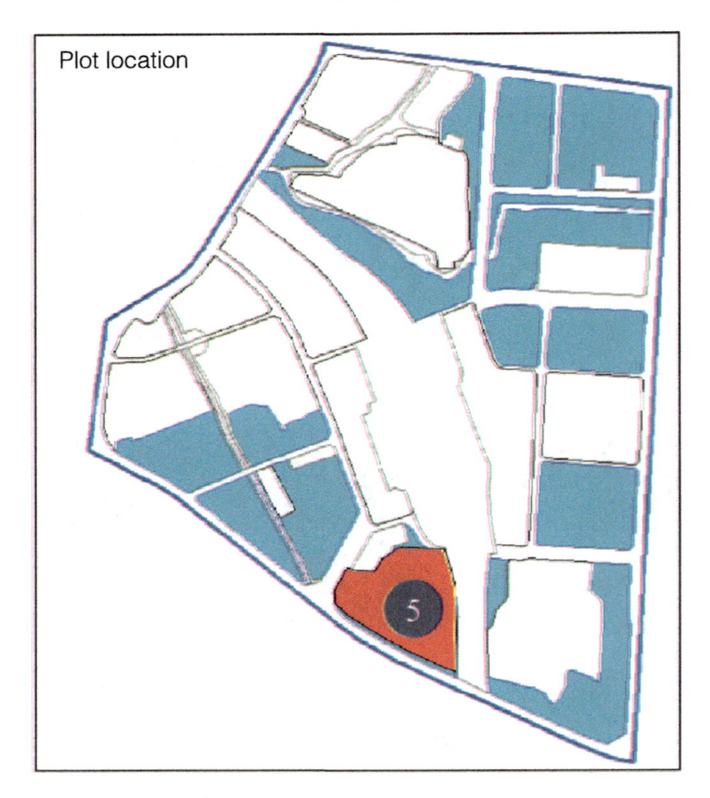

Fig. 6.28 Location of exemplary block for redevelopment (#5)

Fig. 6.29 Conception of functional arrangement based on the comparison of land use applicability

Table 6.17 Final evaluation of land for redevelopment

Lot number	Rating content / points	Feature Description	Retail	Dining	Hotel	Residence	Business office	Administration	Exhibition
1	64	60	61	52	55	50	56	52	42
2	65	64	64	47	58	46	47	47	41
3	64	59	60	50	55	53	46	50	39
4	68	67	64	45	57	49	43	46	41
5	64	61	62	61	66	65	62	53	44
6	53	49	62	46	61	61	44	47	44
7	57	56	66	34	67	63	57	52	52
8	57	55	67	42	65	63	56	53	48
9	62	63	62	40	67	62	51	37	42
10	42	52	38	41	41	38	39	36	39
11	58	60	57	52	64	63	59	57	47
12	51	52	50	42	39	43	38	39	35
13	60	60	62	55	58	61	63	59	50
14	59	57	64	52	58	64	65	62	52

HSR stations and the development and redevelopment of the area surrounding the updated stations involve issues such as how to analyze the overall urban space and the demand of surrounding residents and passengers; the approach to compile continuous planning at various levels; and the recognition of the planning capability and clear administrative competence of development managers.

This book takes 22 cities along the Beijing-Shanghai HSR, the first HSR track to operate in China, as the starting point for its analysis and shows that the station location (relationship between the station and the existing built-up urban area) and urban fiscal capacity produce a significant impact upon the development and con-

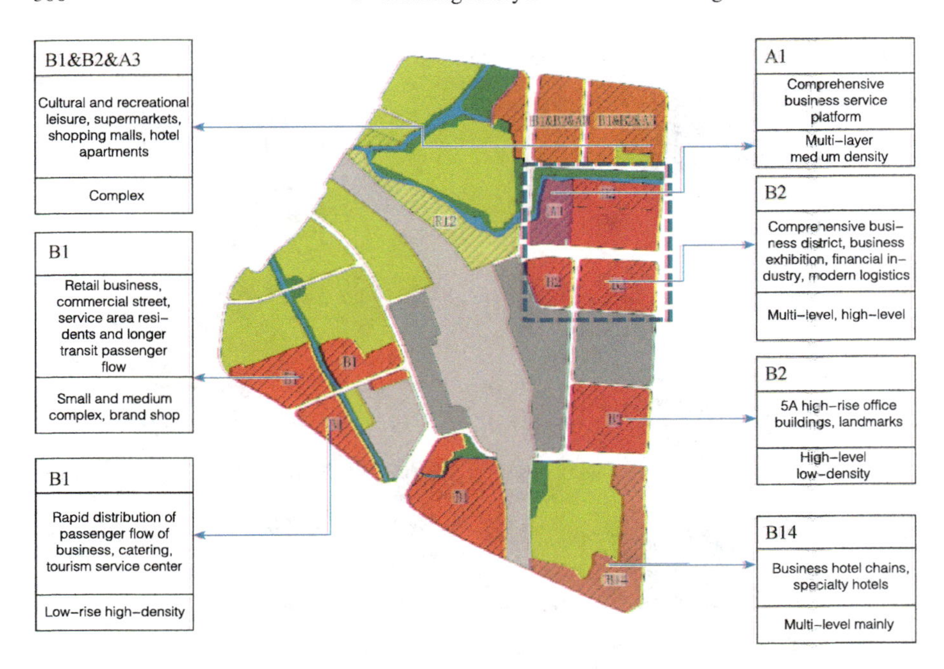

Fig. 6.30 Refinement of functional arrangement

struction of the new town around a HSR station. To further analyze the planning and development of the HSR, a comparative case study was conducted. Two cities, Wuxi and Changzhou, with similar station locations and urban fiscal capacity but different development effects, were selected. The study conducted the comparative analysis based on the exchange value and use value framework in the well-known theory of spatial political economics known as the *growth machine theory*. With the opportunity to study the planning compilation process of development brought by the upgrading of the Wuhu HSR station as an example, this book exemplifies the layer-by-layer analysis method of the planning of the area surrounding the HSR station. It can be easily seen that the planning and development of the new town, or the area surrounding the HSR station in the built-up area, are rooted in multiple complicated and reciprocal planning decision-making processes. The planners must analyze the demand and respond accordingly in the design of space. From functional arrangement to space design, from planning idea to decision implementation, the deduction and promotion of each step decide the final space production as a chain effect and butterfly effect.

The interaction between planning and development implementation displays a relationship among various behavioral subjects. Improvement of market status brings a change in the relationship among behavioral subjects in the planning and development implementation process. A plan that lacks an understanding of the implementation process will not be able to effectively guide the development. Meanwhile, the complexity and uncertainty of the implementation process require the flexible adjustment and adaptation of the plan. An understanding of the relationship between the two can guarantee that the planning compilation is reasonable and feasible. In this way, the local adjustment of the implementation process can avoid producing negative external effects and exerting a negative influence on the overall development. By focusing specifically on the interaction process of the two, we can understand important issues, including how the plan is implemented, what factors lead to planning revision in the implementation process and what requirements the implementation reveals for the plan, and then establish the plan. After answering those questions, the transformation and establishment of a flexible implementation mechanism can be achieved. The planning and development process of the new cities around the HSR stations of Wuxi and Changzhou displays the importance of planning with a clear sequence and consistency, creating implementation mechanisms with timely feedback and creating an implementation agency with stability and clear ownership boundaries for urban development.

The planning and development of the new town around a HSR station involve input and distribution from many public resources. Generally, the process may be tortuous during the time of planning and implementation and may induce a waste of public resources (such as land). The new town and areas surrounding HSR stations have become important strategic platforms for local economic development in China and are influencing and will further deeply exert an impact on the urbanization process of China, producing significant positive and negative effects. Currently, for such new cities, problems ranging from a large planning scale to land lying idle after the construction of infrastructure are in urgent need of resolution. The transformation development advocated in the *12th Five-Year Plan* will move from an outer extension to a full utilization of the existing space. The new cities and areas surrounding HSR stations are the key focus for the planning of the existing space. First, the land intended for first-class development in the new town of HSRs is in need of investment from secondary developers. Second, the updated station has the power to drive urban renewal in the surrounding area. The promotion of the development of an area with the help of a HSR station depends on the quality of the planning and development implementation mechanisms.

Among the theories that can guide the planning and design of areas surrounding the stations, the transit-oriented development (TOD) theory is the spatial model with the most far-reaching influence in the new urbanism of the USA. Compared with conventional TOD (subway stations, bus stations, etc.), the HSR station has a stronger spatial radiation ability, but at the same time, it can be much more difficult to integrate it with the existing urban framework. The HSR station is the portal of the urban linkage area and has export-oriented attributes. The coming and going of large numbers of people may bring investment opportunities and tourism consumption but may also remove resources and talent. This pattern depends on the overall urban attractiveness and regional competitiveness of the city where a station is located. However, in essence, a high density and mixed configuration, as well as a pedestrian-friendly spatial design advocated by TOD, are still very important design principles of the areas surrounding HSR stations. The station land is owned by the former Ministry of Railways, but the area surrounding the station is owned by the local government. Due to this ownership barrier and conflict, the HSR station is built as a pure traffic station, failing to realize the goal of high-density development. Meanwhile, it is difficult to promote a mixed configuration of land use under the existing land use management system, such as taking the regulatory plan as the basic plan and development management tool. The relatively rigid and single land use arrangement calls for a more flexible and scientific planning compilation and management system. Additionally, the pedestrian-friendly space that should be designed around the stations is not established because the grand square around the station is regarded as a necessary spatial accessory for dispersing people. The area surrounding the HSR station is far from a pleasant pedestrian environment thanks to a square on a large scale, a lack of facilities, roads without shade trees, and a loss of the sense of enclosure from the surrounding buildings. With the process of cooperation between the railway corporation and local governments, a change from the administration-oriented decisions of the past to an emphasis on the enhancement of the land exchange value and use value can be expected, which will be beneficial to the realization of the TOD model for HSR stations.

For the spatial arrangement of the new town and the area surrounding the HSR station, besides the three principles of high density, mixed use and pedestrian priority advocated in the TOD model, the spatial model can be refined on the basis of the well-known *three-layer model* (Fig. 6.31). The model proposed in this book divides the new town and the area surrounding the HSR station into two spatial levels, namely the core area of the traffic hub and the urban functional area (Fig. 6.32). The core area of the traffic hub refers to the scope defined by ten minutes' walk from the station, namely a scope with a radius of 1000–1500 m, with the station as the center. In this scope, the functions are mainly traffic services and other core functions that rely on the traffic flow, including regional financial businesses, tourism distribution services, specialized market coordination platforms, hotels and retail commercial facilities. The service target is the HSR passenger, including investors, tourists and regional commuters. Therefore, the results of the survey of passengers' demand for space are of great importance for the planning of the core area. The urban functional area is located at the periphery, referring to a radius of 3000–4000 m from the station and covering an area of 30–50 km^2. The functional configuration of this area will be performed through the close integration of this area with the overall urban situation, and the relevant analysis of the scope of service facilities based on the GIS can be made. In this way, the urban functions that are missing in this area will be defined, and the necessity of building sufficient specific facilities will be confirmed. Meanwhile, the peripheral area should meet the demands of the existing and future residents, and the spatial demand survey for this group will help the planners define the demand intensity of the functions and facilities. It is possible to establish facilities that can exert a regional impact in the core area of the traffic hub and the urban functional area, such as a theme park that is a regional attraction and an enterprise office area with a focus on an industrial use complemented by other urban industrial chains or value chains. To match the local subway lines, subway stations and the configuration of regional facilities in a banded, oval or grid formation can be planned to fit the original circular layer structure. In this way, the structure can be adapted to fit better

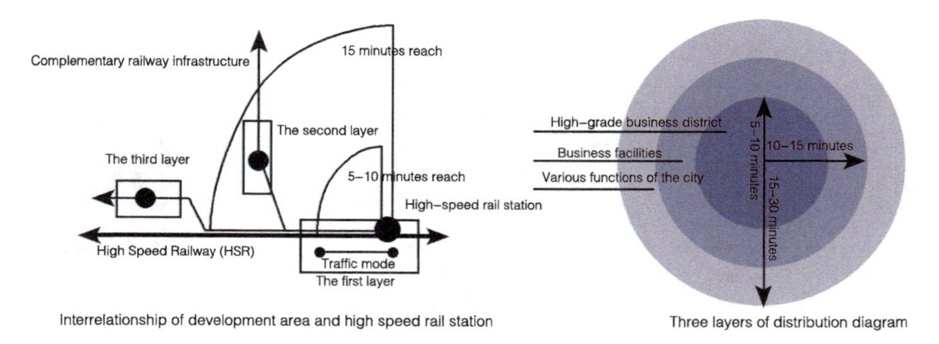

Fig. 6.31 Three-layer model for the area surrounding the HSR station. *Source* Preimus H. HST-Railway stations as dynamic nodes in urban networks [C]//China Planning Network (CPN) 3rd Annual Conference Beijing June. 2006

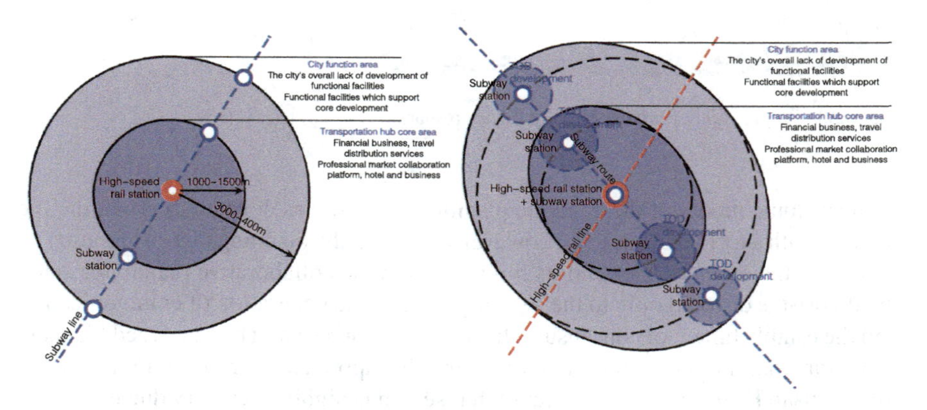

Fig. 6.32 Layer structure of the area surrounding the HSR station

the development of the area surrounding the specific HSR station. Planning for the area surrounding a HSR station, or a larger spatial scope around it, can be regarded as the spatial planning of a new town around a HSR station. With a facility service area calculated using GIS and a questionnaire survey for special groups as the basis, and the analysis and conception of regional resources as a driving force, HSR as an important regional traffic facility can fully play a leading role in urban development.

Urban theory and planning theory underwent a parallel development in the past. The former focuses on the law and norms of the city, while the latter focuses on understanding and constructing the role and methods of planning and planners in urban development (Fig. 6.33). In an article titled *Planning Theory and City* published in the 25th issue of *Planning Education and Research* in 2005, Professor Susan Fainstein proposed the unfeasibility of separating urban theory from planning theory. She believed that planning theory generally ignores the city itself and instead focuses on providing the practice of planning with normative standards. For example,

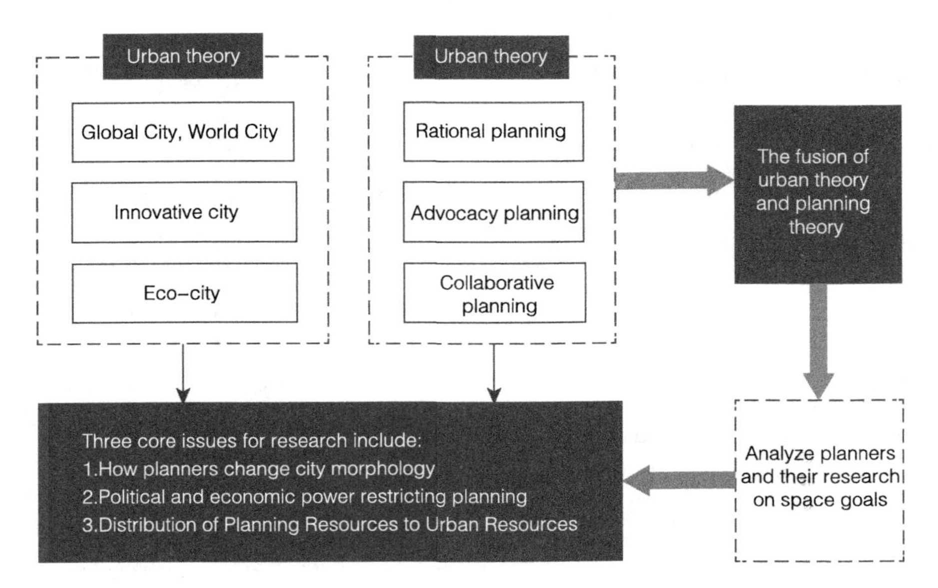

Fig. 6.33 Progression of urban planning theory research

in the rational model of the past, the planning theorist was devoted to providing the planner with a technical strategy; however, currently, the planning theorist focuses on exploring the methodology and its limitations. In the collaborative planning model, the theorist is devoted only to the role of planners, the generation of communication and the establishment of consensus. However, certain research has analyzed planners and their spatial target at the same time, and this approach represents an increasing trend. Susan Fainstein believed that such research is highly necessary due to the historical origin and reason for the existence of planning. In looking back at the origin and evolution of planning and development, we see that the whole process takes the city as the core. The initial motivation of planning and development lies in the criticism of the industrial city and the aspiration to redesign the city according to certain rules. For Howard's garden city mode, Osman's Paris or Burnham's city beautiful movement, the planning is oriented toward creating the expected city. This origin is a logical way to apply the selected model to the urban area in a top-down manner. The evolving planning is devoted to eliminating government prejudices and the preferences of personal interests and changing the dependence on the ultimate goal of the urban blueprint. The planning approach that pursues more democratic participation gradually changes its emphasis from the occupation dominated by design to the social science and is rooted in the understanding and expectations of the city. Meanwhile, urban planning should understand its urban agency and power environment, and the value of urban planning is determined through contact and comparison with urban theory. Neo-Marxism proposed in the 1970s that the structural framework restricts planning ability and especially that the changes wrought by planning do not serve capital. The environment of urban planning is the field of the interwoven various powers of the city. One of the ways to understand such an environment is to verify the planning results and compare them with such urban ideas of similar cities. In addition, planning theory should consider the conditions under which human activities can create a better city for all citizens. To answer this question, we should pay attention to the interactive contact between the planning process and the results. Meanwhile, we also should explore the characteristics of the so-called better city to identify the strategy of realizing a better city and the main barriers. Then, we should understand the relationship among the specific urban development history, urban economic base and social structure and master the related systems of policy-making and decision-making. Therefore, planning theory is both explanatory and normative and is inseparable from urban theory. The research approach of closely linking urban theory with planning theory, proposing the corresponding research problems and forming the relevant analytical framework is the focus of this book.

The HSR is being vigorously constructed, and HSR stations are also continuously being constructed or upgraded. Although several years have passed since the author began to pay attention to the planning and development of the areas surrounding HSR stations in 2009, the HSR effect and the planning and development of the new town around the HSR station still need longer-term evaluation. The planning and implementation of the areas surrounding various stations are ongoing, with the space users entering successively, but the overall driving and promotion of the development of the city still need time. The urban planning discipline requires the implementation of more empirical studies to establish quantitative models or conduct case analyses to identify the real problems in reality and principles behind the simple phenomenon. The empirical research on the planning and development of the new town and the area surrounding the HSR station provides a chance to understand the difficulties and dilemmas in such planning and similar types of development. It displays how the planner changes the spatial form of the area surrounding the station, demonstrates the administrative mechanisms and economic power of the planning that restricts implementation and development and reveals the benefits of urban resource distribution to a certain extent. This book has a restricted view of the planning and development of the new town around the HSR, and the subject is worth exploring in further empirical studies.

Bibliography

Boyle P, Zhou J (2011) Governing urban developments around high-speed train stations: experiences from four European cities. Urban Plan Int 26(3):27–34

Cervero R, Jin M (2009) Rail and property development in Hong Kong: experiences and extensions. Urban Studies 46(10):2019–2043

Cervero R, Landis J (1997) Twenty years of the bay area rapid transit system: land-use and development impacts. Transp Res Part A Policy & Practice 31(4):309–333

Dan I (2009) Large redevelopment initiatives, housing values and gentrification: the case of the Atlanta Beltline. Urban Stud 46(8):1723–1745

Handy S (2005) Smart growth and the transportation-land use connection: what does the research tell us? Int Reg Sci Rev 28(2):146–167

He N, Gu BN (1998) Analysis of the role of urban rail transit in land use. Urban Mass Transit 1(4):32–36

Hess DB, Almeida TM (2007) Impact of proximity to light rail rapid transit on station—area property values in Buffalo, New York. Urban Stud 44(5):1041–1068

Lauria M (1996) Reconstructing urban regime theory: regulating urban politics in a global economy. Sage Publications

Li QL, Ma XD, Zhu CG et al (2007) Analysis of urban spatial structure evolution of Yancheng: based on GIS. Geogr Geo-Info Sci 23(3):69–73

Li LL, Zhang GH, Cao YL (2007) Techical method for planning and adjusting land use around rail transit stations: a case study of Suzhou. UTC J 5(1):30–36

Lin S, Feng L (2011) The social and economic influence of Japanese HSR Construction. Stud Urban and Reg Plan (3):132–156

Liu JL, Zeng XG (2004) Urban rail transit and land use integrated planning based on quantitative analysis. J China Railway Soc 26(3):13–19

Luo PF, Xu YL, Zhang NN (2004) Study on the impacts of regional accessibility of high speed railway: a case study of Nanjing to Shanghai region. Econ Geogr 24(3):407–411

Ma AK, Cao RL, Zhang PG et al (2008) Impacts of the expansion of the railway network on reginal accessibility: a case study of the urban agglomeration along with the Jiaoji railway. J Shandong Normal Univ (Natural Science) 23(2):89–93

Ryan S (1999) Property values and transportation facilities: finding the transportation—land use connection. J Plan Lit 13:412–427

Wang L, Liu G (2007) The evolution of public-private partnership: American downtown redevelopment in the second half of the twentieth century. Urban Plan Int 22(4):21–26

9789811369155